D1329648

BIOMEDICAL
SIGNAL ANALYSIS

IEEE Press Series on Biomedical Engineering

The focus of our series is to introduce current and emerging technologies to biomedical and electrical engineering practitioners, researchers, and students. This series seeks to foster interdisciplinary biomedical engineering education to satisfy the needs of the industrial and academic areas. This requires an innovative approach that overcomes the difficulties associated with the traditional textbooks and edited collections.

Metin Akay, *Series Editor*
Dartmouth College

Books in the IEEE Press Series on Biomedical Engineering

BIOMEDICAL SIGNAL ANALYSIS

A Case-Study Approach

Rangaraj M. Rangayyan

Univeristy of Calgary
Calgary, Alberta, Canada

IEEE Engineering in Medicine
and Biology Society, *Sponsor*

IEEE Press Series on Biomedical Engineering
Metin Akay, *Series Editor*

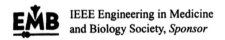

IEEE Press

JOHN WILEY & SONS, INC.

For ordering and customer service, call 1-800-CALL-WILEY.

Library of Congress Cataloging in Publication Data is available.

ISBN 0-471-20811-6

10 9 8 7 6 5 4 3 2

Books of Related Interest from IEEE Press

RANDOM PROCESSES FOR IMAGE AND SIGNAL PROCESSING
Edward R. Dougherty
An SPIE Press book published in cooperation with IEEE Press
A volume in the SPIE/IEEE Series on Imaging Science & Engineering
1999   Hardcover   616 pp   IEEE Order No. PC5747   ISBN 0-7803-3495-7

UNDERSTANDING NEURAL NETWORKS AND FUZZY LOGIC: Basic Concepts & Applications
Stamatios V. Kartalopoulos
A volume in the IEEE Press Understanding & Technology Series
1996   Softcover   232 pp   IEEE Order No. PP5591   ISBN 0-7803-1128-0

Dedication

Mátr dévô bhava
Pitr dévô bhava
Áchárya dévô bhava

Look upon your mother as your God
Look upon your father as your God
Look upon your teacher as your God

— from the sacred Vedic hymns of the *Taittireeya Upanishad* of India.

This book is dedicated to the fond memory of
my mother Srimati Padma Srinivasan Rangayyan
and my father Sri Srinivasan Mandayam Rangayyan,
and to all of my teachers,
in particular, Professor Ivaturi Surya Narayana Murthy.

Preface

Background and Motivation

The establishment of the clinical electrocardiograph (ECG) by the Dutch physician Willem Einthoven in 1903 marked the beginning of a new era in medical diagnostic techniques, including the entry of electronics into health care. Since then, electronics, and subsequently computers, have become integral components of biomedical signal analysis systems, performing a variety of tasks from data acquisition and preprocessing for removal of artifacts to feature extraction and interpretation. Electronic instrumentation and computers have been applied to investigate a host of biological and physiological systems and phenomena, such as the electrical activity of the cardiovascular system, the brain, the neuromuscular system, and the gastric system; pressure variations in the cardiovascular system; sound and vibration signals from the cardiovascular, the musculo-skeletal, and the respiratory systems; and magnetic fields of the brain, to name a few.

The primary step in investigations of physiological systems requires the development of appropriate sensors and instrumentation to transduce the phenomenon of interest into a measurable electrical signal. The next step of analysis of the signals, however, is not always an easy task for a physician or life-sciences specialist. The clinically relevant information in the signal is often masked by noise and interference, and the signal features may not be readily comprehensible by the visual or auditory systems of a human observer. Heart sounds, for example, have most of their energy at or below the threshold of auditory perception of most humans; the interference patterns of a surface electromyographic (EMG) signal are too complex to permit visual analysis. Some repetitious or attention-demanding tasks, such as on-line monitoring of the ECG of a critically ill patient with cardiac rhythm problems, could be uninteresting and tiring for a human observer. Furthermore, the variability present in a given type of signal from one subject to another, and the inter-observer variability inherent in subjective analysis performed by physicians or analysts make consistent understanding or evaluation of any phenomenon difficult, if not impossible. These factors created the need not only for improved instrumentation, but also for the development of methods for objective analysis via signal processing algorithms implemented in electronic hardware or on computers.

Processing of biomedical signals, until a few years ago, was mainly directed toward filtering for removal of noise and power-line interference; spectral analysis to understand the frequency characteristics of signals; and modeling for feature representation and parameterization. Recent trends have been toward quantitative or objective analysis of physiological systems and phenomena via signal analysis. The field of biomedical signal analysis has advanced to the stage of practical application of signal processing and pattern analysis techniques for efficient and improved non-invasive diagnosis, on-line monitoring of critically ill patients, and rehabilitation and sensory aids for the handicapped. Techniques developed by engineers are gaining wider acceptance by practicing clinicians, and the role of engineering in diagnosis and treatment is gaining much-deserved respect.

The major strength in the application of computers in biomedical signal analysis lies in the potential use of signal processing and modeling techniques for quantitative or objective analysis. Analysis of signals by human observers is almost always accompanied by perceptual limitations, inter-personal variations, errors caused by fatigue, errors caused by the very low rate of incidence of a certain sign of abnormality, environmental distractions, and so on. The interpretation of a signal by an expert bears the weight of the experience and expertise of the analyst; however, such analysis is almost always subjective. Computer analysis of biomedical signals, if performed with the appropriate logic, has the potential to add objective strength to the interpretation of the expert. It thus becomes possible to improve the diagnostic confidence or accuracy of even an expert with many years of experience. This approach to improved health care could be labeled as *computer-aided diagnosis*.

Developing an algorithm for biomedical signal analysis, however, is not an easy task; quite often, it might not even be a straightforward process. The engineer or computer analyst is often bewildered by the variability of features in biomedical signals and systems, which is far higher than that encountered in physical systems or observations. Benign diseases often mimic the features of malignant diseases; malignancies may exhibit a characteristic pattern, which, however, is not always guaranteed to appear. Handling all of the possibilities and degrees of freedom in a biomedical system is a major challenge in most applications. Techniques proven to work well with a certain system or set of signals may not work in another seemingly similar situation.

The Problem-solving Approach

The approach I have taken in presenting material in this book is primarily that of development of algorithms for problem solving. Engineers are often said to be (with admiration, I believe) problem solvers. However, the development of a problem statement and gaining of a good understanding of the problem could require a significant amount of preparatory work. I have selected a logical series of problems, from the many case-studies I have encountered in my research work, for presentation in the book. Each chapter deals with a certain type of a problem with biomedical signals. Each chapter begins with a statement of the problem, followed immediately with a few illustrations of the problem with real-life case-studies and the associated signals. Signal processing, modeling, or analysis techniques are then presented, starting with relatively simple "textbook" methods, followed by more sophisticated research approaches directed at the specific problem. Each chapter concludes with one or more applications to significant and practical problems. The book is illustrated copiously with real-life biomedical signals and their derivatives.

The methods presented in the book are at a fairly high level of technical sophistication. A good background in signal and system analysis [1, 2, 3] as well as probability, random variables, and stochastic processes [4, 5, 6, 7, 8, 9] is required, in order to follow the procedures and analysis. Familiarity with systems theory and transforms such as the Laplace and Fourier, the latter in both continuous and discrete versions, will be assumed. We will not be getting into details of the transducers and instrumentation techniques essential for biomedical signal acquisition [10, 11, 12, 13]; instead, we will be studying the problems present in the signals after they have been acquired, concentrating on how to solve the problems. Concurrent or prior study of the physiological phenomena associated with the signals of specific interest, with a clinical textbook, is strongly recommended.

Intended Readership

The book is directed at engineering students in their final year of undergraduate studies or in their graduate studies. Electrical Engineering students with a rich background in signals and systems [1, 2, 3] will be well prepared for the material in the book. Students in other engineering disciplines, or in computer science, physics, mathematics, or geophysics should also be able to appreciate the material in the book. A course on digital signal processing or digital filters [14] would form a useful link, but a capable student without this topic may not face much difficulty.

Practicing engineers, computer scientists, information technologists, medical physicists, and data-processing specialists working in diverse areas such as telecommunications, seismic and geophysical applications, biomedical applications, and hospital information systems may find the book useful in their quest to learn advanced techniques for signal analysis. They could draw inspiration from other applications of signal processing or analysis, and satisfy their curiosity regarding computer applications in medicine and computer-aided medical diagnosis.

Teaching and Learning Plan

The book starts with an illustrated introduction to biomedical signals in Chapter 1. Chapter 2 continues the introduction, but with emphasis on the analysis of multiple channels of related signals. This part of the book may be skipped in the teaching plan for a course if the students have had a previous course on biomedical signals and instrumentation. In such a case, the chapters should be studied as review material in order to get oriented toward the examples to follow in the book.

Chapter 3 deals exclusively with filtering for removal of artifacts as an important precursive step before signal analysis. Basic properties of systems and transforms as well as signal processing techniques are reviewed and described as and when required. The chapter is written so as to facilitate easy comprehension by those who have had a basic course on signals, systems, and transforms [1, 2, 3]. The emphasis is on the application to particular problems in biomedical signal analysis, and not on the techniques themselves. A large number of illustrations are included to provide a visual impression of the problem and the effectiveness of the various filtering methods described.

Chapter 4 presents techniques particularly useful in the detection of events in biomedical signals. Analysis of waveshape and waveform complexity of events and components of signals is the focus of Chapter 5. Techniques for frequency-domain characterization of biomedical signals and systems are presented in Chapter 6. A number of diverse examples are provided in these chapters. Attention is directed to the characteristics of the problems one faces in analyzing and interpreting biomedical signals, rather than to any specific diagnostic application with particular signals.

The material in the book up to and including Chapter 6 will provide more than adequate material for a one-semester (13-week) course at the senior (fourth-year) engineering level. My own teaching experience indicates that this material will require about 36 hours of lectures, augmented with about 12 hours of tutorials (problem-solving sessions) and 10 laboratory sessions.

Modeling biomedical signal-generating processes and systems for parametric representation and analysis is the subject of Chapter 7. Chapter 8 deals with the analysis of nonstationary signals. The topics in these chapters are of higher mathematical complexity than suitable for undergraduate courses. Some sections may be selected and included in a first course on biomedical signal analysis if there is particular interest in these topics. Otherwise, the two chapters could be left for self-study by those in need of the techniques, or included in an advanced course.

Chapter 9 presents the final aspect of biomedical signal analysis, and provides an introduction to pattern classification and diagnostic decision. Although this topic is advanced in nature and could form a graduate-level course on its own, the material is introduced so as to draw the entire exercise of biomedical signal analysis to its concluding stage of diagnostic decision. It is recommended that a few sections from this chapter be included even in a first course on biomedical signal analysis so as to give the students a flavor of the end result.

The topic of data compression has deliberately been left out of the book. Advanced topics such as nonlinear dynamics, time-frequency distributions, wavelet-based analysis, chaos, and fractals are not covered in the book. Adaptive filters and nonstationary signal analysis techniques are introduced in the book, but deserve more attention, depth, and breadth. These topics will form the subjects of a follow-up book that I intend to write.

Each chapter includes a number of study questions and problems to facilitate preparation for tests and examinations. A number of laboratory exercises are also provided at the end of each chapter, which could be used to formulate hands-on exercises with real-life signals. Data files related to the problems and exercises at the end of each chapter are available at the site

ftp://ftp.ieee.org/uploads/press/rangayyan/

MATLAB programs to read the data are also provided where required.

It is strongly recommended that the first one or two laboratory sessions in the course be visits to a local hospital, health sciences center, or clinical laboratory to view biomedical signal acquisition and analysis in a practical (clinical) setting. Signals acquired from fellow students and professors could form interesting and motivating material for laboratory exercises, and should be used to supplement the data files provided. A few workshops by physiologists, neuroscientists, and cardiologists should also be included in the course so as to provide the students with a non-engineering perspective on the subject.

Practical experience with real-life signals is a key element in understanding and appreciating biomedical signal analysis. This aspect could be difficult and frustrating at times, but provides professional satisfaction and educational fun!

RANGARAJ MANDAYAM RANGAYYAN

Calgary, Alberta, Canada
September, 2001

About the Author

Rangaraj (Raj) Mandayam Rangayyan was born in Mysore, Karnataka, India, on 21 July 1955. He received the Bachelor of Engineering degree in Electronics and Communication in 1976 from the University of Mysore at the People's Education Society College of Engineering, Mandya, Karnataka, India, and the Ph.D. degree in Electrical Engineering from the Indian Institute of Science, Bangalore, Karnataka, India, in 1980. He was with the University of Manitoba, Winnipeg, Manitoba, Canada, from 1981 to 1984.

He is, at present, a Professor with the Department of Electrical and Computer Engineering (and an Adjunct Professor of Surgery and Radiology) at the University of Calgary, Calgary, Alberta, Canada. His research interests are in the areas of digital signal and image processing, biomedical signal analysis, medical imaging and image analysis, and computer vision. His current research projects are on mammographic image enhancement and analysis for computer-aided diagnosis of breast cancer; region-based image processing; knee-joint vibration signal analysis for noninvasive diagnosis of articular cartilage pathology; and analysis of textured images by cepstral filtering and sonification. He has lectured extensively in many countries, including India, Canada, United States, Brazil, Argentina, Uruguay, Chile, United Kingdom, The Netherlands, France, Spain, Italy, Finland, Russia, Romania, Egypt, Malaysia, Thailand, China, and Japan. He has collaborated with many research groups in Brazil, Spain, France, and Romania.

He was an Associate Editor of the *IEEE Transactions on Biomedical Engineering* from 1989 to 1996; the Program Chair and Editor of the Proceedings of the IEEE Western Canada Exhibition and Conference on "Telecommunication for Health Care:

Telemetry, Teleradiology, and Telemedicine", July 1990, Calgary, Alberta, Canada; the Canadian Regional Representative to the Administrative Committee of the IEEE Engineering in Medicine and Biology Society (EMBS), 1990–1993; a Member of the Scientific Program Committee and Editorial Board, International Symposium on Computerized Tomography, Novosibirsk, Siberia, Russia, August 1993; the Program Chair and Co-editor of the *Proceedings of the 15th Annual International Conference of the IEEE EMBS*, October 1993, San Diego, CA; and Program Co-chair, 20th Annual International Conference of the IEEE EMBS, Hong Kong, October 1998.

He is the winner of the 1997 and 2001 Research Excellence Awards of the Department of Electrical and Computer Engineering, and the 1997 Research Award of the Faculty of Engineering, University of Calgary. He was awarded the Killam Resident Fellowship and a Sabbatical Fellowship by the University of Calgary in support of writing this book. He was recognized by the IEEE with the award of the Third Millennium Medal in 2000, and was elected as a Fellow of the IEEE in 2001.

Photo by Trudy Lee.

Acknowledgments

To write a book on my favorite subject of biomedical signal analysis has been a long-cherished ambition of mine. Writing this book has been a major task with many facets: challenging, yet yielding more knowledge; tiring, yet stimulating the thirst to understand and appreciate more; difficult, yet satisfying when a part was brought to a certain stage of completion.

A number of very important personalities have shaped me and my educational background. My mother, Srimati Padma Srinivasan Rangayyan, and my father, Sri Srinivasan Mandayam Rangayyan, encouraged me to keep striving to gain higher levels of education and to set and achieve higher goals all the time. I have been very fortunate to have been taught and guided by a number of dedicated teachers, the most important of them being Professor Ivaturi Surya Narayana Murthy, my Ph.D. supervisor, who introduced me to the topic of this book at the Indian Institute of Science, Bangalore, Karnataka, India. It is with great respect and admiration that I dedicate this book as a humble offering to their spirits.

My basic education was imparted by many influential teachers at Saint Joseph's Convent, Saint Joseph's Indian High School, and Saint Joseph's College in Mandya and Bangalore, Karnataka, India. My engineering education was provided by the People's Education Society College of Engineering, Mandya, affiliated with the University of Mysore. I express my gratitude to all of my teachers.

My association with clinical researchers at the University of Calgary and the University of Manitoba has been invaluable in furthering my understanding of the subject matter of this book. I express my deep gratitude to Cyril Basil Frank, Gordon Douglas Bell, Joseph Edward Leo Desautels, Leszek Hahn, and Reinhard Kloiber of the University of Calgary, and Richard Gordon and George Collins of the University of Manitoba, Winnipeg, Manitoba, Canada.

My understanding and appreciation of the subject of biomedical signal analysis has been boosted by the collaborative research and studies performed with my many graduate students, post-doctoral fellows, research associates, and colleagues. I would like to place on record my gratitude to Sridhar Krishnan, Naga Ravindra Mudigonda, Margaret Hilary Alto, Ricardo José Ferrari, Liang Shen, Roseli de Deus Lopes, Antonio César Germano Martins, Marcelo Knörich Zuffo, Begoña Acha Piñero, Carmen Serrano Gotarredona, Sîlvia Delgado Olabarriaga, Christian Roux, Basel Solaiman, Olivier Menut, Denise Guliato, Mihai Ciuc, Vasile Buzuloiu, Titus Zaharia, Constantin Vertan, Sarah Rose, Salahuddin Elkadiki, Kevin Eng, Nema Mohamed El-Faramawy, Arup Das, Farshad Faghih, William Alexander Rolston, Yiping Shen, Zahra Marjan Kazem Moussavi, Joseph Provine, Hieu Ngoc Nguyen, Djamel Boulfelfel, Tamer Farouk Rabie, Katherine Olivia Ladly, Yuanting Zhang, Zhi-Qiang Liu, Raman Bhalachandra Paranjape, Joseph André Rodrigue Blais, Robert Charles Bray, Gopinath Ramaswamaiah Kuduvalli, Sanjeev Tavathia, William Mark Morrow, Timothy Chi Hung Hon, Subhasis Chaudhuri, Paul Soble, Kirby Jaman, Atam Prakash Dhawan, and Richard Joseph Lehner. In particular, I thank Sridhar and Naga for assisting me in preparing illustrations and examples; Sridhar for permitting me to use sections of his M.Sc. and Ph.D. theses; and Sridhar, Naga, Hilary, and Ricardo for careful proofreading of the drafts of the book. Sections of the book were reviewed by Robert Clark, Martin Paul Mintchev, Sanjay Srinivasan, and Abu Bakarr Sesay, University of Calgary; and Ioan Tăbuş, Tampere Technical University, Tampere, Finland; I express my gratitude to them for their comments and advice.

The book has benefited significantly from illustrations and text provided by a number of researchers worldwide, as identified in the references and permissions cited. I thank them all for enriching the book with their gifts of knowledge and kindness. I thank Bert Unterberger for drafting some of the illustrations in the book.

The research projects that have provided me with the background and experience essential in order to write the material in this book have been supported by many agencies. I thank the Natural Sciences and Engineering Research Council of Canada, the Alberta Heritage Foundation for Medical Research, the Alberta Breast Cancer Foundation, the Arthritis Society of Canada, the Nickle Family Foundation of Calgary, Control Data Corporation, the University of Calgary, the University of Manitoba, and the Indian Institute of Science for supporting my research projects.

I thank the Killam Foundation for awarding me a Resident Fellowship to facilitate work on this book. I gratefully acknowledge support from the Alberta Provincial Biomedical Engineering Graduate Programme, funded by a grant from the Whitaker Foundation, toward student assistantship for preparation of exercises and illustrations for this book and the related course ENEL 563 Biomedical Signal Analysis at the University of Calgary. I am pleased to place on record my gratitude for the generous support from the Department of Electrical and Computer Engineering and the Faculty of Engineering at the University of Calgary in terms of supplies, services, and relief from other duties.

My association with the IEEE Engineering in Medicine and Biology Society (EMBS) in many positions has benefited me considerably in numerous ways. In particular, the period as an Associate Editor of the *IEEE Transactions on Biomedical Engineering* was very rewarding, as it provided me with a wonderful opportunity to work with many leading researchers and authors of scientific articles. I thank IEEE EMBS for lending professional support to my career on many fronts. I am grateful to the IEEE Press, in particular, Metin Akay, Series Editor, IEEE Press Series in Biomedical Engineering, for inviting me to write this book.

Writing this book has been a monumental task, often draining me of all of my energy. The infinite source of inspiration and recharging of my energy has been my family — my wife Mayura, my daughter Vidya, and my son Adarsh. While supporting me with their love and affection, they have had to bear the loss of my time and effort at home. I express my sincere gratitude to my family for their love and support, and record their contribution toward the preparation of this book.

It is my humble hope that this book will assist those who seek to enrich their lives and those of others with the wonderful powers of biomedical signal analysis. Electrical and Computer Engineering is indeed a great field in the service of humanity!

RANGARAJ MANDAYAM RANGAYYAN

Calgary, Alberta, Canada
September, 2001

Contents

Symbols and Abbreviations

Note: Bold-face letters represent the vector or matrix form of the variable in the corresponding italicized letters. Variables or symbols used within limited contexts are not listed: they are described within their contexts. The mathematical symbols listed may stand for other entities or variables in different applications; only the common associations are listed for ready reference.

a_k	autoregressive model or filter coefficients
au	arbitrary units
aV{F, L, R}	augmented ECG leads
A_z	area under the ROC curve
ACF	autocorrelation function
ADC	analog-to-digital converter
AI	aortic insufficiency
AM	amplitude modulation
ANC	adaptive noise cancellation
ANN	artificial neural network
AO	aorta, aortic (valve or pressure)
AP	action potential
AR	interval between atrial activity and the corresponding QRS
AR	autoregressive (model or filter)
ARMA	autoregressive, moving-average (model or filter)
AS	aortic stenosis
ASD	atrial septal defect
AV	atrio-ventricular

A2	aortic component of the second heart sound
b_l	moving-average model or filter coefficients
bpm	beats per minute
C	covariance matrix
C_i	the ith class in a pattern classification problem
C_{xy}	covariance between x and y
CCF	cross-correlation function
CD	compact disk
CNS	central nervous system
CP	carotid pulse
CSD	cross-spectral density, cross-spectrum
CV	coefficient of variation
D	dicrotic notch in the carotid pulse
DAC	digital-to-analog converter
DC	direct current; zero frequency
DFT	discrete Fourier transform
DM	diastolic murmur
DW	dicrotic wave in the carotid pulse
$e(n), E(\omega)$	model or estimation error
ECG	electrocardiogram, electrocardiography
ECoG	electrocorticogram
EEG	electroencephalogram
EGG	electrogastrogram
EM	electromagnetic
EMG	electromyogram
ENG	electroneurogram
ERP	event-related potential
E_x	total energy of the signal x
$E[\ ]$	statistical expectation operator
f	frequency variable, usually in Hertz
f_c	cutoff frequency (usually at $-3\ dB$) of a filter in Hertz
f_s	sampling frequency in Hertz
FF	form factor
FFT	fast Fourier transform
FIR	finite impulse response (filter)
FM	frequency modulation
FN	false negative
FNF	false negative fraction
FP	false positive
FPF	false positive fraction
FT	Fourier transform
GLR	generalized likelihood ratio
$h(t), h(n)$	impulse response of a filter
H	Hermitian (complex-conjugate) matrix transposition
Hg	mercury

$H(s), H(z)$	transfer function of a filter
$H(s)$	Laplace transform of $h(t)$
$H(z)$	z-transform of $h(n)$
$H(\omega)$	frequency response of a filter
$H(\omega)$	Fourier transform of $h(t)$
HR	heart rate
HRV	heart-rate variability
HSS	hypertrophic subaortic stenosis
Hz	Hertz
i	index of a series or discrete-time signal
IFT	inverse Fourier transform
IIR	infinite impulse response (filter)
IPI	inter-pulse interval
j	index of a series or discrete-time signal
j	$\sqrt{-1}$
ln	natural logarithm (base e)
L_{ij}	loss function in pattern classification
LA	left atrium
LMS	least mean squares
LP	linear prediction (model)
LV	left ventricle
m	mean
$\mathbf{m}$	mean vector of a pattern class
mA	milliamperes
mm	millimeter
ms	millisecond
mV	millivolt
M	number of samples
MA	moving average (filter)
MCI	muscle-contraction interference
MI	mitral insufficiency
MMSE	minimum mean-squared error
MPC	minimum-phase correspondent
MR	mitral regurgitation
MS	mitral stenosis
MS	mean-squared
MSE	mean-squared error
MU	motor unit
MUAP	motor unit action potential
MVC	maximal voluntary contraction
nA	nanoamperes
N	number of samples
N	filter order
NPV	negative predictive value
p_k	pole of a model

$p(x)$	probability density function of the random variable x
$p(x\|C_i)$	likelihood function of class C_i or state-conditional PDF of x
ppm	pulses per minute
pps	pulses per second
OAE	oto-acoustic emission
P	atrial contraction wave in the ECG
P	percussion wave in the carotid pulse
P	model order or number of poles
$P(x)$	probability of the event x
$P(C_i\|x)$	posterior probability that the observation x is from class C_i
PCG	phonocardiogram
PDA	patent ductus arteriosus
PDF	probability density function
PFP	patello-femoral pulse trains or signals
PI	pulmonary insufficiency
PLP	posterior leaflet prolapse
PPC	physiological patello-femoral crepitus
PPV	positive predictive value
PQ	isoelectric segment in the ECG before ventricular contraction
PS	pulmonary stenosis
PSD	power spectral density, power spectrum
P2	pulmonary component of the second heart sound
Q	model order or number of zeros
QRS	ventricular contraction wave in the ECG
$r, \mathbf{r}$	reference input to an adaptive filter
$r_j(\mathbf{x})$	average risk or loss in pattern classification
RA	right atrium
REM	rapid eye movement
RF	radio-frequency
RLS	recursive least-squares
RLSL	recursive least-squares lattice
RMS	root mean squared
ROC	receiver operating characteristics
RR	interval between two successive QRS waves in an ECG
RV	right ventricle
s	second
s	Laplace-domain variable
$S(\omega), S(k)$	auto- or cross-spectral density; power spectral density
SA	sino-atrial
SD	standard deviation
SEM	spectral error measure
SEP	somatosensory evoked potential
SL	signal length
SM	systolic murmur
SMUAP	single motor-unit action potential

SNR	signal-to-noise ratio
ST	isoelectric segment in the ECG during ventricular contraction
STFT	short-time Fourier transform
S1	first heart sound
S2	second heart sound
S3	third heart sound
S4	fourth heart sound
S^+	sensitivity of a test
S^-	specificity of a test
t	time variable
T	ventricular relaxation wave in the ECG
T	tidal wave in the carotid pulse
T	sampling interval
T	as a superscript: vector or matrix transposition
T^+	positive test result
T^-	negative test result
TF	time-frequency
TFD	time-frequency distribution
Th	threshold
TI	tricuspid insufficiency
TN	true negative
TNF	true negative fraction
TP	true positive
TPF	true positive fraction
TS	tricuspid stenosis
TSE	total squared error
TV	television
V	Volt
V1 – V6	chest leads for ECG
VAG	vibroarthrogram
VCG	vectorcardiography
VMG	vibromyogram
VSD	ventricular septal defect
w	filter tap weight; weighting function
$\mathbf{w}$	filter weight vector
$x(t), x(n)$	a signal in the time domain; usually denotes input
$\mathbf{x}$	vector representation of the signal $x(n)$
$\mathbf{x}$	a feature vector in pattern classification
$X(f), X(\omega)$	Fourier transform of $x(t)$
$X(k)$	Discrete Fourier transform of $x(n)$
$X(z)$	z-transform of $x(n)$
$X(\tau, \omega)$	short-time Fourier transform or time-frequency distribution of $x(t)$
$y(t), y(n)$	a signal in the time domain; usually denotes output
$\mathbf{y}$	vector representation of the signal $y(n)$
$Y(f), Y(\omega)$	Fourier transform of $y(t)$

$Y(k)$	Discrete Fourier transform of $y(n)$
$Y(z)$	z-transform of $y(n)$
z	the z-transform variable
z^{-1}	unit delay operator in discrete-time systems
z_l	zeros of a system
$\mathbf{z}$	a prototype feature vector in pattern classification
ZCR	zero-crossing rate
ZT	the z-transform
1D	one-dimensional
2D	two-dimensional
3D	three-dimensional
I, II, III	limb leads for ECG
α	an EEG wave
β	an EEG wave
γ	an EEG wave
γ_{xy}	correlation coefficient between x and y
γ_i	reflection coefficient
Γ_{xy}	coherence between x and y
δ	an EEG wave
δ	Dirac delta (impulse) function
ε	total squared error
η	a random variable or noise process
θ	an angle
θ	a threshold
θ	an EEG wave
θ, Θ	cross-correlation function
λ	forgetting factor in the RLS filter
μ	the mean (average) of a random variable
μ	a rhythmic wave in the EEG
μ	step size in the LMS filter
μV	microvolt
μm	micrometer
μs	microsecond
ρ	correlation coefficient
σ	the real part of the Laplace variable s (Neper frequency)
σ	the standard deviation of a random variable
σ^2	the variance of a random variable
τ	a time interval, delay, or shift
ϕ, Φ	autocorrelation
ω	frequency variable in radians per second
Ω	frequency variable in radians per second
$*$	when in-line: convolution
$*$	as a superscript: complex conjugation
$^{-}$	average or normalized version of the variable
$^{\wedge}$	complex cepstrum of the signal, if a function of time

ˆ	complex logarithm of the signal, if a function of frequency
˜	estimate of the variable under the symbol
′, ″	first and second derivatives of the preceding function
∀	for all
∈	belongs to or is in (the set)
| |	absolute value or magnitude of
∠	argument of, angle of

1

Introduction to Biomedical Signals

1.1 THE NATURE OF BIOMEDICAL SIGNALS

Living organisms are made up of many component *systems* — the human body, for example, includes the nervous system, the cardiovascular system, and the musculo-skeletal system, among others. Each system is made up of several subsystems that carry on many *physiological processes*. For example, the cardiac system performs the important task of rhythmic pumping of blood throughout the body to facilitate the delivery of nutrients, as well as pumping blood through the pulmonary system for oxygenation of the blood itself.

Physiological processes are complex phenomena, including nervous or hormonal stimulation and control; inputs and outputs that could be in the form of physical material, neurotransmitters, or information; and action that could be mechanical, electrical, or biochemical. Most physiological processes are accompanied by or manifest themselves as *signals* that reflect their nature and activities. Such signals could be of many types, including biochemical in the form of hormones and neuro-transmitters, electrical in the form of potential or current, and physical in the form of pressure or temperature.

Diseases or defects in a biological system cause alterations in its normal phys-iological processes, leading to *pathological processes* that affect the performance, health, and general well-being of the system. A pathological process is typically associated with signals that are different in some respects from the corresponding normal signals. If we possess a good understanding of a system of interest, it becomes possible to observe the corresponding signals and assess the state of the system. The task is not very difficult when the signal is simple and appears at the outer surface of the body. For example, most infections cause a rise in the temperature of the body, which may be sensed very easily, albeit in a relative and *qualitative* manner, via the palm of one's hand. Objective or *quantitative* measurement of temperature requires an instrument, such as a simple thermometer.

A single measurement x of temperature is a *scalar*, and represents the thermal state of the body at a *particular or single instant of time* t (and a particular position). If we record the temperature continuously in some form, say a strip-chart record, we obtain a *signal as a function of time*; such a signal may be expressed in *continuous-time* or *analog* form as $x(t)$. When the temperature is measured at *discrete* points of time, it may be expressed in *discrete-time* form as $x(nT)$ or $x(n)$, where n is the index or measurement sample number of the array of values, and T represents the uniform interval between the time instants of measurement. A discrete-time signal that can take amplitude values only from a limited list of *quantized* levels is called a *digital* signal; the distinction between discrete-time and digital signals is often ignored.

In intensive-care monitoring, the tympanic (ear drum) temperature may sometimes be measured using an infra-red sensor. Occasionally, when catheters are being used for other purposes, a temperature sensor may also be introduced into an artery or the heart to measure the *core* temperature of the body. It then becomes possible to obtain a continuous measurement of temperature, although only a few samples taken at intervals of a few minutes may be stored for subsequent analysis. Figure 1.1 illustrates representations of temperature measurements as a scalar, an array, and a signal that is a function of time. It is obvious that the graphical representation facilitates easier and faster comprehension of trends in the temperature than the numerical format. Long-term recordings of temperature can facilitate the analysis of temperature-regulation mechanisms [15, 16].

Let us now consider another basic measurement in health care and monitoring: that of blood pressure (BP). Each measurement consists of two values — the systolic pressure and the diastolic pressure. BP is measured in millimeters of mercury ($mm\ of\ Hg$) in clinical practice, although the international standard unit for pressure is the *Pascal*. A single BP measurement could thus be viewed as a *vector* $\mathbf{x} = [x_1, x_2]^T$ with two components: x_1 indicating the systolic pressure and x_2 indicating the diastolic pressure. When BP is measured at a few instants of time, we obtain an array of vectorial values $\mathbf{x}(n)$. In intensive-care monitoring and surgical procedures, a pressure transducer may sometimes be inserted into an artery (along with other intra-arterial or intra-venous devices). It then becomes possible to obtain the arterial systolic and diastolic BP on a continuous-time recording, although the values may be transferred to a computer and stored only at sampled instants of time that are several minutes apart. The signal may then be expressed as a function of time $\mathbf{x}(t)$. Figure 1.2 shows BP measurements as a single two-component vector, as an array, and as a function of time. It is clear that the plot as a function of time facilitates rapid observation of trends in the pressure.

33.5 °C

(a)

Time	08:00	10:00	12:00	14:00	16:00	18:00	20:00	22:00	24:00
°C	33.5	33.3	34.5	36.2	37.3	37.5	38.0	37.8	38.0

(b)

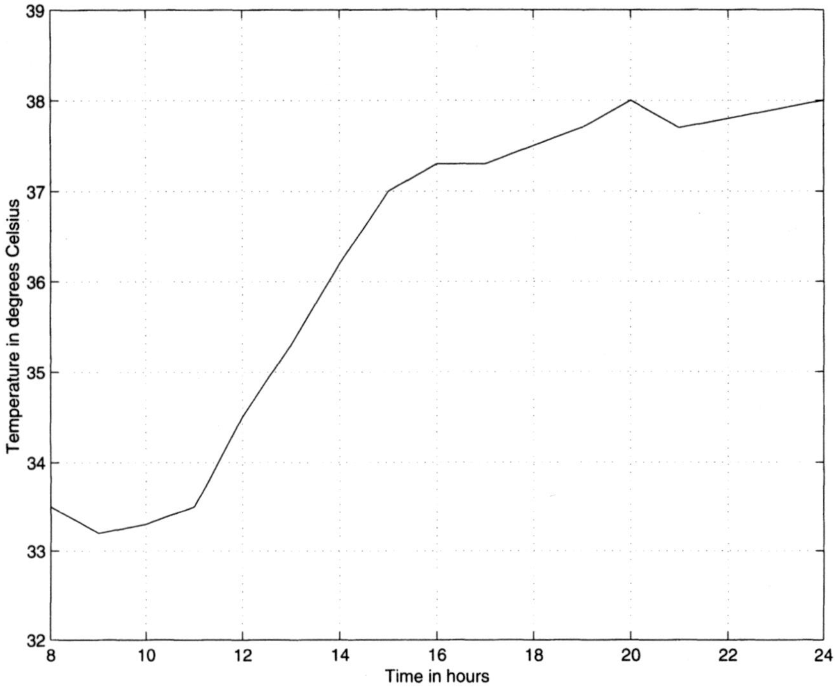

(c)

Figure 1.1 Measurements of the temperature of a patient presented as (a) a scalar with one temperature measurement x at a time instant t; (b) an array $x(n)$ made up of several measurements at different instants of time; and (c) a signal $x(t)$ or $x(n)$. The horizontal axis of the plot represents time in *hours*; the vertical axis gives temperature in *degrees Celsius*. Data courtesy of Foothills Hospital, Calgary.

$$\begin{bmatrix} 122 \\ 66 \end{bmatrix}$$

(a)

Time	08:00	10:00	12:00	14:00	16:00	18:00	20:00	22:00	24:00
Systolic	122	102	108	94	104	118	86	95	88
Diastolic	66	59	60	50	55	62	41	52	48

(b)

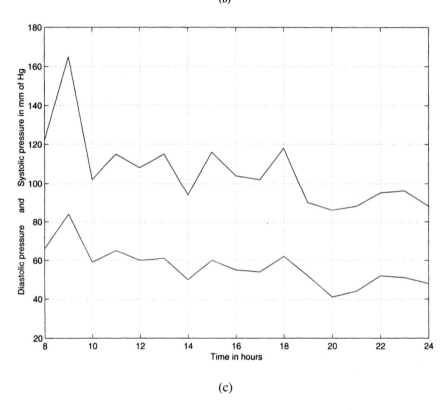

(c)

Figure 1.2 Measurements of the blood pressure of a patient presented as (a) a single pair or vector of systolic and diastolic measurements **x** in *mm of Hg* at a time instant *t*; (b) an array **x**(*n*) made up of several measurements at different instants of time; and (c) a signal **x**(*t*) or **x**(*n*). Note the use of boldface **x** to indicate that each measurement is a vector with two components. The horizontal axis of the plot represents time in *hours*; the vertical axis gives the systolic pressure (upper trace) and the diastolic pressure (lower trace) in *mm of Hg*. Data courtesy of Foothills Hospital, Calgary.

1.2 EXAMPLES OF BIOMEDICAL SIGNALS

The preceding example of body temperature as a signal is a rather simple example of a *biomedical signal*. Regardless of its simplicity, we can appreciate its importance and value in the assessment of the well-being of a child with a fever or that of a critically ill patient in a hospital. The origins and nature of a few other biomedical signals of various types are described in the following subsections, with brief indications of their usefulness in diagnosis. Further detailed discussions on some of the signals will be provided in the context of their analysis for various purposes in the chapters that follow.

1.2.1 The action potential

The action potential (AP) is the electrical signal that accompanies the mechanical contraction of a single cell when stimulated by an electrical current (neural or external) [10, 17, 18, 19, 20, 21]. It is caused by the flow of sodium (Na^+), potassium (K^+), chloride (Cl^-), and other ions across the cell membrane. The action potential is the basic component of all bioelectrical signals. It provides information on the nature of physiological activity at the single-cell level. Recording an action potential requires the isolation of a single cell, and microelectrodes with tips of the order of a few micrometers to stimulate the cell and record the response [10].

Resting potential: Nerve and muscle cells are encased in a semi-permeable membrane that permits selected substances to pass through while others are kept out. Body fluids surrounding cells are conductive solutions containing charged atoms known as ions. In their resting state, membranes of excitable cells readily permit the entry of K^+ and Cl^- ions, but effectively block the entry of Na^+ ions (the permeability for K^+ is 50–100 times that for Na^+). Various ions seek to establish a balance between the inside and the outside of a cell according to charge and concentration. The inability of Na^+ to penetrate a cell membrane results in the following [17]:

- Na^+ concentration inside the cell is far less than that outside.

- The outside of the cell is more positive than the inside of the cell.

- To balance the charge, additional K^+ ions enter the cell, causing higher K^+ concentration inside the cell than outside.

- Charge balance cannot be reached due to differences in membrane permeability for the various ions.

- A state of equilibrium is established with a potential difference, with the inside of the cell being negative with respect to the outside.

A cell in its resting state is said to be *polarized*. Most cells maintain a *resting potential* of the order of -60 to -100 mV until some disturbance or stimulus upsets the equilibrium.

Depolarization: When a cell is excited by ionic currents or an external stimulus, the membrane changes its characteristics and begins to allow Na^+ ions to enter the cell. This movement of Na^+ ions constitutes an ionic current, which further reduces the membrane barrier to Na^+ ions. This leads to an avalanche effect: Na^+ ions rush into the cell. K^+ ions try to leave the cell as they were in higher concentration inside the cell in the preceding resting state, but cannot move as fast as the Na^+ ions. The net result is that the inside of the cell becomes positive with respect to the outside due to an imbalance of K^+ ions. A new state of equilibrium is reached after the rush of Na^+ ions stops. This change represents the beginning of the *action potential*, with a peak value of about $+20$ mV for most cells. An excited cell displaying an action potential is said to be *depolarized*; the process is called *depolarization*.

Repolarization: After a certain period of being in the depolarized state the cell becomes polarized again and returns to its resting potential via a process known as *repolarization*. Repolarization occurs by processes that are analogous to those of depolarization, except that instead of Na^+ ions, the principal ions involved in repolarization are K^+ ions [19]. Membrane depolarization, while increasing the permeability for Na^+ ions, also increases the permeability of the membrane for K^+ ions via a specific class of ion channels known as voltage-dependent K^+ channels. Although this may appear to be paradoxical at first glance, the key to the mechanism for repolarization lies in the time-dependence and voltage-dependence of the membrane permeability changes for K^+ ions compared with that for Na^+ ions. The permeability changes for K^+ during depolarization occur considerably more slowly than those for Na^+ ions, hence the initial depolarization is caused by an inrush of Na^+ ions. However, the membrane permeability changes for Na^+ spontaneously decrease near the peak of the depolarization, whereas those for K^+ ions are beginning to increase. Hence, during repolarization, the predominant membrane permeability is for K^+ ions. Because K^+ concentration is much higher inside the cell than outside, there is a net efflux of K^+ from the cell, which makes the inside more negative, thereby effecting repolarization back to the resting potential.

It should be noted that the voltage-dependent K^+ permeability change is due to a distinctly different class of ion channels than those that are responsible for setting the resting potential. A mechanism known as the $Na^+ - K^+$ pump extrudes Na^+ ions in exchange for transporting K^+ ions back into the cell. However, this transport mechanism carries very little current in comparison with ion channels, and therefore makes a minor contribution to the repolarization process. The $Na^+ - K^+$ pump is essential for resetting the $Na^+ - K^+$ balance of the cell, but the process occurs on a longer time scale than the duration of an action potential.

Nerve and muscle cells repolarize rapidly, with an action potential duration of about 1 ms. Heart muscle cells repolarize slowly, with an action potential duration of $150 - 300$ ms.

The action potential is always the same for a given cell, regardless of the method of excitation or the intensity of the stimulus beyond a threshold: this is known as the *all-or-none* or all-or-nothing phenomenon. After an action potential, there is a period during which a cell cannot respond to any new stimulus, known as the *absolute refractory period* (about 1 ms in nerve cells). This is followed by a *relative refractory period* (several *ms* in nerve cells), when another action potential may be triggered by a much stronger stimulus than in the normal situation.

Figure 1.3 shows action potentials recorded from individual rabbit ventricular and atrial myocytes (muscle cells) [19]. Figure 1.4 shows a ventricular myocyte in its relaxed and fully contracted states. The tissues were first incubated in digestive enzymes, principally collagenase, and then dispersed into single cells using gentle mechanical agitation. The recording electrodes were glass patch pipettes; a whole-cell, current-clamp recording configuration was used to obtain the action potentials. The cells were stimulated at low rates (once per 8 *s*); this is far less than physiological rates. Moreover, the cells were maintained at 20° *C*, rather than body temperature. Nevertheless, the major features of the action potentials shown are similar to those recorded under physiological conditions.

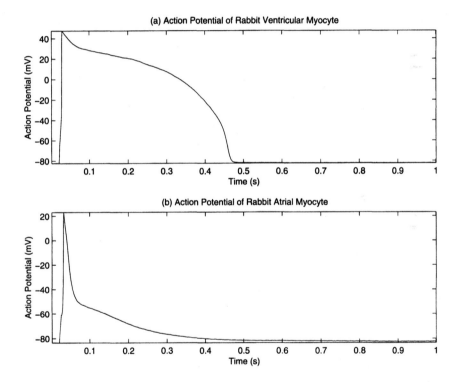

Figure 1.3   Action potentials of rabbit ventricular and atrial myocytes. Data courtesy of R. Clark, Department of Physiology and Biophysics, University of Calgary.

The resting membrane potential of the cells (from 0 to 20 *ms* in the plots in Figure 1.3) is about −83 *mV*. A square pulse of current, 3 *ms* in duration and 1 *nA* in amplitude, was passed through the recording electrode and across the cell membrane, causing the cell to depolarize rapidly. The ventricular myocyte exhibits a depolarized potential of about +40 *mV*; it then slowly declines back to the resting potential level over an interval of about 500 *ms*. The initial, rapid depolarization of

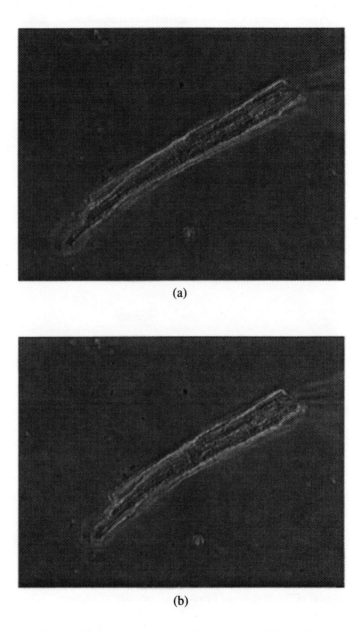

(a)

(b)

Figure 1.4   A single ventricular myocyte (of a rabbit) in its (a) relaxed and (b) fully contracted states. The length of the myocyte is approximately 25 μm. The tip of the glass pipette, faintly visible at the upper-right end of the myocyte, is approximately 2 μm wide. Images courtesy of R. Clark, Department of Physiology and Biophysics, University of Calgary.

the atrial cell is similar to that of the ventricular cell, but does not overshoot zero membrane potential as much as the ventricular action potential; repolarization occurs much more quickly than is the case for the ventricular cell.

Propagation of an action potential: An action potential propagates along a muscle fiber or an unmyelinated nerve fiber as follows [22]: Once initiated by a stimulus, the action potential propagates along the whole length of a fiber without decrease in amplitude by progressive depolarization of the membrane. Current flows from a depolarized region through the intra-cellular fluid to adjacent inactive regions, thereby depolarizing them. Current also flows through the extra-cellular fluids, through the depolarized membrane, and back into the intra-cellular space, completing the local circuit. The energy to maintain conduction is supplied by the fiber itself.

Myelinated nerve fibers are covered by an insulating sheath of *myelin*. The sheath is interrupted every few millimeters by spaces known as the *nodes of Ranvier*, where the fiber is exposed to the interstitial fluid. Sites of excitation and changes of membrane permeability exist only at the nodes, and current flows by jumping from one node to the next in a process known as *saltatory conduction*.

1.2.2 The electroneurogram (ENG)

The ENG is an electrical signal observed as a stimulus and the associated nerve action potential propagate over the length of a nerve. It may be used to measure the velocity of propagation (or conduction velocity) of a stimulus or action potential in a nerve [10]. ENGs may be recorded using concentric needle electrodes or silver – silver-chloride electrodes ($Ag - AgCl$) at the surface of the body.

Conduction velocity in a peripheral nerve may be measured by stimulating a motor nerve and measuring the related activity at two points that are a known distance apart along its course. In order to minimize muscle contraction and other undesired effects, the limb is held in a relaxed posture and a strong but short stimulus is applied in the form of a pulse of about $100\ V$ amplitude and $100 - 300\ \mu s$ duration [10]. The difference in the latencies of the ENGs recorded over the associated muscle gives the conduction time. Knowing the separation distance between the stimulus sites, it is possible to determine the conduction velocity in the nerve [10]. ENGs have amplitudes of the order of $10\ \mu V$ and are susceptible to power-line interference and instrumentation noise.

Figure 1.5 illustrates the ENGs recorded in a nerve conduction velocity study. The stimulus was applied to the ulnar nerve. The ENGs were recorded at the wrist (marked "Wrist" in the figure), just below the elbow (BElbow), and just above the elbow (AElbow) using surface electrodes, amplified with a gain of 2, 000, and filtered to the bandwidth $10 - 10, 000\ Hz$. The three traces in the figure indicate increasing latencies with respect to the stimulus time point, which is the left margin of the plots. The responses shown in the figure are normal, indicate a BElbow – Wrist latency of $3.23\ ms$, and lead to a nerve conduction velocity of $64.9\ m/s$.

Typical values of propagation rate or nerve conduction velocity are [22, 10, 23]:

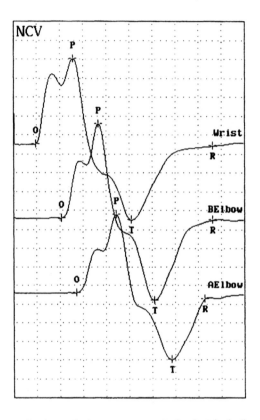

Figure 1.5 Nerve conduction velocity measurement via electrical stimulation of the ulnar nerve. The grid boxes represent 3 *ms* in width and 2 μV in height. AElbow: above the elbow. BElbow: below the elbow. O: onset. P: Peak. T: trough. R: recovery of base-line. Courtesy of M. Wilson and C. Adams, Alberta Children's Hospital, Calgary.

- $45 - 70 \ m/s$ in nerve fibers;

- $0.2 - 0.4 \ m/s$ in heart muscle;

- $0.03 - 0.05 \ m/s$ in time-delay fibers between the atria and ventricles.

Neural diseases may cause a decrease in conduction velocity.

1.2.3 The electromyogram (EMG)

Skeletal muscle fibers are considered to be twitch fibers because they produce a mechanical twitch response for a single stimulus and generate a propagated action potential. Skeletal muscles are made up of collections of *motor units* (MUs), each of which consists of an anterior horn cell (or motoneuron or motor neuron), its axon, and all muscle fibers innervated by that axon. A motor unit is the smallest muscle unit that can be activated by volitional effort. The constituent fibers of a motor unit are activated synchronously. Component fibers of a motor unit extend lengthwise in loose bundles along the muscle. In cross-section, the fibers of a given motor unit are interspersed with the fibers of other motor units [22, 10, 24]. Figure 1.6 (top panel) illustrates a motor unit in schematic form [24].

Large muscles for gross movement have hundreds of fibers per motor unit; muscles for precise movement have fewer fibers per motor unit. The number of muscle fibers per motor nerve fiber is known as the *innervation ratio*. For example, it has been estimated that the platysma muscle (of the neck) has $1,826$ large nerve fibers controlling $27,100$ muscle fibers with $1,096$ motor units and an innervation ratio of 25, whereas the first dorsal interosseus (finger) muscle has 199 large nerve fibers and $40,500$ muscle fibers with 119 motor units and an innervation ratio of 340 [22]. The mechanical output (contraction) of a muscle is the net result of stimulation and contraction of several of its motor units.

When stimulated by a neural signal, each motor unit contracts and causes an electrical signal that is the summation of the action potentials of all of its constituent cells. This is known as the *single-motor-unit action potential* (SMUAP, or simply MUAP), and may be recorded using needle electrodes inserted into the muscle region of interest. Normal SMUAPs are usually biphasic or triphasic, $3 - 15 \ ms$ in duration, $100 - 300 \ \mu V$ in amplitude, and appear with frequency in the range of $6 - 30/s$ [10, 22]. The shape of a recorded SMUAP depends upon the type of the needle electrode used, its positioning with respect to the active motor unit, and the projection of the electrical field of the activity onto the electrodes. Figure 1.7 illustrates simultaneous recordings of the activities of a few motor units from three channels of needle electrodes [25]. Although the SMUAPs are biphasic or triphasic, the same SMUAP displays variable shape from one channel to another. (*Note:* The action potentials in Figure 1.3 are monophasic; the first two SMUAPs in Channel 1 in Figure 1.7 are biphasic, and the third SMUAP in the same signal is triphasic.)

The shape of SMUAPs is affected by disease. Figure 1.8 illustrates SMUAP trains of a normal subject and those of patients with neuropathy and myopathy. Neuropathy causes slow conduction and/or desynchronized activation of fibers, and a polyphasic

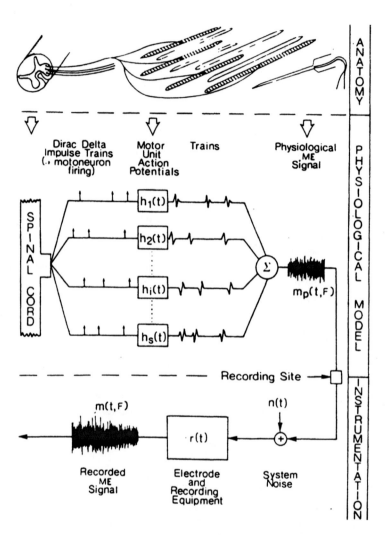

Figure 1.6   Schematic representation of a motor unit and model for the generation of EMG signals. Top panel: A motor unit includes an anterior horn cell or motor neuron (illustrated in a cross-section of the spinal cord), an axon, and several connected muscle fibers. The hatched fibers belong to one motor unit; the non-hatched fibers belong to other motor units. A needle electrode is also illustrated. Middle panel: The firing pattern of each motor neuron is represented by an impulse train. Each system $h_i(t)$ shown represents a motor unit that is activated and generates a train of SMUAPs. The net EMG is the sum of several SMUAP trains. Bottom panel: Effects of instrumentation on the EMG signal acquired. The observed EMG is a function of time t and muscular force produced F. Reproduced with permission from C.J. de Luca, Physiology and mathematics of myoelectric signals, *IEEE Transactions on Biomedical Engineering*, 26:313–325, 1979. ©IEEE.

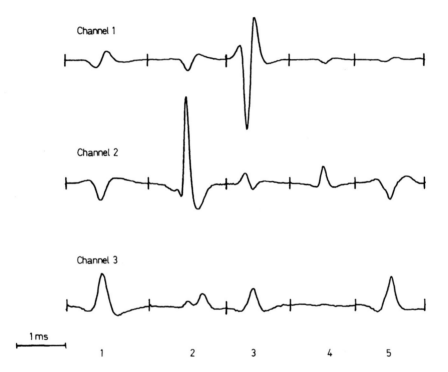

Figure 1.7 SMUAP trains recorded simultaneously from three channels of needle electrodes. Observe the different shapes of the same SMUAPs projected onto the axes of the three channels. Three different motor units are active over the duration of the signals illustrated. Reproduced with permission from B. Mambrito and C.J. de Luca, Acquisition and decomposition of the EMG signal, in *Progress in Clinical Neurophysiology*, Volume 10: Computer-aided Electromyography, Editor: J.E. Desmedt, pp 52–72, 1983. ©S. Karger AG, Basel, Switzerland.

SMUAP with an amplitude larger than normal. The same motor unit may be observed to fire at higher rates than normal before more motor units are recruited. Myopathy involves loss of muscle fibers in motor units, with the neurons presumably intact. *Splintering* of SMUAPs occurs due to asynchrony in activation as a result of patchy destruction of fibers (e.g., in muscular dystrophy), leading to polyphasic SMUAPs. More motor units may be observed to be recruited at low levels of effort.

Gradation of muscular contraction: Muscular contraction levels are controlled in two ways:

- *Spatial recruitment,* by activating new motor units with increasing effort; and

- *Temporal recruitment,* by increasing the frequency of discharge (firing rate) of each motor unit with increasing effort.

Motor units are activated at different times and at different frequencies causing asynchronous contraction. The twitches of individual motor units sum and fuse to form tetanic contraction and increased force. Weak volitional effort causes motor units to fire at about $5 - 15$ *pps* (pulses per second). As greater tension is developed, an *interference pattern* EMG is obtained, with the constituent and active motor units firing in the range of $25 - 50$ *pps*. Grouping of MUAPs has been observed as fatigue develops, leading to decreased high-frequency content and increased amplitude in the EMG [24].

Spatio-temporal summation of the MUAPs of all of the active motor units gives rise to the EMG of the muscle. EMG signals recorded using surface electrodes are complex signals including interference patterns of several MUAP trains and are difficult to analyze. An EMG signal indicates the level of activity of a muscle, and may be used to diagnose neuromuscular diseases such as neuropathy and myopathy.

Figure 1.9 illustrates an EMG signal recorded from the crural diaphragm of a dog using fine-wire electrodes sewn in-line with the muscle fibers and placed 10 *mm* apart [26]. The signal represents one period of breathing (inhalation being the active part as far as the muscle and EMG are concerned). It is seen that the overall level of activity in the signal increases during the initial phase of inhalation. Figure 1.10 shows the early parts of the same signal on an expanded time scale. SMUAPs are seen at the beginning stages of contraction, followed by increasingly complex interference patterns of several MUAPs.

Signal-processing techniques for the analysis of EMG signals will be discussed in Sections 5.2.4, 5.6, 5.9, 5.10, 7.2.1, and 7.3.

1.2.4   The electrocardiogram (ECG)

The ECG is the electrical manifestation of the contractile activity of the heart, and can be recorded fairly easily with surface electrodes on the limbs or chest. The ECG is perhaps the most commonly known, recognized, and used biomedical signal. The rhythm of the heart in terms of beats per minute (*bpm*) may be easily estimated by counting the readily identifiable waves. More important is the fact that the ECG

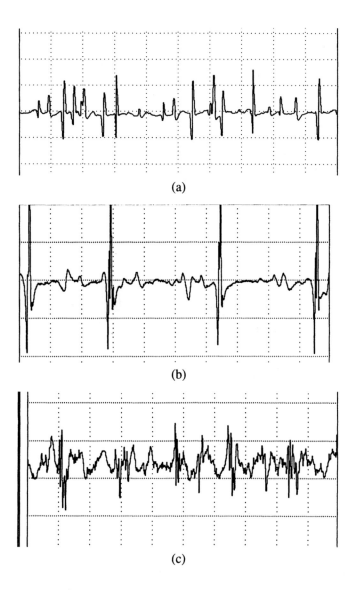

(a)

(b)

(c)

Figure 1.8 Examples of SMUAP trains. (a) From the right deltoid of a normal subject, male, 11 years; the SMUAPs are mostly biphasic, with duration in the range $3 - 5$ ms. (b) From the deltoid of a six-month-old male patient with brachial plexus injury (neuropathy); the SMUAPs are polyphasic and large in amplitude (800 μV), and the same motor unit is firing at a relatively high rate at low-to-medium levels of effort. (c) From the right biceps of a 17-year-old male patient with myopathy; the SMUAPs are polyphasic and indicate early recruitment of more motor units at a low level of effort. The signals were recorded with gauge 20 needle electrodes. The width of each grid box represents a duration of **20** ms; its height represents an amplitude of 200 μV. Courtesy of M. Wilson and C. Adams, Alberta Children's Hospital, Calgary.

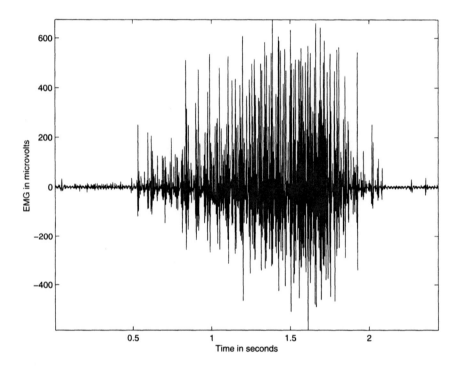

Figure 1.9 EMG signal recorded from the crural diaphragm muscle of a dog using implanted fine-wire electrodes. Data courtesy of R.S. Platt and P.A. Easton, Department of Clinical Neurosciences, University of Calgary.

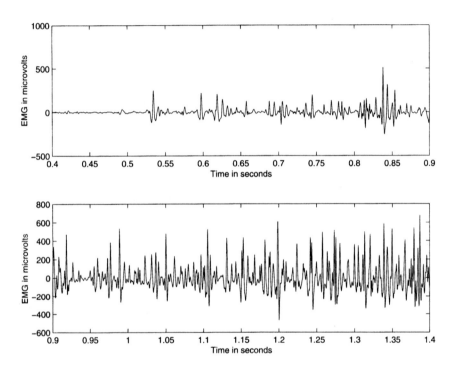

Figure 1.10 The initial part of the EMG signal in Figure 1.9 shown on an expanded time scale. Observe the SMUAPs at the initial stages of contraction, followed by increasingly complex interference patterns of several MUAPs. Data courtesy of R.S. Platt and P.A. Easton, Department of Clinical Neurosciences, University of Calgary.

waveshape is altered by cardiovascular diseases and abnormalities such as myocardial ischemia and infarction, ventricular hypertrophy, and conduction problems.

The heart: The heart is a four-chambered pump with two atria for collection of blood and two ventricles for pumping out of blood. Figure 1.11 shows a schematic representation of the four chambers and the major vessels connecting to the heart. The resting or filling phase of a cardiac chamber is called *diastole*; the contracting or pumping phase is called *systole*.

The right atrium (or auricle, RA) collects impure blood from the superior and inferior vena cavae. During atrial contraction, blood is passed from the right atrium to the right ventricle (RV) through the tricuspid valve. During ventricular systole, the impure blood in the right ventricle is pumped out through the pulmonary valve to the lungs for purification (oxygenation).

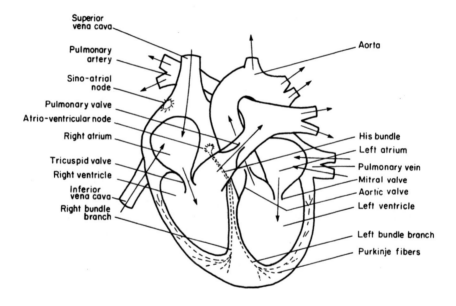

Figure 1.11 Schematic representation of the chambers, valves, vessels, and conduction system of the heart.

The left atrium (LA) receives purified blood from the lungs, which is passed on during atrial contraction to the left ventricle (LV) via the mitral valve. The left ventricle is the largest and most important cardiac chamber. The left ventricle contracts the strongest among the cardiac chambers, as it has to pump out the oxygenated blood through the aortic valve and the aorta against the pressure of the rest of the vascular system of the body. Due to the higher level of importance of contraction of the ventricles, the terms systole and diastole are applied to the ventricles by default.

The heart rate (HR) or cardiac rhythm is controlled by specialized pacemaker cells that form the sino-atrial (SA) node located at the junction of the superior vena cava and the right atrium [23]. The firing rate of the SA node is controlled by impulses from the autonomous and central nervous systems leading to the delivery of the neurotransmitters acetylcholine (for vagal stimulation, causing a reduction in heart rate) or epinephrine (for sympathetic stimulation, causing an increase in the heart rate). The normal (resting) heart rate is about 70 *bpm*. The heart rate is lower during sleep, but abnormally low heart rates below 60 *bpm* during activity could indicate a disorder called *bradycardia*. The instantaneous heart rate could reach values as high as 200 *bpm* during vigorous exercise or athletic activity; a high resting heart rate could be due to illness, disease, or cardiac abnormalities, and is termed *tachycardia*.

The electrical system of the heart: Co-ordinated electrical events and a specialized conduction system intrinsic and unique to the heart play major roles in the rhythmic contractile activity of the heart. The SA node is the basic, natural cardiac pacemaker that triggers its own train of action potentials. The action potential of the SA node propagates through the rest of the heart, causing a particular pattern of excitation and contraction (see Figure 1.12). The sequence of events and waves in a cardiac cycle is as follows [23]:

1. The SA node fires.

2. Electrical activity is propagated through the atrial musculature at comparatively low rates, causing slow-moving depolarization (contraction) of the atria. This results in the P wave in the ECG (see Figure 1.13). Due to the slow contraction of the atria and their small size, the P wave is a slow, low-amplitude wave, with an amplitude of about $0.1 - 0.2\ mV$ and a duration of about $60 - 80\ ms$.

3. The excitation wave faces a propagation delay at the atrio-ventricular (AV) node, which results in a normally iso-electric segment of about $60 - 80\ ms$ after the P wave in the ECG, known as the PQ segment. The pause assists in the completion of the transfer of blood from the atria to the ventricles.

4. The His bundle, the bundle branches, and the Purkinje system of specialized conduction fibers propagate the stimulus to the ventricles at a high rate.

5. The wave of stimulus spreads rapidly from the apex of the heart upwards, causing rapid depolarization (contraction) of the ventricles. This results in the QRS wave of the ECG — a sharp biphasic or triphasic wave of about $1\ mV$ amplitude and $80\ ms$ duration (see Figure 1.13).

6. Ventricular muscle cells possess a relatively long action potential duration of $300 - 350\ ms$ (see Figure 1.3). The plateau portion of the action potential causes a normally iso-electric segment of about $100 - 120\ ms$ after the QRS, known as the ST segment.

7. Repolarization (relaxation) of the ventricles causes the slow T wave, with an amplitude of $0.1 - 0.3\ mV$ and duration of $120 - 160\ ms$ (see Figure 1.13).

Any disturbance in the regular rhythmic activity of the heart is termed *arrhythmia*. Cardiac arrhythmia may be caused by irregular firing patterns from the SA node, or

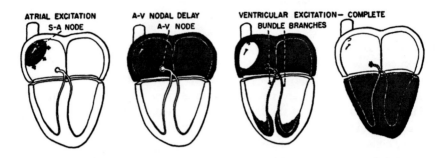

Figure 1.12 Propagation of the excitation pulse through the heart. Reproduced with permission from R.F. Rushmer, *Cardiovascular Dynamics*, 4th edition, ©W.B. Saunders, Philadelphia, PA, 1976.

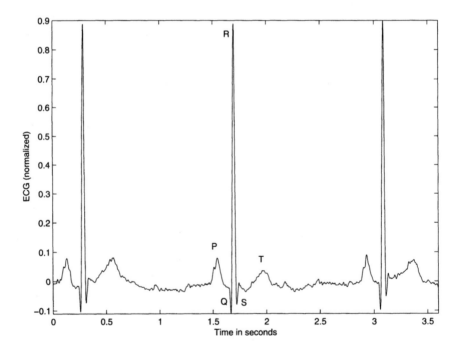

Figure 1.13 A typical ECG signal (male subject of age 24 years). (*Note:* Signal values are not calibrated, that is, specified in physical units, in many applications. As is the case in this plot, signal values in plots in this book are in arbitrary or normalized units unless specified.)

by abnormal and additional pacing activity from other parts of the heart. Many parts of the heart possess inherent rhythmicity and pacemaker properties; for example, the SA node, the AV node, the Purkinje fibers, atrial tissue, and ventricular tissue. If the SA node is depressed or inactive, any one of the above tissues may take over the role of the pacemaker or introduce *ectopic* beats. Different types of abnormal rhythm (arrhythmia) result from variations in the site and frequency of impulse formation. Premature ventricular contractions (PVCs) caused by ectopic foci on the ventricles upset the regular rhythm and may lead to ventricular dissociation and fibrillation — a state of disorganized contraction of the ventricles independent of the atria — resulting in no effective pumping of blood and possibly death. The waveshapes of PVCs are usually very different from that of the normal beats of the same subject due to the different conduction paths of the ectopic impulses and the associated abnormal contraction events. Figure 1.14 shows an ECG signal with a few normal beats and two PVCs. (See Figure 9.5 for an illustration of ventricular bigeminy, where every second pulse from the SA node is replaced by a PVC with a full compensatory pause.)

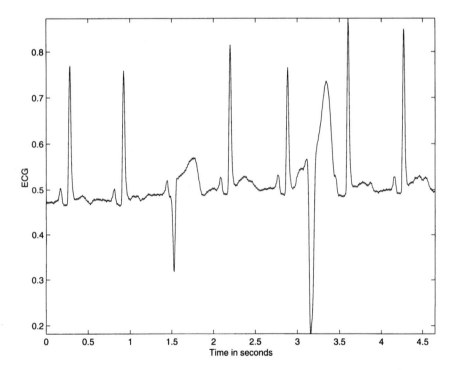

Figure 1.14 ECG signal with PVCs. The third and sixth beats are PVCs. The first PVC has blocked the normal beat that would have appeared at about the same time instant, but the second PVC has not blocked any normal beat triggered by the SA node. Data courtesy of G. Groves and J. Tyberg, Department of Physiology and Biophysics, University of Calgary.

The QRS waveshape is affected by conduction disorders; for example, bundle-branch block causes a widened and possibly jagged QRS. Figure 1.15 shows the ECG signal of a patient with right bundle-branch block. Observe the wider-than-normal QRS complex, which also displays a waveshape that is significantly different from the normal QRS waves. Ventricular hypertrophy (enlargement) could also cause a wider-than-normal QRS.

The ST segment, which is normally iso-electric (flat and in line with the PQ segment) may be elevated or depressed due to myocardial ischemia (reduced blood supply to a part of the heart muscles due to a block in the coronary arteries) or due to myocardial infarction (dead myocardial tissue incapable of contraction due to total lack of blood supply). Many other diseases cause specific changes in the ECG waveshape: the ECG is a very important signal that is useful in heart-rate (rhythm) monitoring and the diagnosis of cardiovascular diseases.

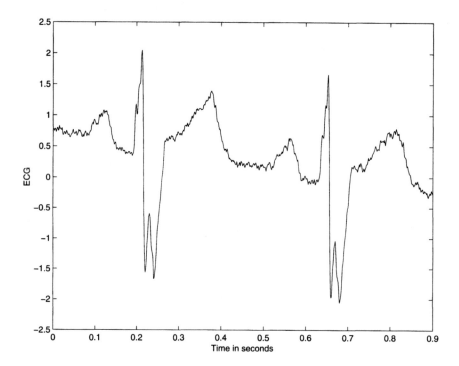

Figure 1.15 ECG signal of a patient with right bundle-branch block and hypertrophy (male patient of age 3 months). The QRS complex is wider than normal, and displays an abnormal, jagged waveform due to desynchronized contraction of the ventricles. (The signal also has a base-line drift, which has not been corrected for.)

ECG signal acquisition: In clinical practice, the standard 12-channel ECG is obtained using four limb leads and chest leads in six positions [23, 27]. The right leg is used to place the reference electrode. The left arm, right arm, and left leg are used to get leads I, II, and III. A combined reference known as *Wilson's central terminal* is formed by combining the left arm, right arm, and left leg leads, and is used as the reference for chest leads. The *augmented* limb leads known as aVR, aVL, and aVF (aV for the augmented lead, R for the right arm, L for the left arm, and F for the left foot) are obtained by using the exploring electrode on the limb indicated by the lead name, with the reference being Wilson's central terminal without the exploring limb lead.

Figure 1.16 shows the directions of the axes formed by the six limb leads. The hypothetical equilateral triangle formed by leads I, II, and III is known as *Einthoven's triangle*. The center of the triangle represents Wilson's central terminal. Schematically, the heart is assumed to be placed at the center of the triangle. The six leads measure projections of the three-dimensional (3D) cardiac electrical vector onto the axes illustrated in Figure 1.16. The six axes sample the $0° - 180°$ range in steps of approximately $30°$. The projections facilitate viewing and analysis of the electrical activity of the heart and from different perspectives in the frontal plane.

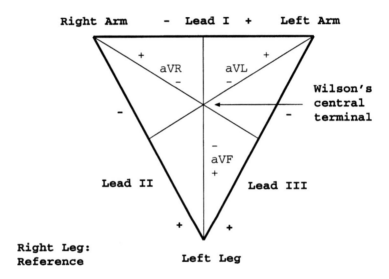

Figure 1.16 Einthoven's triangle and the axes of the six ECG leads formed by using four limb leads.

The six chest leads (written as V1 – V6) are obtained from six standardized positions on the chest [23] with Wilson's central terminal as the reference. The positions for placement of the precordial (chest) leads are indicated in Figure 1.17. The V1 and V2 leads are placed at the fourth intercostal space just to the right and left of the sternum, respectively. V4 is recorded at the fifth intercostal space at the left midclavicular line. The V3 lead is placed half-way between the V2 and V4 leads. The V5 and V6 leads are located at the same level as the V4 lead, but at the anterior axillary line and the midaxillary line, respectively. The six chest leads permit viewing the cardiac electrical vector from different orientations in a cross-sectional plane: V5 and V6 are most sensitive to left ventricular activity; V3 and V4 depict septal activity best; V1 and V2 reflect well activity in the right-half of the heart.

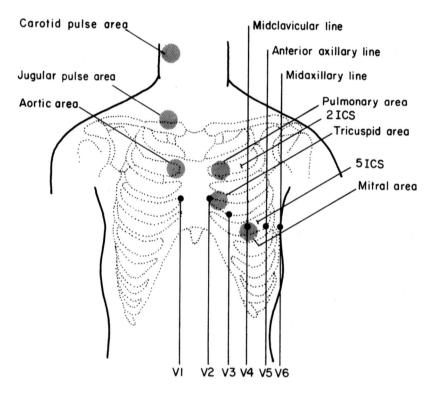

Figure 1.17   Positions for placement of the precordial (chest) leads V1 – V6 for ECG, auscultation areas for heart sounds, and pulse transducer positions for the carotid and jugular pulse signals. ICS: intercostal space.

In spite of being redundant, the 12-lead system serves as the basis of the standard clinical ECG. Clinical ECG interpretation is mainly empirical, based on experimental knowledge. A compact and efficient system has been proposed for *vectorcardiography* or VCG [28, 23], where loops inscribed by the 3D cardiac electrical vector in three mutually orthogonal planes, namely, the frontal, horizontal, and sagittal planes, are plotted and analyzed. Regardless, the 12-lead scalar ECG is the most commonly used procedure in clinical practice.

As the external ECG is a projection of the internal 3D cardiac electrical vector, the external recordings are not unique. Some of the lead inter-relationships are [23, 27]:

- II = I + III

- aVL = (I − III) / 2.

Some of the important features of the standard clinical ECG are:

- A rectangular calibration pulse of 1 mV amplitude and 200 ms duration is applied to produce a pulse of 1 cm height on the paper plot.

- The paper speed used is 25 mm/s, resulting in a graphical scale of 0.04 s/mm or 40 ms/mm. The calibration pulse width will then be 5 mm.

- The ECG signal peak value is normally about 1 mV.

- The amplifier gain used is 1,000.

- Clinical ECG is usually filtered to a bandwidth of about 0.05 − 100 Hz, with a recommended sampling rate of 500 Hz for diagnostic ECG. Distortions in the shape of the calibration pulse may indicate improper filter settings or a poor signal acquisition system.

- ECG for heart-rate monitoring could use a reduced bandwidth 0.5 − 50 Hz.

- High-resolution ECG requires a greater bandwidth of 0.05 − 500 Hz.

Figure 1.18 shows the 12-lead ECG of a normal male adult. The system used to obtain the illustration records three channels at a time: leads I, II, II; aVR, aVL, aVF; V1, V2, V3; and V4, V5, V6 are recorded in the three available channels simultaneously. Other systems may record one channel at a time. Observe the changing shape of the ECG waves from one lead to another. A well-trained cardiologist will be able to deduce the 3D orientation of the cardiac electrical vector by analyzing the waveshapes in the six limb leads. Cardiac defects, if any, may be localized by analyzing the waveshapes in the six chest leads.

Figure 1.19 shows the 12-lead ECG of a patient with right bundle-branch block with secondary repolarization changes. The increased QRS width and distortions in the QRS shape indicate the effects of asynchronous activation of the ventricles due to the bundle-branch block.

Signal-processing techniques to filter ECG signals will be presented in Sections 3.2, 3.3, 3.4, 3.5, and 3.8. Detection of ECG waveforms will be discussed

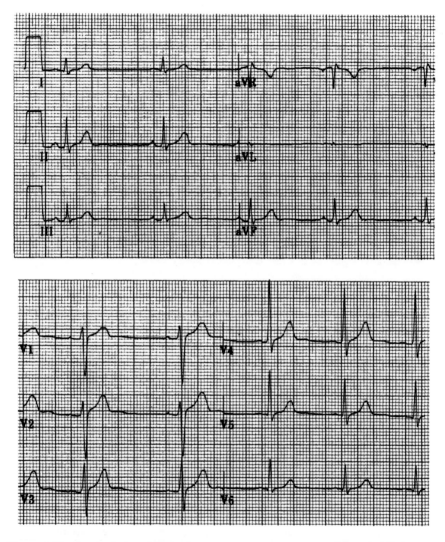

Figure 1.18   Standard 12-lead ECG of a normal male adult. Courtesy of E. Gedamu and L.B. Mitchell, Foothills Hospital, Calgary.

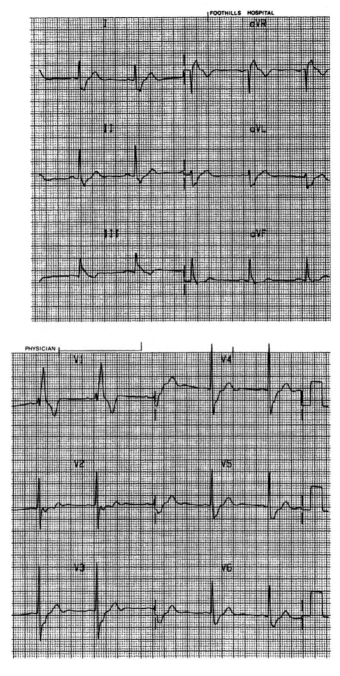

Figure 1.19 Standard 12-lead ECG of a patient with right bundle-branch block. Courtesy of L.B. Mitchell, Foothills Hospital, Calgary.

in Sections 4.2.1, 4.3.2, 4.7, and 4.9. Analysis of ECG waveform shape and classi-fication of beats will be dealt with in Sections 5.2.1, 5.2.2, 5.2.3, 5.4, 5.7, 5.8, 9.2.1, and 9.12. Analysis of heart-rate variability will be described in Sections 7.2.2, 7.8, and 8.9. Reviews of computer applications in ECG analysis have been published by Jenkins [29, 30] and Cox et al. [31].

1.2.5 The electroencephalogram (EEG)

The EEG (popularly known as *brain waves*) represents the electrical activity of the brain [32, 33, 34]. A few important aspects of the organization of the brain are as follows: The main parts of the brain are the cerebrum, the cerebellum, the brain stem (including the midbrain, pons medulla, and the reticular formation), and the thalamus (between the midbrain and the hemispheres). The cerebrum is divided into two hemispheres, separated by a longitudinal fissure across which there is a large connective band of fibers known as the corpus callosum. The outer surface of the cerebral hemispheres, known as the cerebral cortex, is composed of neurons (grey matter) in convoluted patterns, and separated into regions by fissures (sulci). Beneath the cortex lie nerve fibers that lead to other parts of the brain and the body (white matter).

Cortical potentials are generated due to excitatory and inhibitory post-synaptic potentials developed by cell bodies and dendrites of pyramidal neurons. Physiological control processes, thought processes, and external stimuli generate signals in the corresponding parts of the brain that may be recorded at the scalp using surface electrodes. The scalp EEG is an average of the multifarious activities of many small zones of the cortical surface beneath the electrode.

In clinical practice, several channels of the EEG are recorded simultaneously from various locations on the scalp for comparative analysis of activities in different regions of the brain. The International Federation of Societies for Electroencephalography and Clinical Neurophysiology has recommended the $10 - 20$ system of electrode placement for clinical EEG recording [32], which is schematically illustrated in Figure 1.20. The name $10 - 20$ indicates the fact that the electrodes along the midline are placed at $10, 20, 20, 20, 20$, and 10% of the total nasion – inion distance; the other series of electrodes are also placed at similar fractional distances of the corresponding reference distances [32]. The inter-electrode distances are equal along any antero-posterior or transverse line, and electrode positioning is symmetrical. EEG signals may be used to study the nervous system, monitoring of sleep stages, biofeedback and control, and diagnosis of diseases such as epilepsy.

Typical EEG instrumentation settings used are lowpass filtering at $75\ Hz$, and paper recording at $100\ \mu V/cm$ and $30\ mm/s$ for $10 - 20$ minutes over $8 - 16$ si-multaneous channels. Monitoring of sleep EEG and detection of transients related to epileptic seizures may require multichannel EEG acquisition over several hours. Spe-cial EEG techniques include the use of needle electrodes, naso-pharyngeal electrodes, recording the electrocorticogram (ECoG) from an exposed part of the cortex, and the use of intracerebral electrodes. Evocative techniques for recording the EEG include initial recording at rest (eyes open, eyes closed), hyperventilation (after breathing at

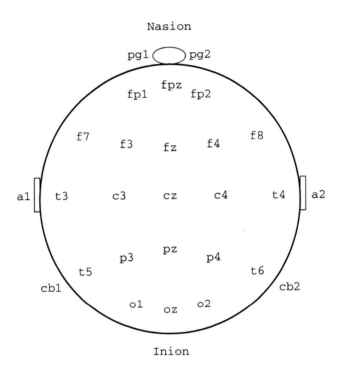

Figure 1.20 The 10 − 20 system of electrode placement for EEG recording [32]. Notes regarding channel labels: pg– naso-pharyngeal, a– auricular (ear lobes), fp– pre-frontal, f– frontal, p– pareital, c– central, o– occipital, t– temporal, cb– cerebellar, z– midline, odd numbers on the left, even numbers on the right of the subject.

20 respirations per minute for 2 – 4 minutes), photic stimulation (with 1 – 50 flashes of light per second), auditory stimulation with loud clicks, sleep (different stages), and pharmaceuticals or drugs.

EEG signals exhibit several patterns of rhythmic or periodic activity. (*Note:* The term *rhythm* stands for different phenomena or events in the ECG and the EEG.) The commonly used terms for EEG frequency (f) bands are:

- Delta (δ): $0.5 \leq f < 4\ Hz$;

- Theta (θ): $4 \leq f < 8\ Hz$;

- Alpha (α): $8 \leq f \leq 13\ Hz$; and

- Beta (β): $f > 13\ Hz$.

Figure 1.21 illustrates traces of EEG signals with the rhythms listed above.

EEG rhythms are associated with various physiological and mental processes [33, 34]. The alpha rhythm is the principal resting rhythm of the brain, and is common in wakeful, resting adults, especially in the occipital area with bilateral synchrony. Auditory and mental arithmetic tasks with the eyes closed lead to strong alpha waves, which are suppressed when the eyes are opened (that is, by a visual stimulus); see Figure 1.21(e) [32].

The alpha wave is replaced by slower rhythms at various stages of sleep. Theta waves appear at the beginning stages of sleep; delta waves appear at deep-sleep stages. High-frequency beta waves appear as background activity in tense and anxious subjects. The depression or absence of the normal (expected) rhythm in a certain state of the subject could indicate abnormality. The presence of delta or theta (slow) waves in a wakeful adult would be considered to be abnormal. Focal brain injury and tumors lead to abnormal slow waves in the corresponding regions. Unilateral depression (left – right asymmetry) of a rhythm could indicate disturbances in cortical pathways. Spikes and sharp waves could indicate the presence of epileptogenic regions in the corresponding parts of the brain.

Figure 1.22 shows an example of eight channels of the EEG recorded simultaneously from the scalp of a subject. All channels display high levels of alpha activity. Figure 1.23 shows 10 channels of the EEG of a subject with spike-and-wave complexes. Observe the distinctly different waveshape and sharpness of the spikes in Figure 1.23 as compared to the smooth waves in Figure 1.22. EEG signals also include spikes, transients, and other waves and patterns associated with various nervous disorders (see Figure 4.1 and Section 4.2.4). Detection of events and rhythms in EEG signals will be discussed in Sections 4.4, 4.5, and 4.6. Spectral analysis of EEG signals will be dealt with in Sections 6.4.3 and 7.5.2. Adaptive segmentation of EEG signals will be described in Section 8.2.2, 8.5, and 8.7.

1.2.6   Event-related potentials (ERPs)

The term *event-related potential* is more general than and preferred to the term *evoked potential*, and includes the ENG or the EEG in response to light, sound,

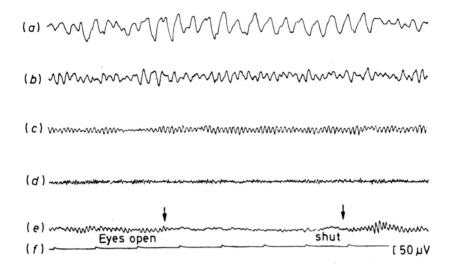

Figure 1.21 From top to bottom: (a) delta rhythm; (b) theta rhythm; (c) alpha rhythm; (d) beta rhythm; (e) blocking of the alpha rhythm by eye opening; (f) 1 *s* time markers and 50 *μV* marker. Reproduced with permission from R. Cooper, J.W. Osselton, and J.C. Shaw, *EEG Technology*, 3rd Edition, 1980. ©Butterworth Heinemann Publishers, a division of Reed Educational & Professional Publishing Ltd., Oxford, UK.

electrical, or other external stimuli. Short-latency ERPs are predominantly dependent upon the physical characteristics of the stimulus, whereas longer-latency ERPs are predominantly influenced by the conditions of presentation of the stimuli.

Somatosensory evoked potentials (SEPs) are useful for noninvasive evaluation of the nervous system from a peripheral receptor to the cerebral cortex. Median nerve short-latency SEPs are obtained by placing stimulating electrodes about $2 - 3$ *cm* apart over the median nerve at the wrist with electrical stimulation at $5 - 10$ *pps*, each stimulus pulse being of duration less than 0.5 *ms* with an amplitude of about 100 *V* (producing a visible thumb twitch). The SEPs are recorded from the surface of the scalp. The latency, duration, and amplitude of the response are measured.

ERPs and SEPs are weak signals typically buried in ongoing activity of associated systems. Examples of ERPs are provided in Figures 3.2 and 3.12. Signal-to-noise ratio (SNR) improvement is usually achieved by synchronized averaging and filtering, which will be described in Section 3.3.1.

1.2.7 The electrogastrogram (EGG)

The electrical activity of the stomach consists of rhythmic waves of depolarization and repolarization of its constituent smooth muscle cells [35, 36, 37]. The activity originates in the mid-corpus of the stomach, with intervals of about 20 *s* in humans. The waves of activity are always present and are not directly associated

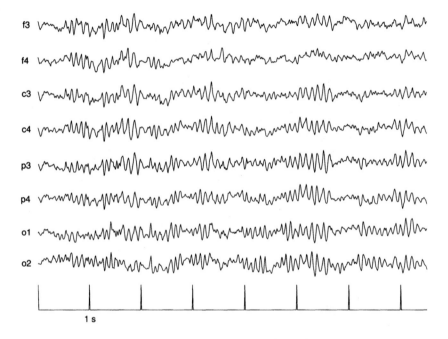

Figure 1.22 Eight channels of the EEG of a subject displaying alpha rhythm. See Figure 1.20 for details regarding channel labels. Data courtesy of Y. Mizuno-Matsumoto, Osaka University Medical School, Osaka, Japan.

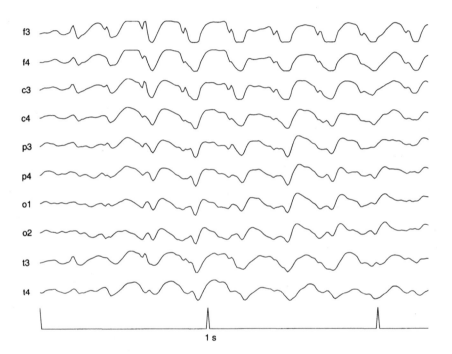

Figure 1.23 Ten channels of the EEG of a subject displaying spike-and-wave complexes. See Figure 1.20 for details regarding channel labels. Data courtesy of Y. Mizuno-Matsumoto, Osaka University Medical School, Osaka, Japan. Note that the time scale is expanded compared to that of Figure 1.22.

with contractions; they are related to the spatial and temporal organization of gastric contractions.

External (cutaneous) electrodes can record the signal known as the electrogastrogram (EGG). Chen et al. [38] used the following procedures to record cutaneous EGG signals. With the subject in the supine position and remaining motionless, the stomach was localized by using a 5 MHz ultrasound transducer array, and the orientation of the distal stomach was marked on the abdominal surface. Three active electrodes were placed on the abdomen along the antral axis of the stomach with an inter-electrode spacing of 3.5 cm. A common reference electrode was placed 6 cm away in the upper right quadrant. Three bipolar signals were obtained from the three active electrodes in relation to the common reference electrode. The signals were amplified and filtered to the bandwidth of $0.02 - 0.3$ Hz with 6 $dB/octave$ transition bands, and sampled at 2 Hz.

The surface EGG is believed to reflect the overall electrical activity of the stomach, including the electrical control activity and the electrical response activity. Chen et al. [38] indicated that gastric dysrhythmia or arrhythmia may be detected via analysis of the EGG. Other researchers suggest that the diagnostic potential of the signal has not yet been established [35, 36]. Accurate and reliable measurement of the electrical activity of the stomach requires implantation of electrodes within the stomach [39], which limits its practical applicability.

1.2.8 The phonocardiogram (PCG)

The heart sound signal is perhaps the most traditional biomedical signal, as indicated by the fact that the stethoscope is the primary instrument carried and used by physicians. The PCG is a vibration or sound signal related to the contractile activity of the cardiohemic system (the heart and blood together) [23, 40, 41, 42, 43, 44], and represents a recording of the heart sound signal. Recording of the PCG signal requires a transducer to convert the vibration or sound signal into an electronic signal: microphones, pressure transducers, or accelerometers may be placed on the chest surface for this purpose. The normal heart sounds provide an indication of the general state of the heart in terms of rhythm and contractility. Cardiovascular diseases and defects cause changes or additional sounds and murmurs that could be useful in their diagnosis.

The genesis of heart sounds: It is now commonly accepted that the externally recorded heart sounds are not caused by valve leaflet movements *per se*, as earlier believed, but by vibrations of the whole cardiovascular system triggered by pressure gradients [23]. The cardiohemic system may be compared to a fluid-filled balloon, which, when stimulated at any location, vibrates as a whole. Externally, however, heart sound components are best heard at certain locations on the chest individually, and this localization has led to the concept of *secondary sources* on the chest related to the well-known auscultatory areas: the mitral, aortic, pulmonary, and tricuspid areas [23]. The standard auscultatory areas are indicated in Figure 1.17. The mitral area is near the apex of the heart. The aortic area is to the right of the sternum, in the second right-intercostal space. The tricuspid area is in the fourth intercostal space near the right sternal border. The pulmonary area lies at the left parasternal line in the second or third left-intercostal space [23].

A normal cardiac cycle contains two major sounds — the first heart sound (S1) and the second heart sound (S2). Figure 1.24 shows a normal PCG signal, along with the ECG and carotid pulse tracings. S1 occurs at the onset of ventricular contraction, and corresponds in timing to the QRS complex in the ECG signal.

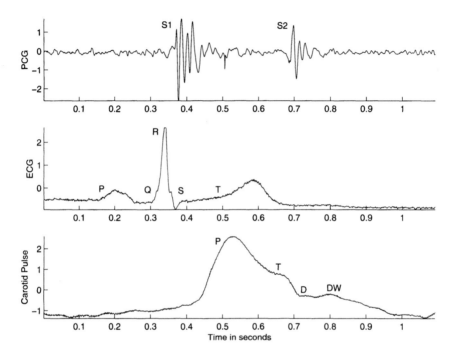

Figure 1.24 Three-channel simultaneous record of the PCG, ECG, and carotid pulse signals of a normal male adult.

The initial vibrations in S1 occur when the first myocardial contractions in the ventricles move blood toward the atria, sealing the atrio-ventricular (AV — mitral and tricuspid) valves (see Figure 1.25). The second component of S1 begins with abrupt tension of the closed AV valves, decelerating the blood. Next, the semilunar (aortic and pulmonary) valves open and the blood is ejected out of the ventricles. The third component of S1 may be caused by oscillation of blood between the root of the aorta and the ventricular walls. This is followed by the fourth component of S1, which may be due to vibrations caused by turbulence in the ejected blood flowing rapidly through the ascending aorta and the pulmonary artery.

Following the systolic pause in the PCG of a normal cardiac cycle, the second sound S2 is caused by the closure of the semilunar valves. While the primary vibrations occur in the arteries due to deceleration of blood, the ventricles and atria also vibrate, due to transmission of vibrations through the blood, valves, and the valve rings. S2 has two components, one due to closure of the aortic valve (A2)

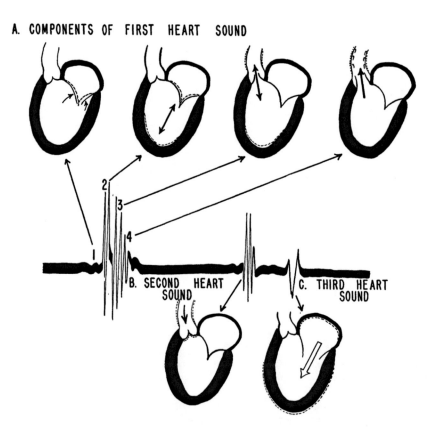

Figure 1.25 Schematic representation of the genesis of heart sounds. Only the left portion of the heart is illustrated as it is the major source of the heart sounds. The corresponding events in the right portion also contribute to the sounds. The atria do not contribute much to the heart sounds. Reproduced with permission from R.F. Rushmer, *Cardiovascular Dynamics*, 4th edition, ©W.B. Saunders, Philadelphia, PA, 1976.

and another due to closure of the pulmonary valve (P2). The aortic valve normally closes before the pulmonary valve, and hence A2 precedes P2 by a few milliseconds. Pathologic conditions could cause this gap to widen, or may also reverse the order of occurrence of A2 and P2. The A2 – P2 gap is also widened in normal subjects during inspiration. (*Note:* The PCG signal in Figure 1.24 does not show the A2 and P2 components separately.)

In some cases a third heart sound (S3) may be heard, corresponding to sudden termination of the ventricular rapid-filling phase. Because the ventricles are filled with blood and their walls are relaxed during this part of diastole, the vibrations of S3 are of very low frequency. In late diastole, a fourth heart sound (S4) may be heard sometimes, caused by atrial contractions displacing blood into the distended ventricles. In addition to these sounds, valvular clicks and snaps are occasionally heard.

Heart murmurs: The intervals between S1 and S2, and S2 and S1 of the next cycle (corresponding to ventricular systole and diastole, respectively) are normally silent. Murmurs, which are caused by certain cardiovascular defects and diseases, may occur in these intervals. Murmurs are high-frequency, noise-like sounds that arise when the velocity of blood becomes high as it flows through an irregularity (such as a constriction or a baffle). Typical conditions in the cardiovascular system that cause turbulence in blood flow are valvular stenosis and insufficiency. A valve is said to be stenosed when, due to the deposition of calcium or other reasons, the valve leaflets are stiffened and do not open completely, and thereby cause an obstruction or baffle in the path of the blood being ejected. A valve is said to be insufficient when it cannot close effectively and causes reverse leakage or regurgitation of blood through a narrow opening.

Systolic murmurs (SM) are caused by conditions such as ventricular septal defect (VSD — essentially a hole in the wall between the left ventricle and the right ventricle), aortic stenosis (AS), pulmonary stenosis (PS), mitral insufficiency (MI), and tricuspid insufficiency (TI). Semilunar valvular stenosis (aortic stenosis, pulmonary stenosis) causes an obstruction in the path of blood being ejected during systole. AV valvular insufficiency (mitral insufficiency, tricuspid insufficiency) causes regurgitation of blood to the atria during ventricular contraction.

Diastolic murmurs (DM) are caused by conditions such as aortic or pulmonary insufficiency (AI, PI), and mitral or tricuspid stenosis (MS, PS). Other conditions causing murmurs are atrial septal defect (ASD), patent ductus arteriosus (PDA), as well as certain physiological or functional conditions that cause increased cardiac output or blood velocity.

Features of heart sounds and murmurs, such as intensity, frequency content, and timing, are affected by many physical and physiological factors such as the recording site on the thorax, intervening thoracic structures, left ventricular contractility, position of the cardiac valves at the onset of systole, the degree of the defect present, the heart rate, and blood velocity. For example, S1 is loud and delayed in mitral stenosis; right bundle-branch block causes wide splitting of S2; left bundle-branch block results in reversed splitting of S2; acute myocardial infarction causes a pathologic S3; and severe mitral regurgitation (MR) leads to an increased S4 [40, 41, 42, 43, 44].

Although murmurs are noise-like events, their features aid in distinguishing between different causes. For example, aortic stenosis causes a diamond-shaped midsystolic murmur, whereas mitral stenosis causes a decrescendo – crescendo type diastolic – presystolic murmur. Figure 1.26 illustrates the PCG, ECG, and carotid pulse signals of a patient with aortic stenosis; the PCG displays the typical diamond-shaped murmur in systole.

Recording PCG signals: PCG signals are normally recorded using piezoelectric contact sensors that are sensitive to displacement or acceleration at the skin surface. The PCG signals illustrated in this section were obtained using a Hewlett Packard HP21050A transducer, which has a nominal bandwidth of $0.05 - 1,000 \ Hz$. The carotid pulse signals shown in this section were recorded using the HP21281A pulse transducer, which has a nominal bandwidth of $0 - 100 \ Hz$. PCG recording is normally performed in a quiet room, with the patient in the supine position with the head resting on a pillow. The PCG transducer is placed firmly on the desired position on the chest using a suction ring and/or a rubber strap.

Use of the ECG and carotid pulse signals in the analysis of PCG signals will be described in Sections 2.2.1, 2.2.2, and 2.3. Segmentation of the PCG based on events detected in the ECG and carotid pulse signals will be discussed in Section 4.10. A particular type of synchronized averaging to detect A2 in S2 will be the topic of Section 4.11. Spectral analysis of the PCG and its applications will be presented in Sections 6.2.1, 6.4.5, 6.6, and 7.10. Parametric modeling and detection of S1 and S2 will be described in Sections 7.5.2 and 7.9. Modeling of sound generation in stenosed coronary arteries will be discussed in Section 7.7.1. Adaptive segmentation of PCG signals with no other reference signal will be explored in Section 8.8.

1.2.9 The carotid pulse (CP)

The carotid pulse is a pressure signal recorded over the carotid artery as it passes near the surface of the body at the neck. It provides a pulse signal indicating the variations in arterial blood pressure and volume with each heart beat. Because of the proximity of the recording site to the heart, the carotid pulse signal closely resembles the morphology of the pressure signal at the root of the aorta; however, it cannot be used to measure absolute pressure [41]. The carotid pulse is a useful adjunct to the PCG and can assist in the identification of S2 and its components.

The carotid pulse rises abruptly with the ejection of blood from the left ventricle to the aorta, reaching a peak called the percussion wave (P, see Figure 1.24). This is followed by a plateau or a secondary wave known as the tidal wave (T), caused by a reflected pulse returning from the upper body. Next, closure of the aortic valve causes a notch known as the dicrotic notch (D). The dicrotic notch may be followed by the dicrotic wave (DW, see Figure 1.24) due to a reflected pulse from the lower body [41]. The carotid pulse trace is affected by valvular defects such as mitral insufficiency and aortic stenosis [41]; however, it is not commonly used in clinical diagnosis.

The carotid pulse signals shown in this section were recorded using the HP21281A pulse transducer, which has a nominal bandwidth of $0 - 100 \ Hz$. The carotid pulse

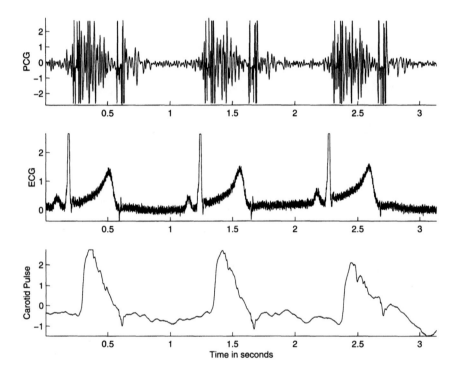

Figure 1.26   Three-channel simultaneous record of the PCG, ECG, and carotid pulse signals of a patient (female, 11 years) with aortic stenosis. Note the presence of the typical diamond-shaped systolic murmur and the split nature of S2 in the PCG.

signal is usually recorded with the PCG and ECG signals. Placement of the carotid pulse transducer requires careful selection of a location on the neck as close to the carotid artery as possible, where the pulse is felt the strongest, usually by a trained technician (see Figure 1.17).

Details on intervals that may be measured from the carotid pulse and their use in segmenting the PCG will be presented in Sections 2.2.2 and 2.3. Signal-processing techniques for the detection of the dicrotic notch will be described in Section 4.3.3. Use of the dicrotic notch for segmentation of PCG signals will be explored in Sections 4.10 and 4.11. Application of the carotid pulse to averaging of PCG spectra in systole and diastole will be proposed in Section 6.4.5.

1.2.10   Signals from catheter-tip sensors

For very specific and close monitoring of cardiac function, sensors placed on catheter tips may be inserted into the cardiac chambers. It then becomes possible to acquire several signals such as left ventricular pressure, right atrial pressure, aortic (AO) pressure, and intracardiac sounds [43, 44]. While these signals provide valuable and accurate information, the procedures are invasive and are associated with certain risks.

Figures 1.27 and 1.28 illustrate multi-channel aortic, left ventricular, and right ventricular pressure recordings from a dog using catheter-tip sensors. The ECG signal is also shown. Observe in Figure 1.27 that the right ventricular and left ventricular pressures increase exactly at the instant of each QRS complex. The aortic pressure peaks slightly after the increase in the left ventricular pressure. The notch (incisura) in the aortic pressure signal is due to closure of the aortic valve. (The same notch propagates through the vascular system and appears as the dicrotic notch in the carotid pulse signal.) The left ventricular pressure range ($10 - 110\ mm\ of\ Hg$) is much larger than the right ventricular pressure range ($5 - 25\ mm\ of\ Hg$). The aortic pressure range is limited to the vascular BP range of $80 - 120\ mm\ of\ Hg$.

The signals in Figure 1.28 display the effects of PVCs. Observe the depressed ST segment in the ECG signal in the figure, likely due to myocardial ischemia. (It should be noted that the PQ and ST segments of the ECG signal in Figure 1.27 are iso-electric, even though the displayed values indicate a non-zero level. On the other hand, in the ECG in Figure 1.28, the ST segment stays below the corresponding iso-electric PQ segment.) The ECG complexes appearing just after the $2\ s$ and $3\ s$ markers are PVCs arising from different ectopic foci, as evidenced by their markedly different waveforms. Although the PVCs cause a less-than-normal increase in the left ventricular pressure, they do not cause a rise in the aortic pressure, as no blood is effectively pumped out of the left ventricle during the ectopic beats.

1.2.11   The speech signal

Human beings are social creatures by nature, and have an innate need to communicate. We are endowed with the most sophisticated vocal system in nature. The speech signal

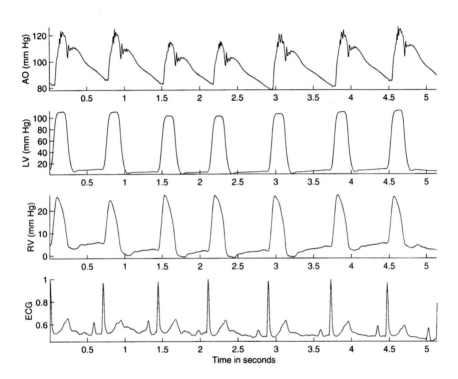

Figure 1.27 Normal ECG and intracardiac pressure signals from a dog. AO represents aortic pressure near the aortic valve. Data courtesy of R. Sas and J. Tyberg, Department of Physiology and Biophysics, University of Calgary.

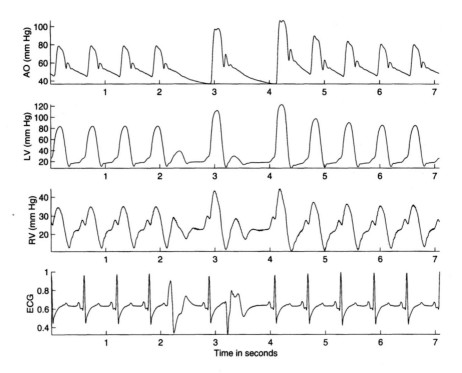

Figure 1.28   ECG and intracardiac pressure signals from a dog with PVCs. Data courtesy of R. Sas and J. Tyberg, Department of Physiology and Biophysics, University of Calgary.

is an important signal, although it is more commonly considered as a communication signal than a biomedical signal. However, the speech signal can serve as a diagnostic signal when speech and vocal-tract disorders need to be investigated [45].

Speech sounds are produced by transmitting puffs of air from the lungs through the vocal tract (as well as the nasal tract for certain sounds) [46]. The vocal tract starts at the vocal cords or glottis in the throat and ends at the lips and the nostrils. The shape of the vocal tract is varied to produce different types of sound units or *phonemes* which, when concatenated, form speech. In essence, the vocal tract acts as a filter that modulates the spectral characteristics of the input puffs of air. It is evident that the system is dynamic, and that the filter, and therefore the speech signal produced, have time-varying characteristics, that is, they are nonstationary (see Section 3.1.2).

Speech sounds may be classified mainly as voiced, unvoiced, and plosive sounds [46]. Voiced sounds involve the participation of the glottis: air is forced through the vocal cords held at a certain tension. The result is a series of quasi-periodic pulses of air which is passed through the vocal tract. The input to the vocal tract may be treated as an impulse train that is almost periodic. Upon convolution with the impulse response of the vocal tract, which is held steady at a certain configuration for the duration of the voiced sound desired, a quasi-periodic signal is produced with a characteristic waveshape that is repeated. All vowels are voiced sounds. Figure 1.29 shows the speech signal of the word "safety" spoken by a male. Figure 1.30 shows, in the upper trace, a portion of the signal corresponding to the /E/ sound (the letter "a" in the word). The quasi-periodic nature of the signal is evident. Features of interest in voiced signals are the pitch (average interval between the repetitions of the vocal-tract impulse response or basic wavelet) and the resonance or formant frequencies of the vocal-tract system.

An unvoiced sound (or fricative) is produced by forcing a steady stream of air through a narrow opening or constriction formed at a specific position along the vocal tract. The result is a turbulent signal that appears almost like random noise. In fact, the input to the vocal tract is a broadband random signal, which is filtered by the vocal tract to yield the desired sound. Fricatives are unvoiced sounds, as they do not involve any activity (vibration) of the vocal cords. The phonemes /S/, /SH/, /Z/, and /F/ are examples of fricatives. The lower trace in Figure 1.30 shows a portion of the signal corresponding to the /S/ sound in the word "safety". The signal has no identifiable structure, and appears to be random (see also Figures 3.1, 3.3, and 3.4, as well as Section 3.1.2). The transfer function of the vocal tract, as evidenced by the spectrum of the signal itself, would be of interest in analyzing a fricative.

Plosives, also known as stops, involve complete closure of the vocal tract, followed by an abrupt release of built-up pressure. The phonemes /P/, /T/, /K/, and /D/ are examples of plosives. The sudden burst of activity at about 1.1 s in Figure 1.29 illustrates the plosive nature of /T/. Plosives are hard to characterize as they are transients; their properties are affected by the preceding phoneme as well. For more details on the speech signal, see Rabiner and Schafer [46].

Signal-processing techniques for extraction of the vocal-tract response from voiced speech signals will be described in Section 4.8.3. Frequency-domain characteristics of speech signals will be illustrated in Section 7.6.3 and 8.4.1.

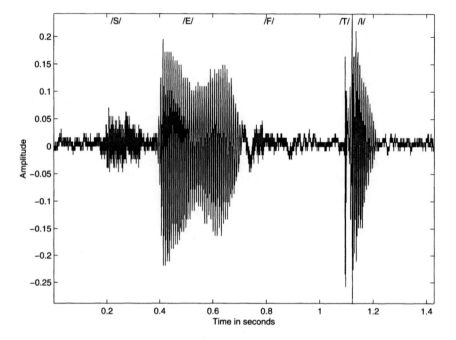

Figure 1.29   Speech signal of the word "safety" uttered by a male speaker. Approximate time intervals of the various phonemes in the word are /S/: $0.2 - 0.35$ *s*; /E/: $0.4 - 0.7$ *s*; /F/: $0.75 - 0.95$ *s*; /T/: transient at 1.1 *s*; /I/: $1.1 - 1.2$ *s*. Background noise is also seen in the signal before the beginning and after the termination of the speech, as well as during the stop interval before the plosive /T/.

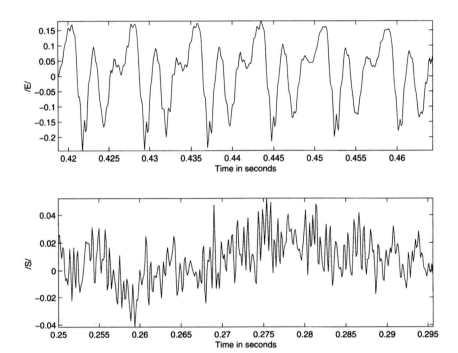

Figure 1.30 Segments of the signal in Figure 1.29 on an expanded scale to illustrate the quasi-periodic nature of the voiced sound /E/ in the upper trace, and the almost-random nature of the fricative /S/ in the lower trace.

1.2.12   The vibromyogram (VMG)

The VMG is the direct mechanical manifestation of contraction of a skeletal muscle, and is a vibration signal that accompanies the EMG. The signal has also been named as the sound-, acoustic-, or phono-myogram. Muscle sounds or vibrations are related to the change in dimensions (contraction) of the constituent muscle fibers (see Figure 1.4), and may be recorded using contact microphones or accelerometers (such as the Dytran 3115A accelerometer, Dytran, Chatsworth, CA) placed on the muscle surface [47, 48]. The frequency and intensity of the VMG have been shown to vary in direct proportion to the contraction level. The VMG, along with the EMG, may be useful in studies related to neuromuscular control, muscle contraction, athletic training, and biofeedback. VMG signal analysis, however, is not as well established or popular as EMG analysis.

Simultaneous analysis of the VMG and EMG signals will be discussed in Section 2.2.5. Adaptive cancellation of the VMG from knee-joint vibration signals will be the topic of Sections 3.6.2, 3.6.3, and 3.10. Analysis of muscle contraction using the VMG will be described in Section 5.10.

1.2.13   The vibroarthrogram (VAG)

The knee joint: As illustrated in Figure 1.31, the knee joint is formed between the femur, the patella, and the tibia. The knee joint is the largest articulation in the human body that can effectively move from $0°$ extension to $135°$ flexion, together with $20°$ to $30°$ rotation of the flexed leg on the femoral condyles. The joint has four important features: (1) a joint cavity, (2) articular cartilage, (3) a synovial membrane, and (4) a fibrous capsule [49, 50]. The knee joint is known as a synovial joint, as it contains a lubricating substance called the synovial fluid. The patella (knee cap), a sesamoid bone, protects the joint, and is precisely aligned to slide in the groove (trochlea) of the femur during leg movement. The knee joint is made up of three compartments: (1) the patello-femoral, (2) the lateral tibio-femoral, and (3) the medial tibio-femoral compartments. The patello-femoral compartment is classified as a synovial gliding joint and the tibio-femoral as a synovial hinge joint [51]. The anterior and posterior cruciate ligaments as well as the lateral and medial ligaments bind the femur and tibia together, give support to the knee joint, and limit movement of the joint. The various muscles around the joint help in the movement of the joint and contribute to its stability.

The knee derives its physiological movement and its typical rolling – gliding mechanism of flexion and extension from its six degrees of freedom: three in translation and three in rotation. The translations of the knee take place on the anterior – posterior, medial – lateral, and proximal – distal axes. The rotational motion consists of flexion – extension, internal – external rotation, and abduction – adduction.

Although the tibial plateaus are the main load-bearing structures in the knee, the cartilage, menisci, and ligaments also bear loads. The patella aids knee extension by lengthening the lever arm of the quadriceps muscle throughout the entire range of motion, and allows a better distribution of compressive stresses on the femur [52].

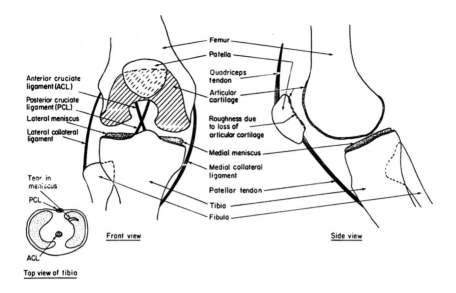

Figure 1.31   Front and side views of the knee joint (the two views are not mutually orthogonal). The inset shows the top view of the tibia with the menisci.

Articular cartilage: Two types of cartilage are present in the knee joint: the *articular cartilage*, which covers the ends of bones, and the wedge-shaped fibrocartilaginous structure called the *menisci*, located between the femur and the tibia [53]. The shock-absorbing menisci are composed of the medial meniscus and the lateral meniscus, which are two crescent-shaped plates of fibrocartilage that lie on the articular surface of the tibia.

The articular surfaces of the knee joint are the large curved condyles of the femur, the flattened condyles (medial and lateral plateaus) of the tibia, and the facets of the patella. There are three types of articulation: an intermediate articulation between the patella and the femur, and lateral and medial articulation between the femur and the tibia. The articular surfaces are covered by cartilage, like all the major joints of the body. Cartilage is vital to joint function because it protects the underlying bone during movement. Loss of cartilage function leads to pain, decreased mobility, and in some instances, deformity and instability.

Knee-joint disorders: The knee is the most commonly injured joint in the body. Arthritic degeneration of injured knees is a well-known phenomenon, and is known to result from a variety of traumatic causes. Damage to the stabilizing ligaments of the knee, or to the shock-absorbing fibrocartilage pads (the menisci) are two of the most common causes of deterioration of knee-joint surfaces. Impact trauma to the articular cartilage surfaces themselves could lead to surface deterioration and secondary osteoarthritis.

Non-traumatic conditions of the knee joint include the extremely common idiopathic condition known as chondromalacia patella (soft cartilage of the patella), in which articular cartilage softens, fibrillates, and sheds off the undersurface of the patella. Similarly, the meniscal fibrocartilage of the knee can apparently soften, which could possibly lead to degenerative tears and secondary changes in the regional hyaline surfaces.

Knee-joint sounds: Considerable noise is often associated with degeneration of knee-joint surfaces. The VAG is the vibration signal recorded from a joint during movement (articulation) of the joint. Normal joint surfaces are smooth and produce little or no sound, whereas joints affected by osteoarthritis and other degenerative diseases may have suffered cartilage loss and produce grinding sounds. Detection of knee-joint problems via the analysis of VAG signals could help avoid unnecessary exploratory surgery, and also aid better selection of patients who would benefit from surgery [54, 55, 56, 57, 58, 59, 60]. The VAG signal, however, is not yet well understood, and is a difficult signal to analyze due to its complex nonstationary characteristics.

Further details on the VAG signal will be provided in Sections 2.2.6, 3.2.6, and 8.2.3. Modeling of a specific type of VAG signal known as patello-femoral crepitus will be presented in Sections 7.2.4, 7.3, and 7.7.2. Adaptive filtering of the VAG signal to remove muscle-contraction interference will be described in Sections 3.6.2, 3.6.3, and 3.10. Adaptive segmentation of VAG signals into quasi-stationary segments will be illustrated in Sections 8.6.1 and 8.6.2. The role of VAG signal analysis in the detection of articular cartilage diseases will be discussed in Section 9.13.

1.2.14   Oto-acoustic emission signals

The oto-acoustic emission (OAE) signal represents the acoustic energy emitted by the cochlea either spontaneously or in response to an acoustic stimulus. The discovery of the existence of this signal indicates that the cochlea not only receives sound but also produces acoustic energy [61]. The OAE signal could provide objective information on the micromechanical activity of the preneural or sensory components of the cochlea that are distal to the nerve-fiber endings. Analysis of the OAE signal could lead to improved noninvasive investigative techniques to study the auditory system. The signal may also assist in screening of hearing function and in the diagnosis of hearing impairment.

1.3   OBJECTIVES OF BIOMEDICAL SIGNAL ANALYSIS

The representation of biomedical signals in electronic form facilitates computer processing and analysis of the data. Figure 1.32 illustrates the typical steps and processes involved in computer-aided diagnosis and therapy based upon biomedical signal analysis.

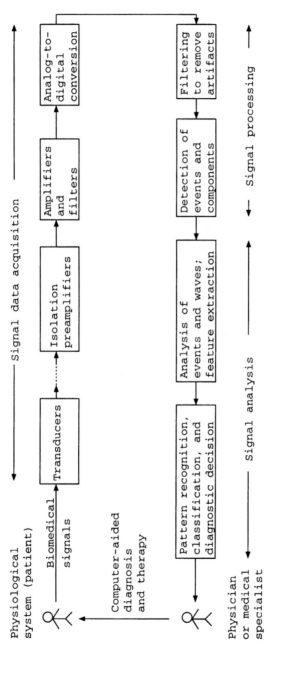

Figure 1.32 Computer-aided diagnosis and therapy based upon biomedical signal analysis.

The major objectives of biomedical instrumentation and signal analysis [17, 13, 10, 11, 12] are:

- *Information gathering* — measurement of phenomena to interpret a system.

- *Diagnosis* — detection of malfunction, pathology, or abnormality.

- *Monitoring* — obtaining continuous or periodic information about a system.

- *Therapy and control* — modification of the behavior of a system based upon the outcome of the activities listed above to ensure a specific result.

- *Evaluation* — objective analysis to determine the ability to meet functional requirements, obtain proof of performance, perform quality control, or quantify the effect of treatment.

Signal acquisition procedures may be categorized as being invasive or noninvasive, and active or passive.

Invasive versus noninvasive procedures: Invasive procedures involve the placement of transducers or other devices inside the body, such as needle electrodes to record MUAPs, or insertion of catheter-tip sensors into the heart via a major artery or vein to record intracardiac signals. Noninvasive procedures are desirable in order to minimize risk to the subject. Recording of the ECG using limb or chest electrodes, the EMG with surface electrodes, or the PCG with microphones or accelerometers placed on the chest are noninvasive procedures.

Note that making measurements or imaging with x-rays, ultrasound, and so on, may be classified as invasive procedures, as they involve penetration of the body with externally administered radiation, even though the radiation is invisible and there is no visible puncturing or invasion of the body.

Active versus passive procedures: Active data acquisition procedures require external stimuli to be applied to the subject, or require the subject to perform a certain activity to stimulate the system of interest in order to elicit the desired response or signal. For example, recording an EMG signal requires contraction of the muscle of interest, say the clenching of a fist; recording the VAG signal from the knee requires flexing of the leg over a certain joint angle range; recording visual ERP signals requires the delivery of flashes of light to the subject. While these stimuli may appear to be innocuous, they do carry risks in certain situations for some subjects: flexing the knee beyond a certain angle may cause pain for some subjects; strobe lights may trigger epileptic seizures in some subjects. The investigator should be aware of such risks, factor them in a *risk – benefit analysis*, and be prepared to manage adverse reactions.

Passive procedures do not require the subject to perform any activity. Recording of the ECG using limb or chest electrodes, the EEG during sleep using scalp-surface electrodes, or the PCG with microphones or accelerometers placed on the chest are passive procedures, but require contact between the subject and the instruments. Note that although the procedure is passive, the system of interest is active under its own natural control in these procedures. Acquiring an image of a subject with reflected natural light (with no flash from the camera) or with the natural infra-red (thermal) emission could be categorized as a passive and non-contact procedure.

Most organizations require ethical approval by specialized committees for experimental procedures involving human or animal subjects, with the aim of minimizing the risk and discomfort to the subject and maximizing the benefits to both the subjects and the investigator.

The human – instrument system: The components of a *human – instrument system* [17, 13, 10, 11, 12] are:

- *The subject or patient:* It is important always to bear in mind that the main purpose of biomedical instrumentation and signal analysis is to provide a certain benefit to the subject or patient. All systems and procedures should be designed so as not to unduly inconvenience the subject, and not to cause any harm or danger. In applying invasive or risky procedures, it is extremely important to perform a risk – benefit analysis and determine if the anticipated benefits of the procedure are worth placing the subject at the risks involved.

- *Stimulus or procedure of activity:* Application of stimuli to the subject in active procedures requires instruments such as strobe light generators, sound generators, and electrical pulse generators. Passive procedures require a standardized protocol of the desired activity to ensure repeatability and consistency of the experiment.

- *Transducers:* electrodes, sensors.

- *Signal-conditioning equipment:* amplifiers, filters.

- *Display equipment:* oscilloscopes, strip-chart or paper recorders, computer monitors, printers.

- *Recording, data processing, and transmission equipment:* analog instrumentation tape recorders, analog-to-digital converters (ADCs), digital-to-analog converters (DACs), digital tapes, compact disks (CDs), diskettes, computers, telemetry systems.

- *Control devices:* power supply stabilizers and isolation equipment, patient intervention systems.

The science of measurement of physiological variables and parameters is known as *biometrics.* Some of the aspects to be considered in the design, specification, or use of biomedical instruments [17, 13, 10, 11, 12] are:

- *Isolation of the subject or patient* — of paramount importance so that the subject is not placed at the risk of electrocution.

- *Range of operation* — the minimum to maximum values of the signal or parameter being measured.

- *Sensitivity* — the smallest signal variation measurable. This determines the resolution of the system.

- *Linearity* — desired over at least a portion of the range of operation. Any nonlinearity present may need to be corrected for at later stages of signal processing.

- *Hysteresis* — a lag in measurement due to the direction of variation of the entity being measured. Hysteresis may add a bias to the measurement, and should be corrected for.

- *Frequency response* — represents the variation of sensitivity with frequency. Most systems encountered in practice exhibit a lowpass behavior, that is, the sensitivity of the system decreases as the frequency of the input signal increases. Signal restoration techniques may be required to compensate reduced high-frequency sensitivity.

- *Stability* — an unstable system could preclude repeatability and consistency of measurements.

- *Signal-to-noise ratio (SNR)* — power-line interference, grounding problems, thermal noise, and so on, could compromise the quality of the signal being acquired. A good understanding of the signal-degrading phenomena present in the system is necessary in order to design appropriate filtering and correction procedures.

- *Accuracy* — includes the effects of errors due to component tolerance, movement, or mechanical errors; drift due to changes in temperature, humidity, or pressure; reading errors due to, for example, parallax; and zeroing or calibration errors.

1.4   DIFFICULTIES ENCOUNTERED IN BIOMEDICAL SIGNAL ACQUISITION AND ANALYSIS

In spite of the long history of biomedical instrumentation and its extensive use in health care and research, many practical difficulties are encountered in biomedical signal acquisition, processing, and analysis [17, 13, 10, 11, 12]. The characteristics of the problems, and hence their potential solutions, are unique to each type of signal. Particular attention should be paid to the following issues.

Accessibility of the variables to measurement: Most of the systems and organs of interest, such as the cardiovascular system and the brain, are located well within the body (for good reasons!). While the ECG may be recorded using limb electrodes, the signal so acquired is but a projection of the true 3D cardiac electrical vector of the heart onto the axis of the electrodes. Such a signal may be sufficient for rhythm monitoring, but could be inadequate for more specific analysis of the cardiac system such as atrial electrical activity. Accessing the atrial electrical activity at the source requires insertion of an electrode close to the atrial surface or within the atria.

Similarly, measurement of blood pressure using a pressure cuff over an arm gives an estimate of the brachial arterial pressure. Detailed study of pressure variations within the cardiac chambers or arteries over a cardiac cycle would require insertion of catheters with pressure sensors into the heart. Such invasive procedures provide access to the desired signals at their sources and often provide clear and useful signals, but carry high risks.

The surface EMG includes the interference pattern of the activities of several motor units even at very low levels of muscular contraction. Acquisition of SMUAPs requires access to the specific muscle layer or unit of interest by insertion of fine-wire or needle electrodes. The procedure carries risks of infection and damage to muscle fibers, and causes pain to the subject during muscular activity.

An investigator should assess the system and variables of interest carefully and determine the minimal level of intervention absolutely essential to the data acquisition procedure. The trade-off to be performed is that of integrity and quality of the information acquired versus the pain and risks to the subject.

Variability of the signal source: It is evident from the preceding sections that the various systems that comprise the human body are dynamic systems with several variables. Biomedical signals represent the dynamic activity of physiological systems and the states of their constituent variables. The nature of the processes or the variables could be deterministic or random (stochastic); a special case is that of periodicity or quasi-periodicity.

A normal ECG exhibits a regular rhythm with a readily identifiable waveshape (the QRS complex) in each period, and under such conditions the signal may be referred to as a deterministic and periodic signal. However, the cardiovascular system of a heart patient may not stay in a given state over significant periods and the waveshape and rhythm may vary over time.

The surface EMG is the summation of the MUAPs of the motor units that are active at the given instant of time. Depending upon the level of contraction desired (at the volition of the subject), the number of active motor units varies, increasing with increasing effort. Furthermore, the firing intervals or the firing rate of each motor unit also vary in response to the level of contraction desired, and exhibit stochastic properties. While the individual MUAPs possess readily identifiable and simple monophasic, biphasic, or triphasic waveshapes, the interference pattern of several motor units firing at different rates will appear as an almost random signal with no visually recognizable waves or waveshapes.

The dynamic nature of biological systems causes most signals to exhibit stochastic and nonstationary behavior. This means that signal statistics such as mean, variance, and spectral density change with time. For this reason, signals from a dynamic system should be analyzed over extended periods of time including various possible states of the system, and the results should be placed in the context of the corresponding states.

Inter-relationships and interactions among physiological systems: The various systems that compose the human body are not mutually independent; rather, they are inter-related and interact in various ways. Some of the interactive phenomena are compensation, feedback, cause-and-effect, collateral effects, loading, and take-over of function of a disabled system or part by another system or part. For example, the second heart sound exhibits a split during active inspiration in normal subjects due to reduced intra-thoracic pressure and decreased venous return to the left side of the heart [41] (but not during expiration); this is due to normal physiological processes. However, the second heart sound is split in both inspiration and expiration due to delayed right ventricular contraction in right bundle-branch block, pulmonary valvular stenosis or insufficiency, and other conditions [41]. Ignoring this inter-relationship could lead to misinterpretation of the signal.

Effect of the instrumentation or procedure on the system: The placement of transducers on and connecting a system to instruments could affect the performance or alter the behavior of the system, and cause spurious variations in the parameters being investigated. The experimental procedure or activity required to elicit the signal may lead to certain effects that could alter signal characteristics. This aspect may not always be obvious unless careful attention is paid. For example, the placement of a relatively heavy accelerometer may affect the vibration characteristics of a muscle and compromise the integrity of the vibration or sound signal being measured. Fatigue may set in after a few repetitions of an experimental procedure, and subsequent measurements may not be indicative of the true behavior of the system; the system may need some rest between procedures or their repetitions.

Physiological artifacts and interference: One of the pre-requisites for obtaining a good ECG signal is for the subject to remain relaxed and still with no movement. Coughing, tensing of muscles, and movement of the limbs cause the corresponding EMG to appear as an undesired artifact. In the absence of any movement by the subject, the only muscular activity in the body would be that of the heart. When chest leads are used, even normal breathing could cause the associated EMG of the chest muscles to interfere with the desired ECG. It should also be noted that breathing causes beat-to-beat variations in the RR interval, which should not be mistaken to be sinus arrhythmia. An effective solution would be to record the signal with the subject holding breath for a few seconds. This simple solution does not apply in long-term monitoring of critically ill patients or in recording the ECG of infants; signal-processing procedures would then be required to remove the artifacts.

A unique situation is that of acquiring the ECG of a fetus through surface electrodes placed over the mother's abdomen: the maternal ECG appears as an interference in this situation. No volitional or external control is possible or desirable to prevent the artifact in this situation, which calls for more intelligent adaptive cancellation techniques using multiple channels of various signals [62].

Another example of physiological interference or cross-talk is that of muscle-contraction interference (MCI) in the recording of the knee-joint VAG signal [63]. The rectus femoris muscle is active (contracting) during the swinging movement of the leg required to elicit the joint vibration signal. The VMG of the muscle is propagated to the knee and appears as an interference. Swinging the leg mechanically using a mechanical actuator is a possible solution; however, this represents an unnatural situation, and may cause other sound or vibration artifacts from the machine. Adaptive filtering using multi-channel vibration signals from various points is a feasible solution [63].

Energy limitations: Most biomedical signals are generated at microvolt or millivolt levels at their sources. Recording such signals requires very sensitive transducers and instrumentation with low noise levels. The connectors and cables need to be shielded as well, in order to obviate pickup of ambient electromagnetic (EM) signals. Some applications may require transducers with integrated amplifiers and signal conditioners so that the signal leaving the subject at the transducer level is much stronger than ambient sources of potential interference.

When external stimuli are required to elicit a certain response from a system, the level of the stimulus is constrained due to safety factors and physiological limitations. Electrical stimuli to record the ENG need to be limited in voltage level so as to not cause local burns or interfere with the electrical control signals of the cardiac or nervous systems. Auditory and visual stimuli are constrained by the lower thresholds of detectability and upper thresholds related to frequency response, saturation, or pain.

Patient safety: Protection of the subject or patient from electrical shock or radiation hazards is an unquestionable requirement of paramount importance. The relative levels of any other risks involved should be assessed when a choice is available between various procedures, and analyzed against their relative benefits. Patient safety concerns may preclude the use of a procedure that may yield better signals or results than others, or require modifications to a procedure that may lead to inferior signals. Further signal-processing steps would then become essential in order to improve signal quality or otherwise compensate for the initial loss.

1.5 COMPUTER-AIDED DIAGNOSIS

Physicians, cardiologists, neuroscientists, and health-care technologists are highly trained and skilled practitioners. Why then would we want to use computers or electronic instrumentation for the analysis of biomedical signals? The following points provide some arguments in favor of the application of computers to process and analyze biomedical signals.

- Humans are highly skilled and fast in the analysis of visual patterns and waveforms, but are slow in arithmetic operations with large numbers of values. The ECG of a single cardiac cycle (heart beat) could have up to 200 numerical values; the corresponding PCG up to 2,000. If signals need to be processed to remove noise or extract a parameter, it would not be practical for a person to perform such computation. Computers can perform millions of arithmetic operations per second. It should be noted, however, that recognition of waveforms and images using mathematical procedures typically requires huge numbers of operations that could lead to slow responses in such tasks from low-level computers.

- Humans could be affected by fatigue, boredom, and environmental factors, and are susceptible to committing errors. Long-term monitoring of signals, for example, the heart rate and ECG of a critically ill patient, by a human observer watching an oscilloscope or computer tracing is neither economical nor feasible. A human observer could be distracted by other events in the surrounding areas and may miss short episodes or transients in the signal. Computers, being inanimate but mathematically accurate and consistent machines, can be designed to perform computationally specific and repetitive tasks.

- Analysis by humans is usually subjective and qualitative. When comparative analysis is required between the signal of a subject and another or a standard pattern, a human observer would typically provide a qualitative response. For example, if the QRS width of the ECG is of interest, a human observer may remark that the QRS of the subject is wider than the reference or normal. More specific or objective comparison to the accuracy of the order of a few milliseconds would require the use of electronic instrumentation or a computer. Derivation of quantitative or numerical features from signals with large numbers of samples would certainly demand the use of computers.

- Analysis by humans is subject to inter-observer as well as intra-observer variations (with time). Given that most analyses performed by humans are based upon qualitative judgment, they are liable to vary with time for a given observer, or from one observer to another. The former could also be due to lack of diligence or due to inconsistent application of knowledge, and the latter due to variations in training and level of understanding. Computers can apply a given procedure repeatedly and whenever recalled in a consistent manner. It is further possible to encode the knowledge (to be more specific, the logic) of many experts into a single computational procedure, and thereby enable a computer with the collective intelligence of several human experts in the area of interest.

- Most biomedical signals are fairly slow (lowpass) signals, with their bandwidth limited to a few tens to a few thousand Hertz. Typical sampling rates for digital processing of biomedical signals therefore range from $100\ Hz$ to $10 - 20\ kHz$. Sampling rates as above facilitate *on-line, real-time* analysis of biomedical signals with even low-end computers. Note that the term "real-time analysis" may be used to indicate the processing of each sample of the signal before the next sample arrives, or the processing of an epoch or episode such as an ECG beat before the next one is received in its entirety in a buffer. Heart-rate monitoring of critically ill patients would certainly demand real-time ECG analysis. However, some applications do not require on-line, real-time analysis: for example, processing a VAG signal to diagnose cartilage degeneration, and analysis of a long-term ECG record obtained over several hours using an ambulatory system do not demand immediate attention and results. In such cases, computers could be used for *off-line* analysis of pre-recorded signals with sophisticated signal-processing and time-consuming modeling techniques. The speed required for real-time processing and the computational complexities of modeling techniques in the case of off-line applications both would rule out the possibility of performance of the tasks by humans.

One of the important points to note in the above discussion is that *quantitative analysis* becomes possible by the application of computers to biomedical signals. The logic of medical or clinical diagnosis via signal analysis could then be *objectively* encoded and *consistently* applied in routine or repetitive tasks. However, it should be emphasized at this stage that the end-goal of biomedical signal analysis should be seen as computer-*aided* diagnosis and not automated diagnosis. A physician or medical specialist typically uses a significant amount of information in addition to signals and measurements, including the general physical appearance and mental state of the patient, family history, and socio-economic factors affecting the patient, many of which are not amenable to quantification and logistic rule-based processes. Biomedical signals are, at best, indirect indicators of the state of the patient; most cases lack a direct or unique signal – pathology relationship [31]. The results of signal analysis need to be integrated with other clinical signs, symptoms, and information by a physician. Above all, the *intuition* of the specialist plays an important role in arriving at the final diagnosis. For these reasons, and keeping in mind the realms of practice of various licensed and regulated professions, liability, and legal factors, the final diagnostic decision is best left to the physician or medical specialist. It is expected that quantitative and objective analysis facilitated by the application of computers to biomedical signal analysis will lead to a more accurate diagnostic decision by the physician.

On the importance of quantitative analysis:

"When you can measure what you are speaking about, and express it in numbers, you know something about it; but when you cannot measure it, when you cannot express it in numbers, your knowledge is of a meager and unsatisfactory kind: it may be the beginning of knowledge, but you have scarcely, in your thoughts, advanced to the stage of *science*."

— *Lord Kelvin (William Thomson, 1824 – 1907) [64]*

On assumptions made in quantitative analysis:

"Things do not in general run around with their measure stamped on them like the capacity of a freight car; it requires a certain amount of investigation to discover what their measures are ... What most experimenters take for granted before they begin their experiments is infinitely more interesting than any results to which their experiments lead."

— *Norbert Wiener (1894 – 1964)*

1.6 REMARKS

We have taken a general look at the nature of biomedical signals in this chapter, and seen a few signals illustrated for the purpose of gaining familiarity with their typical appearance and features. Specific details of the characteristics of the signals and their processing or analysis will be dealt with in subsequent chapters.

We have also stated the objectives of biomedical instrumentation and signal analysis. Some practical difficulties that arise in biomedical signal investigation were discussed in order to draw attention to the relevant practical issues. The suitability and desirability of the application of computers for biomedical signal analysis were discussed, with emphasis on objective and quantitative analysis toward the end-goal of computer-aided diagnosis. The remaining chapters will deal with specific techniques and applications.

1.7 STUDY QUESTIONS AND PROBLEMS

(*Note:* Some of the questions may require background preparation with other sources on the ECG (for example, Rushmer [23]), the EMG (for example, Goodgold and Eberstein [22]), and biomedical instrumentation (for example, Webster [10].)

1. Give two reasons to justify the use of electronic instruments and computers in medicine.

2. State any two objectives of using biomedical instrumentation and signal analysis.

3. Distinguish between open-loop and closed-loop monitoring of a patient.

4. List three common types or sources of artifact in a biomedical instrument.

5. A nerve cell has an action potential of duration 10 *ms* including the refractory period. What is the maximum rate (in pulses per second) at which this cell can transmit electrical activity?

6. Consider a myocardial cell with an action potential of duration 300 *ms* including its refractory period. What is the maximum rate at which this cell can be activated (fired) into contraction?

7. Distinguish between spatial and temporal recruitment of motor units to obtain increasing levels of muscular activity.

8. Consider three motor units with action potentials (SMUAPs) that are of different biphasic and triphasic shapes. Consider the initial stages of contraction of the related muscle. Draw three plots of the net EMG of the three motor units for increasing levels of contraction with the spatial and temporal recruitment phenomena invoked individually and in combination. Assume low levels of contraction and that the SMUAPs do not overlap.

9. Draw a typical ECG waveform over one cardiac cycle indicating the important component waves, their typical durations, and the typical intervals between them. Label each wave or interval with the corresponding cardiac event or activity.

10. Draw the waveform corresponding to two cycles of a typical ECG signal and indicate the following waves and periods: (a) the P, QRS, and T waves; (b) the RR interval; (c) atrial contraction; (d) atrial relaxation; (e) ventricular contraction; and (f) ventricular relaxation.

11. Explain why the P and T waves are low-frequency signals whereas the QRS complex is a high-frequency signal. Include diagrams of action potentials and an ECG waveform in your reasoning.

12. Explain the reasons for widening of the QRS complex in the case of certain cardiac diseases.

13. Give two examples that call for the use of electronic instruments and/or computers in ECG analysis.

14. A heart patient has a regular SA node pulse (firing) pattern and an irregular ectopic focus. Over a period of 10 s, the SA node was observed to fire regularly at $t = 0, 1, 2, 3, 4, 5, 6, 7, 8$, and 9 s. The ectopic focus was observed to fire at $t = 1.3, 2.8, 6.08$, and 7.25 s.

    Draw two impulse sequences corresponding to the firing patterns of the SA node and the ectopic focus. Draw a schematic waveform of the resulting ECG of the patient. Explain the source of each beat (SA node or ectopic focus) and give reasons.

15. A patient has ventricular bigeminy, where every second pulse from the SA node is replaced by a premature ventricular ectopic beat with a full compensatory pause. (See Figure 9.5 for an illustration of bigeminy.) The SA-node firing rate is regular at 80 beats a minute, and each ectopic beat precedes the blocked SA node pulse by 100 ms.

    (a) Draw a schematic trace of the ECG for 10 beats, marking the time scale in detail.

    (b) Draw a histogram of the RR intervals for the ECG trace.

    (c) What is the average RR interval computed over the 10 beats?

16. Draw a typical PCG (heart sound signal) waveform over one cardiac cycle indicating the important component waves, their typical durations, and the typical intervals between them. Label each wave or interval with the corresponding cardiac event or activity.

17. Give two examples that require the application of electronic instruments and/or computers in EEG analysis.

18. Distinguish between ECG rhythms and EEG rhythms. Sketch one example of each.

1.8 LABORATORY EXERCISES AND PROJECTS

1. Visit an ECG, EMG, or EEG laboratory in your local hospital or health sciences center. View a demonstration of the acquisition of a few biomedical signals. Request a specialist in a related field to explain how he or she would interpret the signals. Volunteer to be the experimental subject and experience first-hand a biomedical signal acquisition procedure!

2. Set up an ECG acquisition system and study the effects of the following conditions or actions on the quality and nature of the signal: loose electrodes; lack of electrode gel; the subject holding his/her breath or breathing freely during the recording procedure; and the subject coughing, talking, or squirming during signal recording.

3. Using a stethoscope, listen to your own heart sounds and those of your friends. Examine the variability of the sounds with the site of auscultation. Study the effects of heavy breathing and speaking by the subject as you are listening to the heart sound signal.

4. Record speech signals of vowels (/A/, /I/, /U/, /E/, /O/), diphthongs (/EI/, /OU/), fricatives (/S/, /F/), and plosives (/T/, /P/), as well as words with all three types of sounds (for example, safety, explosive, hearty, heightened, house). You may be able to perform this experiment with the microphone on your computer workstation. Study the waveform and characteristics of each signal.

2

Analysis of Concurrent, Coupled, and Correlated Processes

The human body is a complex integration of a number of biological systems with several ongoing physiological, functional, and possibly pathological processes. Most biological processes within a body are not independent of one another; rather, they are mutually correlated and bound together by physical or physiological control and communication phenomena. Analyzing any single process without due attention to others that are concurrent, coupled, or correlated with the process may provide only partial information and pose difficulties in the comprehension of the process. The problem, then, is how do we recognize the existence of concurrent, coupled, and correlated phenomena? How do we obtain the corresponding signals and identify the correlated features? Unfortunately, there is no simple or universal rule to apply to this problem.

Ideally, an investigator should explore the system or process of interest from all possible angles and use multidisciplinary approaches to identify several potential sources of information. The signals so obtained may be electrical, mechanical, biochemical, or physical, among the many possibilities, and may exhibit interrelationships confounded by peculiarities of transduction, time delays, multipath transmission or reflection, waveform distortions, and filtering effects that may need to be accounted for in their simultaneous analysis. Events or waves in signals of interest may be nonspecific and difficult to identify and analyze. How could we exploit the concurrency, coupling, and correlation present between processes or related signals to better understand a system?

2.1 PROBLEM STATEMENT

Determine the correspondences, correlation, and inter-relationships present be-
tween concurrent signals related to a common underlying physiological system
or process, and identify their potential applications.

The statement above represents, of necessity at this stage of the discussion, a
rather vague and generic problem. The case-studies and applications presented in the
following sections provide a few illustrative examples dealing with specific systems
and problems. Signal processing techniques for the various tasks identified in the
case-studies will be developed in chapters that follow. Note that the examples cover
a diverse range of systems, processes, and signals. The specific problem of your
interest will very likely not be directly related to any of the case-studies presented
here. It is expected that a study of the examples provided will expand the scope of
your analytical skills and lead to improved solution of your specific case.

2.2 ILLUSTRATION OF THE PROBLEM WITH CASE-STUDIES

2.2.1 The electrocardiogram and the phonocardiogram

A clinical ECG record typically includes 12 channels of sequentially or simultaneously recorded signals, and can be used on its own to diagnose many cardiac diseases.
This is mainly due to the simple and readily identifiable waveforms in the ECG, and
the innumerable studies that have firmly established clinical ECG as a standard procedure, albeit as an empirical one. The PCG, on the other hand, is a more complex
signal. PCG waveforms cannot be visually analyzed except for the identification of
gross features such as the presence of murmurs, time delays as in a split S2, and
envelopes of murmurs. An advantage with the PCG is that it may be listened to;
auscultation of heart sounds is more commonly performed than visual analysis of the
PCG signal. However, objective analysis of the PCG requires the identification of
components, such as S1 and S2, and subsequent analysis tailored to the nature of the
components.

Given a run of a PCG signal over several cardiac cycles, visual identification of
S1 and S2 is possible if there are no murmurs between the sounds, and if the heart
rate is low such that the S2 – S1 (of the next beat) interval is longer than the S1 – S2
interval (as expected in normal situations). At high heart rates and with the presence
of murmurs or premature beats, identification of S1 and S2 could be difficult.

Problem: *Identify the beginning of S1 in a PCG signal and extract the heart*
sound signal over one cardiac cycle.

Solution: The ECG and PCG are concurrent phenomena, with the noticeable
difference that the former is electrical while the latter is mechanical (sound or vibration). It is customary to record the ECG with the PCG; see Figures 1.24 and 1.26 for
examples.

The QRS wave in the ECG is directly related to ventricular contraction, as the
summation of the action potentials of ventricular muscle cells (see Section 1.2.4).

As the ventricles contract, the tension in the *chordae tendineae* and the pressure of retrograde flow of blood toward the atria seal the AV valves shut, thereby causing the initial vibrations of S1 [23] (see Section 1.2.8). Thus S1 begins immediately after the QRS complex. Given the nonspecific nature of vibration signals and the various possibilities in the transmission of the heart sounds to the recording site on the chest, detection of S1 on its own is a difficult problem.

As will be seen in Sections 3.3.1, 4.3.1, and 4.3.2, detection of the QRS is fairly easy, given that the QRS is the sharpest wave in the ECG over a cardiac cycle; in fact, the P and T waves may be almost negligible in many ECG records. Thus the QRS complex in the ECG is a reliable indicator of the beginning of S1, and may be used to segment a PCG record into individual cardiac cycles: from the beginning of one QRS (and thereby S1) to the beginning of the next QRS and S1. This method may be applied visually or via signal processing techniques: the former requires no further explanation but will be expanded upon in Section 2.3; the latter will be dealt with in Section 4.10.

2.2.2 The phonocardiogram and the carotid pulse

Identification of the diastolic segment of the PCG may be required in some applications in cardiovascular diagnosis [65]. Ventricular systole ends with the closure of the aortic and pulmonary valves, indicated by the aortic (A2) and pulmonary (P2) components of the second heart sound S2 (see Section 1.2.8). The end of contraction is also indicated by the T wave in the ECG, and S2 appears slightly after the end of the T wave (see Figure 1.24). S2 may be taken to be the end of systole and the beginning of ventricular relaxation or diastole. (*Note:* Shaver et al. [43] and Reddy et al. [44] have included S2 in the part of their article on systolic sounds.) However, as in the case of S1, S2 is also a nonspecific vibrational wave that cannot be readily identified (even visually), especially when murmurs are present.

Given the temporal relationship between the T wave and S2, it may appear that the former may be used to identify the latter. This, however, may not always be possible in practice, as the T wave is often a low-amplitude and smooth wave and is sometimes not recorded at all (see Figure 1.14). ST segment elevation (as in Figure 1.14) or depression (as in Figure 1.28) may make even visual identification of the end of the T wave difficult. Thus the T wave is not a reliable indicator to use for identification of S2.

Problem: *Identify the beginning of S2 in a PCG signal.*

Solution: Given the inadequacy of the T wave as an indicator of diastole, we need to explore other possible sources of information. Closure of the aortic valve is accompanied by deceleration and reversal of blood flow in the aorta. This causes a sudden drop in the blood pressure within the aorta, which is already on a downward slope due to the end of systolic activity. The sudden change in pressure causes an *incisura* or notch in the aortic pressure wave (see Figures 1.27 and 1.28). The aortic pressure signal may be obtained using catheter-tip sensors [43, 44], but the procedure would be invasive. Fortunately, the notch is transmitted through the arterial system, and may be observed in the carotid pulse (see Section 1.2.9) recorded at the neck.

The dicrotic notch D in the carotid pulse signal will bear a delay with respect to the corresponding notch in the aortic pressure signal, but has the advantage of being accessible in a noninvasive manner. (Similar events occur in the pulmonary artery, but provide no externally observable effects.) See Figures 1.24 and 1.26 for examples of three-channel PCG – ECG – carotid pulse recordings that illustrate the D – S2 – T relationships. The dicrotic notch may thus be used as a reliable indicator of the end of systole or beginning of diastole that may be obtained in a noninvasive manner. The average S2 – D delay has been found to be 42.6 ms with a standard deviation of 5 ms [66] (see also Tavel [41]), which should be subtracted from the dicrotic notch position to obtain the beginning of S2.

Signal processing techniques for the detection of the dicrotic notch and segmentation of the PCG will be described in Sections 4.3.3, 4.10, and 4.11.

2.2.3   The ECG and the atrial electrogram

Most studies on the ECG and the PCG pay more attention to ventricular activity than to atrial activity, and even then, more to *left* ventricular activity than to the right. Rhythm analysis is commonly performed using QRS complexes to obtain inter-beat intervals known as RR intervals. Such analysis neglects atrial activity.

Recollect that the AV node introduces a delay between atrial contraction initiated by the SA node impulse and the consequent ventricular contraction. This delay plays a major role in the coordinated contraction of the atria and the ventricles. Certain pathological conditions may disrupt this coordination, and even cause AV dissociation [23]. It then becomes necessary to study atrial activity independent of ventricular activity and establish their association, or lack thereof. Thus the interval between the P wave and the QRS (termed the PR interval) would be a valuable adjunct to the RR interval in rhythm analysis. Unfortunately, the atria, being relatively small chambers with weak contractile activity, cause a small and smooth P wave in the external ECG. Quite often the P wave may not be recorded or seen in the external ECG; see, for example, leads I and V3 – V6 in Figure 1.18.

Problem: *Obtain an indicator of atrial contraction to measure the PR interval.*

Solution: One of the reasons for the lack of specificity of the P wave is the effect of transmission from the atria to the external recording sites. An obvious solution would be to insert electrodes into one of the atria via a catheter and record the signal at the source. This would, of course, constitute an invasive procedure. Jenkins et al. [67, 68, 29, 30] proposed a unique and very interesting procedure to obtain a strong and clear signal of atrial activity: they developed a pill electrode that could be swallowed and lowered through the esophagus to a position close to the left atrium (the bipolar electrode pill being held suspended by wires about 35 cm from the lips). The procedure may or may not be termed invasive, although an object is inserted into the body (and removed after the procedure), as the action required is that of normal swallowing of a tablet-like object. The gain required to obtain a good atrial signal was $2 - 5$ times that used in ECG amplifiers. With a $5 - 100$ Hz bandpass filter, Jenkins et al. obtained an SNR of 10.

Figure 2.1 shows recordings from a normal subject of the atrial electrogram from the pill electrode and an external ECG lead. Atrial contraction is clearly indicated by a sharp spike in the atrial electrogram. Measurement of the PR interval (or the AR interval, as called by Jenkins et al.) now becomes an easy task, with identification of the spike in the atrial electrogram (the "A" wave, as labeled by Jenkins et al.) being easier than identification of the QRS in the ECG.

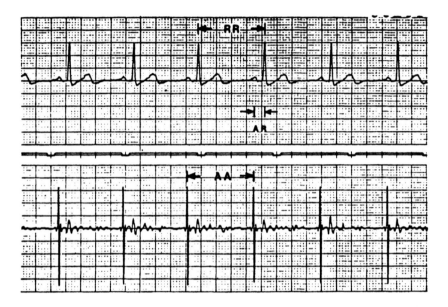

Figure 2.1 Pill-electrode recording of the atrial electrogram (lower tracing) and the external ECG (upper tracing) of a normal subject. The pulse train between the two signals indicates intervals of 1 s. Reproduced with permission from J.M. Jenkins, D. Wu, and R. Arzbaecher, Computer diagnosis of abnormal cardiac rhythms employing a new P-wave detector for interval measurement, *Computers and Biomedical Research*, 11:17–33, 1978. ©Academic Press.

Figure 2.2 shows the atrial electrogram and external ECG of a subject with ectopic beats. The PVCs have no immediately preceding atrial activity. The first PVC has blocked the conduction of the atrial activity occurring immediately after, resulting in a compensatory pause before the following normal beat. The second PVC has not blocked the subsequent atrial wave, but has caused a longer-than-normal AV delay and an aberrant conduction path, which explains the different waveshape of the consequent beat. The third PVC has not affected the timing of the following SA-node-initiated pulse, but has caused a change in waveshape in the resulting QRS-T by altering the conduction path [67, 68, 29, 30].

Jenkins et al. developed a four-digit code for each beat, as illustrated in Figure 2.2. The first digit was coded as

0: abnormal waveshape, or

1: normal waveshape,

as determined by a correlation coefficient computed between the beat being processed and a normal template (see Sections 3.3.1, 4.4.2, and 5.4.1). The remaining three digits encoded the nature of the RR, AR, and AA intervals, respectively, as

0: short,

1: normal, or

2: long.

The absence of a preceding A wave related to the beat being analyzed was indicated by the code x in the fourth digit (in which case the AR interval is longer than the RR interval). Figure 2.2 shows the code for each beat. Based upon the code for each beat, Jenkins et al. were able to develop a computerized method to detect a wide variety of arrhythmia.

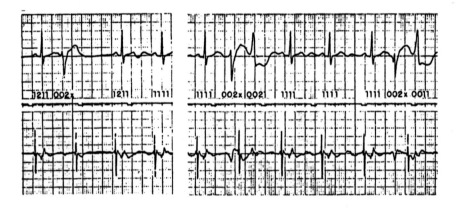

Figure 2.2   Atrial electrogram (lower tracing) and the external ECG (upper tracing) of a subject with ectopic beats. The pulse train between the two signals indicates intervals of 1 s. Reproduced with permission from J.M. Jenkins, D. Wu, and R. Arzbaecher, Computer diagnosis of abnormal cardiac rhythms employing a new P-wave detector for interval measurement, *Computers and Biomedical Research*, 11:17–33, 1978. ©Academic Press.

2.2.4 Cardio-respiratory interaction

The heart rate is affected by normal breathing due to the coupling and interaction existing between the cardiac and respiratory systems [69, 70, 71, 72, 73, 74]. Breathing also affects the transmission of the heart sounds from the cardiac chambers to the chest surface. Durand et al. [75] recorded intracardiac and chest-surface PCG signals and derived the dynamic transfer function of the heart – thorax acoustic system in dogs. Analysis of the synchronization and coupling within the cardio-respiratory system could require sophisticated analysis of several signals acquired simultaneously from the cardiac and respiratory systems [76]. A few techniques for the analysis of heart-rate variability (HRV) based upon RR interval data will be described in Sections 7.2.2, 7.8, and 8.9.

2.2.5 The electromyogram and the vibromyogram

The EMG signal has been studied extensively and the relationship between EMG signal parameters and muscle contraction level has been established [22, 24]. It is known that the EMG root mean-squared (RMS) and mean frequency values increase with increasing muscle contraction until fatigue sets in, at which point both values begin to decrease. In this situation, while the muscle output measured is mechanical contraction (using force or strain transducers), the signal analyzed is electrical in character. A direct mechanical signal related to basic muscle-fiber or motor unit phenomena may be desired in some situations.

Problem: *Obtain a mechanical signal that is a direct indicator of muscle-fiber or motor unit activity to study muscle contraction and force development.*

Solution: The VMG, as introduced in Section 1.2.12, is a vibration signal measured from a contracting muscle. The signal is a direct manifestation of the contraction of muscle fibers, and as such represents mechanical activity at the muscle-fiber or motor-unit level. The VMG signal is the mechanical counterpart and contemporary of the EMG signal. Although no direct relationship has been established between the force outputs of individual motor units and the net force output of the muscle, it has been shown that the RMS and mean frequency parameters of the VMG signal increase with muscle force output, in patterns that parallel those of the EMG. Thus the VMG may be used to quantify muscular contraction [47].

Given the simplicity and noninvasive nature of EMG and VMG measurement, simultaneous analysis of the two signals is an attractive and viable application. Such techniques may find use in biofeedback and rehabilitation [48]. Figure 2.3 shows simultaneous EMG – VMG recodings at two levels of contraction of the rectus femoris muscle [48]. Both signals are interference patterns of several active motor units even at low levels of muscle effort, and cannot be analyzed visually. However, a general increase in the power levels of the signals from the lower effort to the higher effort case may be observed. Signal processing techniques for simultaneous EMG – VMG studies will be described in Section 5.10.

2.2.6 The knee-joint and muscle vibration signals

We saw in Section 1.2.13 that the vibration (VAG) signals produced by the knee joint during active swinging movement of the leg may bear diagnostic information. However, the VMG associated with the rectus femoris muscle that must necessarily be active during extension of the leg could appear as an interference and corrupt the VAG signal [63].

Problem: *Suggest an approach to remove muscle-contraction interference from the knee-joint vibration signal.*

Solution: The VMG interference signal gets transmitted from the source muscle location to the VAG recording position at the skin surface over the patella (knee cap) through the intervening muscles and bones (see Figure 3.11 and Section 3.2.6). Although the interference signal has been found to be of very low frequency (around 10 Hz), the frequency content of the signal varies with muscular effort and knee-joint

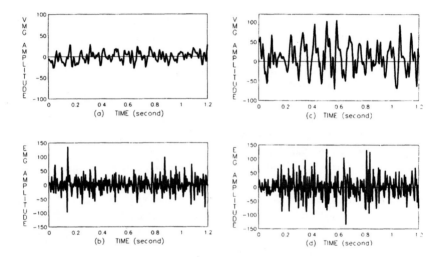

Figure 2.3 Simultaneous EMG – VMG records at two levels of contraction of the rectus femoris muscle. (a) VMG at 40% of the maximal voluntary contraction (MVC) level. (b) EMG at 40% MVC. (c) VMG at 60% MVC. (d) EMG at 60% MVC. Reproduced with permission from Y.T. Zhang, C.B. Frank, R.M. Rangayyan, and G.D. Bell, Relationships of the vibromyogram to the surface electromyogram of the human rectus femoris muscle during voluntary isometric contraction, *Journal of Rehabilitation Research and Development*, 33(4): 395–403, 1996. ©Department of Veterans Affairs.

angle. The rectus femoris muscle and the knee-joint systems are coupled dynamic systems with vibration characteristics that vary with activity level, and hence time; thus simple highpass or bandpass filtering of the VAG signal is not an appropriate solution.

An approach to solve the problem would be to record the VMG signal at the rectus femoris at the same time as the VAG signal of interest is acquired from the patella position. Adaptive filtering and noise cancellation techniques [77, 62, 63] could then be applied, with the VAG signal as the primary input and the VMG signal as the reference input. Assuming that the VMG signal that arrives at the patella is strongly correlated with the VMG signal at the rectus femoris and not correlated with the VAG signal of interest, the adaptive filter should remove the interference and estimate the desired VAG signal. Details of adaptive filters will be provided in Sections 3.6 and 3.10.

2.3 APPLICATION: SEGMENTATION OF THE PCG INTO SYSTOLIC AND DIASTOLIC PARTS

Problem: *Show how the ECG and carotid pulse signals may be used to break a PCG signal into its systolic and diastolic parts.*

Solution: A cardiac cycle may be divided into two important parts based upon ventricular activity: systole and diastole. The systolic part starts with S1 and ends at the beginning of S2; it includes any systolic murmur that may be present in the signal. The diastolic part starts with S2, and ends just before the beginning of the S1 of the next cardiac cycle. (The aortic and pulmonary valves close slightly before the A2 and P2 components of S2. Therefore systole may be considered to have ended just before S2. Although Shaver et al. [43] and Reddy et al. [44] have included S2 in the part of their article on systolic sounds, we shall include S2 in the diastolic part of the PCG.) The diastolic part includes any diastolic murmur that may be present in the signal; it might also include S3 and S4, if present, as well as AV valve-opening snaps, if any.

We saw in Section 2.2.1 that the QRS complex in the ECG may be used as a reliable marker of the beginning of S1. We also saw, in Section 2.2.2, that the dicrotic notch in the carotid pulse may be used to locate the beginning of S2. Thus, if we have both the ECG and carotid pulse signals along with the PCG, it becomes possible to break the PCG into its systolic and diastolic parts.

Figure 2.4 shows three-channel PCG – ECG – carotid pulse signals of a subject with systolic murmur due to aortic stenosis (the same as in Figure 1.26), with the systolic and diastolic parts of the PCG marked in relation to the QRS and D events. The demarcation was performed by visual inspection of the signals in this example. Signal processing techniques to detect the QRS and D waves will be presented in Section 4.3. Adaptive filtering techniques to break the PCG into stationary segments without the use of any other reference signal will be described in Section 8.8.

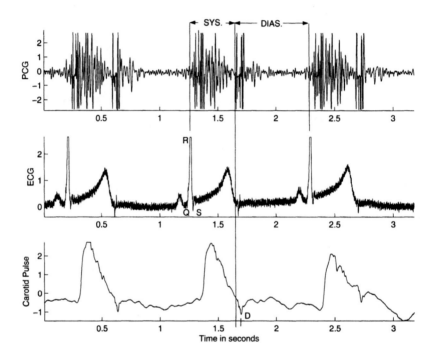

Figure 2.4   Demarcation of the systolic (SYS.) and diastolic (DIAS.) parts of the PCG signal in Figure 1.26 by using the ECG and carotid pulse as reference signals. The QRS complex and the dicrotic notch D are marked on the ECG and carotid pulse signals, respectively.

2.4 REMARKS

This chapter has introduced the notion of using multiple channels of biomedical signals to obtain information on concurrent, coupled, and correlated phenomena with the aim of obtaining an improved understanding of a system or obtaining reference signals for various purposes. The main point to note is that physiological systems are complex systems with multiple variables and outputs that should be studied from various approaches in order to gain multifaceted information.

Some of the problems have been stated in fairly general terms due to the introductory nature of the chapter. Subsequent chapters will present more illustrations of specific problems and applications of the notions gained from this chapter. A number of examples will be provided to illustrate the use of multiple channels of signals to obtain timing information.

2.5 STUDY QUESTIONS AND PROBLEMS

1. A patient has ventricular bigeminy: every second pulse from the SA node is replaced by a premature ventricular ectopic beat (PVC) with a full compensatory pause. (See Figure 9.5 for an illustration of bigeminy.) The SA-node rate is regular at 80 beats a minute, and each ectopic beat precedes the blocked SA-node pulse by 100 ms.

   Draw a schematic three-channel representation of the ECG, the atrial electrogram (or SA-node firing pattern), and the firing pattern of the ectopic focus for 10 beats, marking the time scale in detail. Identify the correspondences and relationships between the activities in the three channels.

2. Draw schematic representations of the ECG, PCG, and carotid pulse signals. Label all waves in the three signals. Identify their common relationships to events in the cardiac cycle.

2.6 LABORATORY EXERCISES AND PROJECTS

(*Note:* The following projects require access to a physiological signal recording laboratory.)

1. Using a multichannel biomedical signal acquisition system, obtain simultaneous recordings of an ECG channel and a signal related to respiration (temperature, airflow, or pressure in the nostril). Study the variations in the RR interval with inspiration and expiration. Repeat the experiment with the subject holding his/her breath during the signal acquisition period.

2. Obtain simultaneous recordings of an ECG lead, the PCG, the carotid pulse, and the pulse at the wrist. Study the temporal correspondences (and delays) between events in the various channels.

3. Record an ECG lead and PCG signals from two or three auscultation areas (mitral, aortic, pulmonary, tricuspid, and apex: see Figure 1.17) simultaneously. Study the variations in the intensities and characteristics of S1 and S2 and their components in the PCGs from the various recording sites.

3

Filtering for Removal of Artifacts

Most biomedical signals appear as weak signals in an environment that is teeming with many other signals of various origins. Any signal other than that of interest could be termed as an interference, artifact, or simply *noise*. The sources of noise could be physiological, the instrumentation used, or the environment of the experiment.

This chapter starts with an introduction to the nature of the artifacts that are commonly encountered in biomedical signals. Several illustrations of signals corrupted by various types of artifacts are provided. Details of the design of filters, spanning a broad range of approaches, from linear time-domain and frequency-domain fixed filters to the optimal Wiener filter to adaptive filters, are then described. The chapter concludes with demonstrations of application of the filters described to ECG and VAG signals.

(*Note:* A good background in signal and system analysis [1, 2, 3] as well as probability, random variables, and stochastic processes [4, 5, 6, 7, 8, 9] is required, in order to follow the procedures and analysis described in this chapter. Familiarity with systems theory and transforms such as the Laplace transform, the Fourier transform in both the continuous and discrete form, and the z-transform will be assumed.)

3.1 PROBLEM STATEMENT

Noise is omnipresent! The problems caused by artifacts in biomedical signals are vast in scope and variety; their potential for degrading the performance of the most sophisticated signal processing algorithms is high. The enormity of the problem of noise removal and its importance are reflected by the size of this chapter and its placement as the first one on signal processing techniques. Let us start with a generic statement of the problem and investigate its nature:

Analyze the various types of artifacts that corrupt biomedical signals and explore filtering techniques to remove them without degrading the signal of interest.

If during an ECG acquisition procedure the subject coughs or squirms, the EMG associated with such activity will pose an interference or artifact. In adult patients, such physiological interference may be minimized by strict instructions and self-control; this solution may, however, not be applicable to infants and children. An intriguing example of physiological interference is that of the mother's ECG appearing along with that of the fetus, with the latter being of interest. No external control is feasible or desirable in this case, and the investigator is forced to develop innovative solutions to extract the signal of interest.

Due to the weak levels of most biomedical signals at their source, high amplification factors of several hundred to several thousand may be required. Electronic noise in the instrumentation amplifiers also gets amplified along with the desired signal. While it is possible to reduce the thermal component of the noise by cooling the devices to very low temperatures, this step may not be practical in most applications; the cost could also be prohibitive. Low-noise power supplies and modern electronic amplifiers with high input impedance, high common-mode rejection ratio, and high power-supply rejection ratio are desirable for the acquisition of biomedical signals [10].

Our environment is filled with EM waves, both natural and man-made. EM waves broadcast by radio and television (TV) stations and those radiated by fluorescent lighting devices, computer monitors, and other systems used in the laboratory or work environment are picked up by cables, devices, and connectors. The $50 \, Hz$ or $60 \, Hz$ power-supply waveform is notorious for the many ways in which it can get mixed with and corrupt the signal of interest. Such interference may be termed as being due to the environment of the experiment. Simple EM shielding of cables and grounding of the chassis of equipment reduce EM and power-supply interference in most cases. Experiments dealing with very weak signals such as ERPs and EEGs may require a wire-mesh-shielded cage to contain the subject and the instruments.

The ECG is a relatively strong signal with a readily identifiable waveform. Most types of interference that affect ECG signals may be removed by bandpass filters. Other signals of less recognizable waveforms and broader bandwidths may not be amenable to simple filtering procedures. In the case of signals such as ERPs or SEPs the noise levels could be much higher than the signal levels, rendering the latter unrecognizable in a single recording. It is important to gain a good understanding of the noise processes involved before one attempts to filter or preprocess a signal.

3.1.1   Random noise, structured noise, and physiological interference

A *deterministic signal* is one whose value at a given instant of time may be computed using a closed-form mathematical function of time, or predicted from a knowledge of a few past values of the signal. A signal that does not meet this condition may be labeled as a *nondeterministic signal* or a random signal.

Test for randomness: Random signals are generally expected to display more excursions about a certain reference level within a specified interval than signals that are predictable. Kendall [78] and Challis and Kitney [79] recommend a test for randomness based upon the number of peaks or troughs in the signal. A peak or a trough is defined by a set of three consecutive samples of the signal, with the central sample being either the maximum or minimum, respectively. As the direction of excursion of the signal changes at peaks and troughs, such points are collectively known as *turning points*. A simple test for a turning point is that the sign of the first-order difference (derivative) at the current sample of the signal be not equal to that at the preceding sample. Given a signal of N samples, the signal may be labeled as being random if the number of turning points is greater than the threshold $\frac{2}{3}(N - 2)$ [78, 79]. In the case of a signal of varying characteristics, that is, a nonstationary signal, the test would have to be conducted using a running window of N samples. The width of the window should be chosen, keeping in mind the shortest duration over which the signal may remain in a given state. The method as above was used by Mintchev et al. [39] to study the dynamics of the level of randomness in EGG signals.

Figure 3.1 illustrates the variation in the number of turning points in a moving window of 50 ms (400 samples with the sampling frequency $f_s = 8\ kHz$) for the speech signal of the word "safety". The threshold for randomness for $N = 400$ according to the rule above is 265. It is seen from the figure that the test indicates that the signal is random for the fricatives /S/ (over the interval of $0.2 - 0.4\ s$, approximately) and /F/ $(0.7 - 0.9\ s)$, and not random for the remaining portions, as expected. (See also Section 1.2.11 and Figures 1.29 and 1.30.)

Random noise: The term *random noise* refers to an interference that arises from a random process such as thermal noise in electronic devices. A random process is characterized by the probability density function (PDF) representing the probabilities of occurrence of all possible values of a random variable. (See Papoulis [4] or Bendat and Piersol [5] for background material on probability, random variables, and stochastic processes.) Consider a random process η that is characterized by the PDF $p_\eta(\eta)$. The mean μ_η of the random process η is given by the first-order moment of the PDF, defined as

$$\mu_\eta = E[\eta] = \int_{-\infty}^{\infty} \eta\, p_\eta(\eta)\, d\eta, \tag{3.1}$$

where $E[\ ]$ represents the *statistical expectation operator*. It is common to assume the mean of a random noise process to be zero.

The mean-squared (MS) value of the random process η is given by the second-order moment of the PDF, defined as

$$E[\eta^2] = \int_{-\infty}^{\infty} \eta^2\, p_\eta(\eta)\, d\eta. \tag{3.2}$$

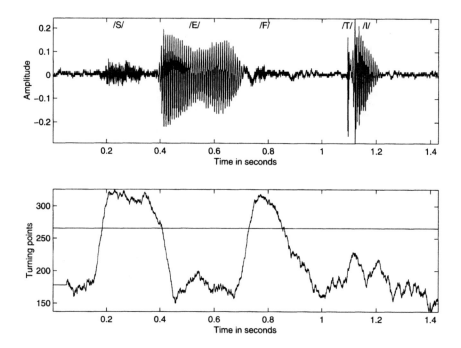

Figure 3.1   Top: Speech signal of the word "safety" uttered by a male speaker. Bottom: Count of turning points in a moving window of 50 ms (400 samples with $f_s = 8\ kHz$). The threshold for randomness for $N = 400$ is 265.

The variance σ_η^2 of the process is defined as the second central moment:

$$\sigma_\eta^2 = E[(\eta - \mu_\eta)^2] = \int_{-\infty}^{\infty} (\eta - \mu_\eta)^2 \, p_\eta(\eta) \, d\eta. \tag{3.3}$$

The square root of the variance gives the standard deviation (SD) σ_η of the process. Note that $\sigma_\eta^2 = E[\eta^2] - \mu_\eta^2$. If the mean is zero, it follows that $\sigma_\eta^2 = E[\eta^2]$, that is, the variance and the MS values are the same.

When the values of a random process η form a time series or a function of time, we have a random signal (or a stochastic process) $\eta(t)$. The statistical measures described above then have physical meanings: the mean represents the DC component, the MS value represents the average power, and the square root of the mean-squared value (the root mean-squared or RMS value) gives the average noise magnitude or level. The measures are useful in calculating the SNR, which is commonly defined as the ratio of the peak-to-peak amplitude range of the signal to the RMS value of the noise, or as the ratio of the average power of the signal to that of the noise.

Observe the use of the same symbol η to represent the random variable, the random process, and the random signal as a function of time. The subscript of the PDF or the statistical parameter derived indicates the random process of concern. The context of the discussion or expression should make the meaning of the symbol clear.

A biomedical signal of interest $x(t)$ may also, for the sake of generality, be considered to be a realization of a random process x. For example, although a normal heart sound signal is heard as the same comforting *lub – dub* sound over every cycle, the corresponding PCG vibration waveforms are not precisely the same from one cycle to another. The PCG signal may be represented as a random process exhibiting certain characteristics *on the average*.

When a (random) signal $x(t)$ is observed in an environment with random noise, the measured signal $y(t)$ may be treated as a realization of another random process y. In most cases the noise is additive, and the observed signal is expressed as

$$y(t) = x(t) + \eta(t). \tag{3.4}$$

Each of the random processes x and y is characterized by its own PDF $p_x(x)$ and $p_y(y)$, respectively.

In most practical applications, the random processes representing a signal of interest and the noise affecting the signal may be assumed to be *statistically independent processes*. Two random processes x and η are said to be statistically independent if their joint PDF $p_{x,\eta}(x, \eta)$ is equal to the product of their individual PDFs given as $p_x(x)p_\eta(\eta)$. It then follows that the first-order moment and second-order central moment of the signals $x(t)$ and $y(t)$ are related as

$$E[y] = \mu_y = E[x] = \mu_x, \tag{3.5}$$

$$E[(y - \mu_y)^2] = \sigma_y^2 = \sigma_x^2 + \sigma_\eta^2, \tag{3.6}$$

where μ represents the mean and σ^2 represents the variance of the random process indicated by the subscript, and it is assumed that $\mu_\eta = 0$.

Ensemble averages: When the PDFs of the random processes of concern are not known, it is common to approximate the statistical expectation operation by averages computed using a collection or *ensemble* of sample observations of the random process. Such averages are known as *ensemble averages*. Suppose we have M observations of the random process x as functions of time: $x_1(t), x_2(t), \ldots, x_M(t)$. We may estimate the mean of the process at a particular instant of time t_1 as

$$\mu_x(t_1) = \lim_{M \to \infty} \frac{1}{M} \sum_{k=1}^{M} x_k(t_1). \tag{3.7}$$

Figure 3.2 illustrates ten sample acquisitions of flash visual ERPs (see also Figure 3.12). The vertical lines at $t = t_1$ and $t = t_2 = t_1 + \tau$ represent the ensemble averaging process at two different instants of time.

The autocorrelation function (ACF) $\phi_{xx}(t_1, t_1 + \tau)$ of a random process x that is a time series is given by

$$\phi_{xx}(t_1, t_1 + \tau) = E[x(t_1)x(t_1 + \tau)] = \int_{-\infty}^{\infty} x(t_1) \, x(t_1 + \tau) \, p_x(x) \, dx, \tag{3.8}$$

which may be estimated as

$$\phi_{xx}(t_1, t_1 + \tau) = \lim_{M \to \infty} \frac{1}{M} \sum_{k=1}^{M} x_k(t_1) \, x_k(t_1 + \tau), \tag{3.9}$$

where τ is the delay parameter. If the signals are complex, one of the functions in the expression above should be conjugated; in this book we shall deal with physiological signals that are always real. The two vertical lines at $t = t_1$ and $t = t_2 = t_1 + \tau$ in Figure 3.2 represent the ensemble averaging process to compute $\phi_{xx}(t_1, t_2)$. The ACF indicates how the values of a signal at a particular instant of time are statistically related to (or have characteristics in common with) values of the same signal at another instant of time.

When dealing with random processes that are observed as functions of time (or stochastic processes), it becomes possible to compute ensemble averages at every point of time. Then, we obtain an averaged function of time $\bar{x}(t)$ as

$$\bar{x}(t) = \mu_x(t) = \frac{1}{M} \sum_{k=1}^{M} x_k(t) \tag{3.10}$$

for all time t. The signal $\bar{x}(t)$ may be used to represent the random process x as a prototype; see the last trace (framed) in Figure 3.2.

Time averages: When we have a sample observation of a random process $x_k(t)$ as a function of time, it is possible to compute *time averages* or *temporal statistics* by integrating along the time axis:

$$\mu_x(k) = \lim_{T \to \infty} \frac{1}{T} \int_{-T/2}^{T/2} x_k(t) \, dt. \tag{3.11}$$

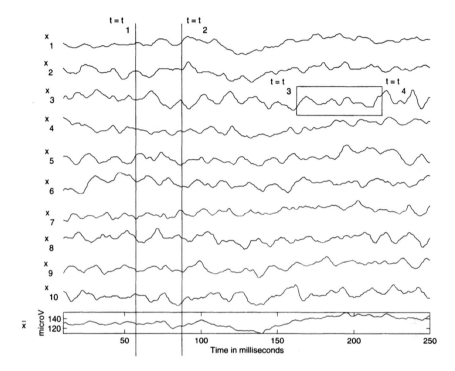

Figure 3.2   Ten sample acquisitions (x_1 to x_{10}) of individual flash visual ERPs from the occipital midline (oz) position of a normal adult male (the author of this book!). The ear lobes were used to form the reference lead (a1a2), and the left forehead was used as the reference (see Figure 1.20). The signals may be treated as ten realizations of a random process in the form of time series or signals. The vertical lines at $t = t_1$ and $t = t_2 = t_1 + \tau$ represent the ensemble averaging process at two different instants of time. The last plot (framed) gives the ensemble average or prototype $\bar{x}(t)$ of the ten individual signals. The horizontal box superimposed on the third trace represents the process of computing temporal statistics over the duration $t = t_3$ to $t = t_4$ of the sample ERP $x_3(t)$. See also Figure 3.12. Data courtesy of L. Alfaro and H. Darwish, Alberta Children's Hospital, Calgary.

The integral would be replaced by a summation in the case of sampled or discrete-time signals. The time-averaged ACF $\phi_{xx}(\tau, k)$ is given by

$$\phi_{xx}(\tau, k) = \lim_{T \to \infty} \frac{1}{T} \int_{-T/2}^{T/2} x_k(t) \, x_k(t+\tau) \, dt. \tag{3.12}$$

(See Section 6.4 for details on estimation of the ACF of finite-length data sequences.) The horizontal box superimposed on the third trace in Figure 3.2 represents the process of computing temporal statistics over the duration $t = t_3$ to $t = t_4$ of the sample ERP $x_3(t)$ selected from the ensemble of ERPs illustrated in the figure.

Random noise may thus be characterized in terms of ensemble and/or temporal statistics. The mean does not play an important role: it is usually assumed to be zero, or may be subtracted out if it is not zero. The ACF plays an important role in the characterization of random processes. The Fourier transform (FT) of the ACF is the power spectral density (PSD) function, which is useful in spectral analysis and filter design.

Covariance and cross-correlation: When two random processes x and y need to be compared, we could compute the covariance between them as

$$C_{xy} = E[(x-\mu_x)(y-\mu_y)] = \int_{-\infty}^{\infty} \int_{-\infty}^{\infty} (x-\mu_x)(y-\mu_y) \, p_{x,y}(x, y) \, dx \, dy, \tag{3.13}$$

where $p_{x,y}(x, y)$ is the joint PDF of the two processes. The covariance parameter may be normalized to get the correlation coefficient, defined as

$$\rho_{xy} = \frac{C_{xy}}{\sigma_x \sigma_y}, \tag{3.14}$$

with $-1 \le \rho_{xy} \le +1$. A high covariance indicates that the two processes have similar statistical variability or behavior. The processes x and y are said to be uncorrelated if $\rho_{xy} = 0$. Two processes that are statistically independent are also uncorrelated; the converse of this property is, in general, not true.

When dealing with random processes x and y that are functions of time, the cross-correlation function (CCF) between them is defined as

$$\theta_{xy}(t_1, t_1 + \tau) = E[x(t_1)y(t_1 + \tau)] = \int_{-\infty}^{\infty} \int_{-\infty}^{\infty} x(t_1) \, y(t_1 + \tau) \, p_{x,y}(x, y) \, dx \, dy. \tag{3.15}$$

Correlation functions are useful in analyzing the nature of variability and spectral bandwidth of signals, as well as for detection of events by template matching. The discussion on random processes will be continued in the next subsection.

Structured noise: Power-line interference at 50 Hz or 60 Hz is an example of structured noise: the typical waveform of the interference is known in advance. It should, however, be noted that the phase of the interfering waveform will not usually be known. Furthermore, the interfering waveform may not be an exact sinusoid; this is indicated by the presence of harmonics of the fundamental 50 Hz or 60 Hz component.

Physiological interference: As we have already noted, the human body is a complex conglomeration of several systems and processes. Several physiological processes could be active at a given instant of time, each one producing many signals of different types. A patient or experimental subject may not be able to exercise control on all physiological processes and systems. The appearance of signals from systems or processes other than those of interest may be termed as physiological interference; several examples are listed below.

- EMG related to coughing, breathing, or squirming affecting the ECG

- EGG interfering with precordial ECG

- Maternal ECG getting added to the fetal ECG of interest

- ECG interfering with the EEG

- Ongoing EEG in ERPs and SEPs

- Breath, lung, or bowel sounds contaminating the heart sounds (PCG)

- Heart sounds getting mixed with breath or lung sounds

- Muscle sound (VMG) interference in joint sounds (VAG)

- Needle-insertion activity appearing at the beginning of a needle-EMG recording

Physiological interference may not be characterized by any specific waveform or spectral content, and is typically dynamic and nonstationary (varying with the level of the activity of relevance and hence with time; see the next subsection for a discussion on stationarity). Thus simple, linear bandpass filters will usually not be effective in removing physiological interference.

3.1.2 Stationary versus nonstationary processes

We saw in the previous subsection that random processes may be characterized in terms of their ensemble and/or temporal statistics. A random process is said to be *stationary in the strict sense* or *strongly stationary* if its ensemble averages of all orders are independent of time, that is, they do not vary with time. In practice, only first-order and second-order averages are used. A random process is said to be *weakly stationary* or *stationary in the wide sense* if its ensemble mean and ACF do not vary with time. Then, from Equations 3.7 and 3.9, we have $\mu_x(t_1) = \mu_x$ and $\phi_{xx}(t_1, t_1 + \tau) = \phi_{xx}(\tau)$. The ACF is now a function of the delay parameter τ only; the PSD of the process does not vary with time.

A stationary process is said to be *ergodic* if the temporal statistics computed are independent of the sample observed; that is, the same result is obtained for any sample observation $x_k(t)$. The time averages in Equations 3.11 and 3.12 are then independent of k: $\mu_x(k) = \mu_x$ and $\phi_{xx}(\tau, k) = \phi_{xx}(\tau)$. All ensemble statistics may be replaced by temporal statistics when analyzing ergodic processes. Ergodic processes are an important type of stationary random processes since their statistics may be computed from a single observation as a function of time. The use of ensemble and temporal averages for noise filtering will be illustrated in Sections 3.3.1 and 3.3.2, respectively.

Signals or processes that do not meet the conditions described above may be, in general, called *nonstationary processes*. A nonstationary process possesses statistics that vary with time. It is readily seen in Figure 1.15 (see also Figure 3.6) that the mean level (base-line) of the signal is varying over the duration of the signal. Therefore, the signal is nonstationary in the mean, a first-order statistical measure. Figure 3.3 illustrates the variance of the speech signal of the word "safety" computed in a moving window of 50 ms (400 samples with $f_s = 8\ kHz$). As the variance changes significantly from one portion of the signal to another, it should be concluded that the signal is nonstationary in its second-order statistics (variance, SD, or RMS). While the speech signal is stationary in the mean, this is not an important characteristic as the mean is typically removed from speech signals. (A DC signal bears no information related to vibration or sound.)

Note that the variance displays a behavior that is almost the opposite of that of the turning points count in Figure 3.1. Variance is sensitive to changes in amplitude, with large swings about the mean leading to large variance values. The procedure to detect turning points examines the presence of peaks and troughs with no consideration of their relative amplitudes; the low-amplitude ranges of the fricatives in the signal have resulted in low variance values, even though their counts of turning points are high.

Most biomedical systems are dynamic and produce nonstationary signals (for example, EMG, EEG, VMG, PCG, VAG, and speech signals). However, a physical or physiological system has limitations in the rate at which it can change its characteristics. This limitation facilitates breaking a signal into segments of short duration (typically a few tens of milliseconds), over which the statistics of interest are not varying, or may be assumed to remain the same. The signal is then referred to as a *quasi-stationary process*; the approach is known as *short-time analysis*. Figure 3.4 illustrates the spectrogram of the speech signal of the word "safety". The spectrogram was computed by computing an array of magnitude spectra of segments of the signal of duration 64 ms; an overlap of 32 ms was permitted between successive segments. It is evident that the spectral characteristics of the signal vary over its duration: the fricatives demonstrate more high-frequency content than the vowels, and also lack formant (resonance) structure. The signal is therefore nonstationary in terms of its PSD; since the PSD is related to the ACF, the signal is also nonstationary in the second-order statistical measure of the ACF.

Further discussion and examples of techniques of this nature will be presented in Sections 8.4.1 and 8.5. Adaptive signal processing techniques may also be designed to detect changes in certain statistical measures of an observed signal; the signal may then be broken into quasi-stationary segments of variable duration that meet the specified conditions of stationarity. Methods for analysis of nonstationary signals will be discussed in Chapter 8. Adaptive segmentation of the EEG, VAG, and PCG signals will be discussed in Sections 8.5, 8.6, 8.7, and 8.8.

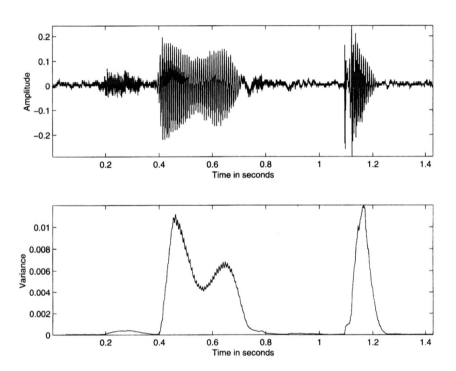

Figure 3.3 Top: Speech signal of the word "safety" uttered by a male speaker. Bottom: Variance computed in a moving window of 50 ms (400 samples with $f_s = 8\ kHz$).

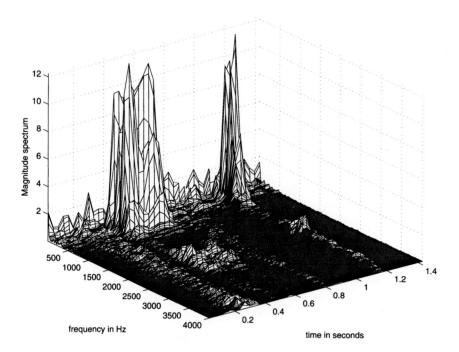

Figure 3.4 Spectrogram of the speech signal of the word "safety" uttered by a male speaker. (The signal is also illustrated in Figures 1.29, 3.1, and 3.3.) Each curve represents the magnitude spectrum of the signal in a moving window of duration 64 ms (512 samples with $f_s = 8\ kHz$), with the window advance interval being 32 ms. The spectrogram is plotted on a linear scale to display better the major differences between the voiced and unvoiced sounds.

Certain systems, such as the cardiac system, normally perform rhythmic operations. The resulting signal, such as the ECG, PCG, or carotid pulse, is then almost periodic, and may be referred to as a *cyclo-stationary signal*. The statistics of the PCG signal vary within the duration of a cardiac cycle, especially when murmurs are present, but repeat themselves at regular intervals. The cyclic repetition of the process facilitates ensemble averaging, using epochs or events extracted from an observation of the signal over many cycles (which is, strictly speaking, a single function of time). Exploitation of the cyclic nature of the ECG signal for synchronized averaging to reduce noise will be illustrated in Section 3.3.1. Application of the same concept to estimate the envelopes of PCG signals will be described in Section 5.5.2. Further extensions of the approach to extract A2 from S2 in PCG signals will be demonstrated in Section 4.11; those to estimate the PSDs of PCG segments in systole and diastole will be presented in Section 6.4.5.

3.2 ILLUSTRATION OF THE PROBLEM WITH CASE-STUDIES

The following case-studies present several examples of various types of interference in biomedical signals of different origins. The aim of this section is to gain familiarity with the various possibilities of interference and their general characteristics. Filtering techniques to remove various types of interference will be described in later sections.

3.2.1 Noise in event-related potentials

An ERP is a signal obtained in response to a stimulus. The response is usually of very small amplitude (of the order of 10 μV), and is buried in ambient EEG activity and noise. The waveform of a single response may be barely recognizable against the background activity. Figure 3.2 shows ten individual flash visual ERP signals. The signals were recorded at the occipital midline (oz) position, with the left and right ear lobes combined to form the reference lead (a1a2). The left forehead was used as the reference. The ERP signals are buried in ongoing EEG and power-line (60 Hz) interference, and cannot be analyzed using the individual acquisitions shown in the figure.

3.2.2 High-frequency noise in the ECG

Figure 3.5 shows a segment of an ECG signal with high-frequency noise. The noise could be due to the instrumentation amplifiers, the recording system, pickup of ambient EM signals by the cables, and so on. The signal illustrated has also been corrupted by power-line interference at 60 Hz and its harmonics, which may also be considered as a part of high-frequency noise relative to the low-frequency nature of the ECG signal.

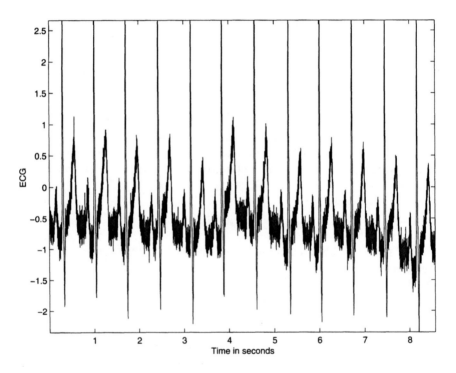

Figure 3.5 ECG signal with high-frequency noise.

3.2.3 Motion artifact in the ECG

Low-frequency artifacts and base-line drift may be caused in chest-lead ECG signals by coughing or breathing with large movement of the chest, or when an arm or leg is moved in the case of limb-lead ECG acquisition. The EGG is a common source of artifact in chest-lead ECG. Poor contact and polarization of the electrodes may also cause low-frequency artifacts. Base-line drift may sometimes be caused by variations in temperature and bias in the instrumentation and amplifiers as well. Figure 3.6 shows an ECG signal with low-frequency artifact. Base-line drift makes analysis of isoelectricity of the ST segment difficult. A large base-line drift may cause the positive or negative peaks in the ECG to be clipped by the amplifiers or the ADC.

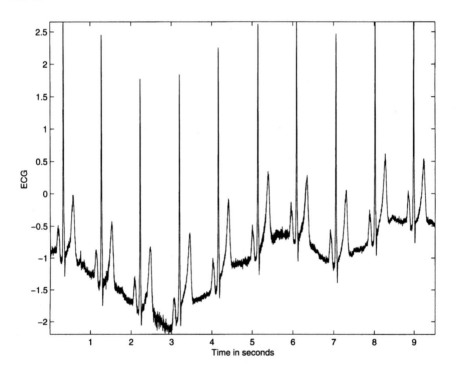

Figure 3.6   ECG signal with low-frequency artifact.

3.2.4 Power-line interference in ECG signals

The most commonly encountered periodic artifact in biomedical signals is the power-line interference at $50\ Hz$ or $60\ Hz$. If the power-line waveform is not a pure sinusoid due to distortions or clipping, harmonics of the fundamental frequency could also appear. Harmonics will also appear if the interference is a periodic waveform that is not a sinusoid (such as rectangular pulses).

Power-line interference may be difficult to detect visually in signals having non-specific waveforms such as the PCG or EMG; however, the interference is easily visible if present on well-defined signal waveforms such as the ECG or carotid pulse signals. In either case, the power spectrum of the signal should provide a clear indication of the presence of power-line interference as an impulse or spike at 50 Hz or 60 Hz; harmonics, if present, will appear as additional spikes at integral multiples of the fundamental frequency.

Figure 3.7 shows a segment of an ECG signal with 60 Hz interference. Observe the regular or periodic structure of the interference, which rides on top of the ECG waves. Figure 3.8 shows the power spectrum of the signal. The periodic interference is clearly displayed as a spike at not only its fundamental frequency of 60 Hz, but also as spikes at 180 Hz and 300 Hz, which represent the third and fifth harmonics, respectively. (The recommended sampling rate for ECG signals is 500 Hz; the higher rate of 1, 000 Hz was used in this case as the ECG was recorded as a reference signal with the PCG. The larger bandwidth also permits better illustration of artifacts and filtering.)

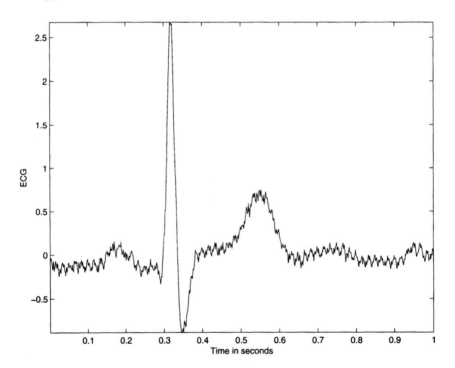

Figure 3.7   ECG signal with power-line (60 Hz) interference.

The bandwidth of interest of the ECG signal, which is usually in the range 0.05 − 100 Hz, includes the 60 Hz component; hence simple lowpass filtering will not be appropriate for removal of power-line interference. Lowpass filtering of the ECG to a bandwidth lower than 60 Hz could smooth and blur the QRS complex as well as

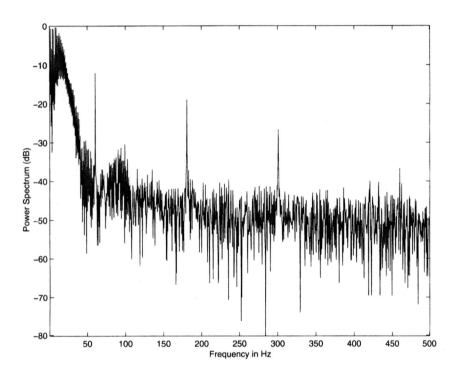

Figure 3.8 Power spectrum of the ECG signal in Figure 3.7 with power-line interference. The spectrum illustrates peaks at the fundamental frequency of 60 Hz as well as the third and fifth harmonics at 180 Hz and 300 Hz, respectively.

affect the PQ and ST segments. The ideal solution would be to remove the 60 Hz component without sacrificing any other component.

3.2.5 Maternal interference in fetal ECG

Figure 3.9 shows an ECG signal recorded from the abdomen of a pregnant woman. Shown also is a simultaneously recorded ECG from the woman's chest. Comparing the two, we see that the abdominal ECG demonstrates multiple peaks (QRS complexes) corresponding to the maternal ECG (occurring at the same time instants as the QRS complexes in the chest lead) as well as several others at weaker levels and a higher repetition rate. The non-maternal QRS complexes represent the ECG of the fetus. Observe that the QRS complex shapes of the maternal ECG from the chest and abdominal leads have different shapes due to the projection of the cardiac electrical vector onto different axes. Given that the two signals being combined have almost the same bandwidth, how would we be able to separate them and obtain the fetal ECG that we would be interested in?

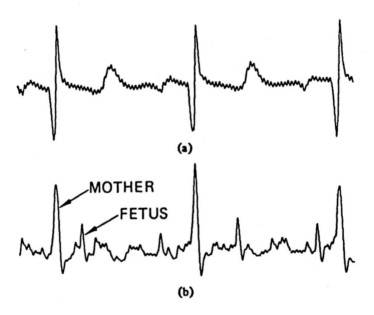

Figure 3.9   ECG signals of a pregnant woman from abdominal and chest leads: (a) chest-lead ECG, and (b) abdominal-lead ECG; the former presents the maternal ECG whereas the latter is a combination of the maternal and fetal ECG signals. (See also Figure 3.58.) Reproduced with permission from B. Widrow, J.R. Glover, Jr., J.M. McCool, J. Kaunitz, C.S. Williams, R.H. Hearn, J.R. Zeidler, E. Dong, Jr., R.C. Goodlin, Adaptive noise cancelling: Principles and applications, *Proceedings of the IEEE*, 63(12):1692–1716, 1975. ©IEEE.

3.2.6 Muscle-contraction interference in VAG signals

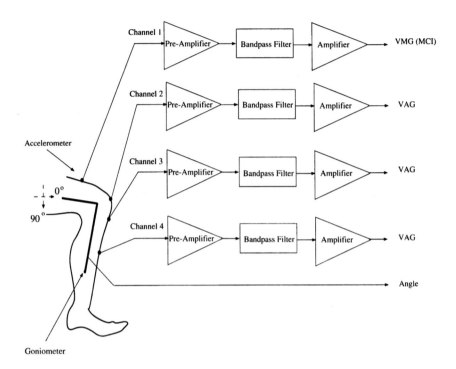

Figure 3.10 Experimental setup to measure VMG and VAG signals at different positions along the leg [63].

Figure 3.10 shows the recording setup used by Zhang et al. [63] to study the possibility of VMG signals appearing as muscle-contraction interference in VAG signals. The left-hand column in Figure 3.11 shows VMG signals recorded using accelerometers placed at the distal rectus femoris (thigh), mid-patella (knee cap), tibial tuberosity, and mid-tibial shaft positions of a subject during isometric contraction of the rectus femoris muscle (with no leg or knee movement). The right-hand column of the figure shows vibration signals recorded at the same positions using the same accelerometers, but during isotonic contraction (swinging movement of the leg). The top signal (a) in the right-hand column indicates the VMG signal generated at the rectus femoris during acquisition of the VAG signals; parts (b) – (d) of the right-hand column show the VAG signals.

VAG signals are difficult to analyze as they have no predefined or recognizable waveforms; it is even more difficult to identify any noise or interference that may be present in VAG signals. The signals shown in Figure 3.11 indicate that a transformed version of the VMG could get added to the VAG, especially during extension of the leg when the rectus femoris muscle is active (the second halves of the VAG signals in parts (b) – (d) of the right-hand column). The left-hand column of VMG signals in Figure 3.11 illustrates that the VMG generated at the distal rectus femoris gets

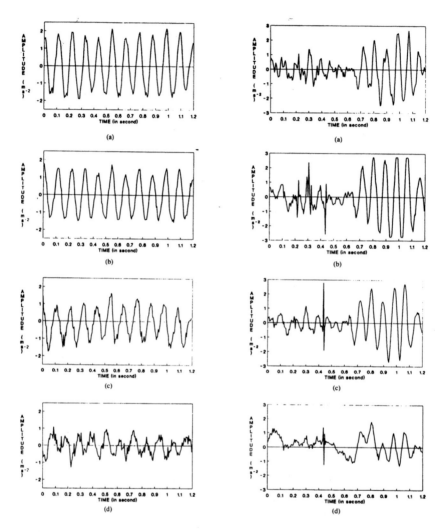

Figure 3.11 Left-hand column: VMG signals recorded simultaneously at (top-to-bottom) (a) the distal rectus femoris, (b) mid-patella, (c) tibial tuberosity, and (d) mid-tibial shaft positions during isometric contraction (no leg or knee movement). Right-hand column: Vibration signals recorded simultaneously at the same positions as above during isotonic contraction (swinging movement of the leg). Observe the muscle-contraction interference appearing in the extension parts (second halves) of each of the VAG signals (plots (b) – (d)) in the right-hand column [63]. The recording setup is shown in Figure 3.10. Reproduced with permission from Y.T. Zhang, R.M. Rangayyan, C.B. Frank, and G.D. Bell, Adaptive cancellation of muscle-contraction interference from knee joint vibration signals, *IEEE Transactions on Biomedical Engineering*, 41(2):181–191, 1994. ©IEEE.

transmitted well down the leg and appears at the other recording positions. It may be observed from the VAG signals in the right-hand column that vibration signals comparable to the VMG are present in the VAG channels (b) – (d) during extension (second halves) but not as prominent in flexion (first halves). Interestingly enough, the knee-joint "crepitus" and click signals that appear in the first half of the VAG signal at the mid-patella position (right (b)) have been transmitted downwards along the leg to the tibial tuberosity (right (c)) and mid-tibial shaft (right (d)) positions farther down the leg, presumably along the tibia, but not upwards to the distal rectus femoris position (right (a)).

It should also be noted that the VAG signal cannot be expected to be the same during the extension and flexion parts of a swing cycle: extension causes more stress or force per unit area on the patello-femoral joint than flexion. Furthermore, the VAG and VMG signals are nonstationary: characteristics of the VAG vary with the quality of the cartilage surfaces that come into contact at different joint angles, while the VMG varies in accordance with the level of contraction of the muscles involved. To make the problem even more difficult, the bandwidths of the two signals overlap in the range of about $0 - 100 \ Hz$. These factors make removal of the VMG or muscle-contraction interference from VAG signals a challenge.

3.2.7 Potential solutions to the problem

Now that we have gained an understanding of a few sources of artifacts in biomedical signals and their nature, we are prepared to look at specific problems and develop effective filtering techniques to solve them. The following sections investigate artifacts of various types and demonstrate increasingly complex signal processing techniques to remove them. The problem statement at the beginning of each section defines the nature of the problem in as general terms as possible, sets the terms and conditions, and defines the scope of the investigation to follow. The solution proposed provides the details of an appropriate filtering technique. Each solution is demonstrated with an illustration of its application. Further examples of application of the techniques studied are provided at the end of the chapter. Comparative evaluation of filtering techniques is also provided where applicable.

A practical problem encountered by an investigator in the field may not precisely match a specific problem considered in this chapter. However, it is expected that the knowledge of several techniques and an appreciation of the results of their application gained from this chapter will help in designing innovative and appropriate solutions to new problems.

3.3 TIME-DOMAIN FILTERS

Certain types of noise may be filtered directly in the time domain using signal processing techniques or digital filters. An advantage of time-domain filtering is that spectral characterization of the signal and noise may not be required (at least in a direct manner). Time-domain processing may also be faster in most cases than frequency-domain filtering.

3.3.1 Synchronized averaging

Problem: *Propose a time-domain technique to remove random noise given the possibility of acquiring multiple realizations of the signal or event of interest.*

Solution: Linear filters fail to perform when the signal and noise spectra overlap. Synchronized signal averaging can separate a repetitive signal from noise without distorting the signal [27, 79]. ERP or SEP epochs may be obtained a number of times by repeated application of the stimulus; they may then be averaged by using the stimulus as a trigger for aligning the epochs. ECG signals may be filtered by detecting the QRS complexes and using their positions to align the waveforms for synchronized averaging. If the noise is random with zero mean and is uncorrelated with the signal, averaging will improve the SNR.

Let $y_k(n)$ represent one realization of a signal, with $k = 1, 2, \ldots, M$ representing the ensemble index, and $n = 1, 2, \ldots, N$ representing the time-sample index. (Some authors use the notation nT, $T = 1/f_s$ being the sampling interval, where f_s is the sampling frequency, to denote the index of a sampled signal; in this book we shall use just n, the sample number.) M is the number of copies (events, epochs, or realizations) of the signal available, and N is the number of time samples in each copy of the signal (event). We may express the observed signal as

$$y_k(n) = x_k(n) + \eta_k(n), \tag{3.16}$$

where $x_k(n)$ represents the original uncorrupted signal and $\eta_k(n)$ represents the noise in the kth copy of the observed signal. Now, if for each instant of time n we add the M copies of the signal, we get

$$\sum_{k=1}^{M} y_k(n) = \sum_{k=1}^{M} x_k(n) + \sum_{k=1}^{M} \eta_k(n); \ n = 1, 2, \ldots, N. \tag{3.17}$$

If the repetitions of the signal are identical and aligned, $\sum_{k=1}^{M} x_k(n) = Mx(n)$. If the noise is random and has zero mean and variance σ_η^2, $\sum_{k=1}^{M} \eta_k(n)$ will tend to zero as M increases, with a variance of $M\sigma_\eta^2$. The RMS value of the noise in the averaged signal is $\sqrt{M}\sigma_\eta$. Thus the SNR of the signal will increase by a factor of $\frac{M}{\sqrt{M}}$ or $\sqrt{M}$. The larger the number of epochs or realizations that are averaged, the better will be the SNR of the result. Note that synchronized averaging is a type of ensemble averaging.

An algorithmic description of synchronized averaging is as follows:

1. Obtain a number of realizations of the signal or event of interest.

2. Determine a reference point for each realization of the signal. This is directly given by the trigger if the signal is obtained by external stimulation (such as

ERPs or SEPs), or may be obtained by detecting the repetitive events in the signal if it is quasi-periodic (such as the QRS complex in the ECG or S1 and S2 in the PCG).

3. Extract parts of the signal corresponding to the events and add them to a buffer. Note that it is possible for the various parts to be of different durations. Alignment of the copies at the trigger point is important; the tail ends of all parts may not be aligned.

4. Divide the result in the buffer by the number of events added.

Figure 3.12 illustrates two single-flash ERPs in the upper two traces. The results of averaging over 10 and 20 flashes are shown in the third and fourth plots, respectively, in the same figure. The averaging process has facilitated identification of the first positivity and the preceding and succeeding troughs (marked on the fourth trace) with certainty; the corresponding features are not reliably seen in the single acquisitions (see also the single-flash ERPs in Figure 3.2). Visual ERPs are analyzed in terms of the latencies of the first major peak or positivity, labeled as P120 due to the fact that the normal expected latency for adults is 120 ms; the trough or negativity before P120, labeled as N80; and the trough following P120, labeled as N145. The N80, P120, and N145 latencies measured from the averaged signal in Trace 4 of Figure 3.12 are 85.7, 100.7, and 117 ms, respectively, which are considered to be within the normal range for adults.

Illustration of application: The upper trace in Figure 3.13 illustrates a noisy ECG signal over several beats. In order to obtain trigger points, a sample QRS complex of 86 ms duration (86 samples at a sampling rate of $1,000\ Hz$) was extracted from the the first beat in the signal and used as a template. Template matching was performed using a normalized correlation coefficient defined as [79]

$$\gamma_{xy}(k) = \frac{\sum_{n=0}^{N-1}[x(n) - \bar{x}][y(n-k) - \bar{y}]}{\sqrt{\sum_{n=0}^{N-1}[x(n) - \bar{x}]^2 \sum_{n=0}^{N-1}[y(n-k) - \bar{y}]^2}}\ , \qquad (3.18)$$

where x is the template, y is the ECG signal, $\bar{x}$ and $\bar{y}$ are the averages of the corresponding signals over the N samples considered, and k is the time index of the signal y at which the template is placed. (Jenkins et al. [67] used a measure similar to $\gamma_{xy}(k)$ but without subtraction of the mean and without the shift parameter k to match segmented ECG cycles with a template.) The lower trace in Figure 3.13 shows $\gamma_{xy}(k)$, where it is seen that the cross-correlation result peaks to values near unity at the locations of the QRS complexes in the signal. Averaging inherent in the cross-correlation formula (over N samples) has reduced the effect of noise on template matching.

By choosing an appropriate threshold, it becomes possible to obtain a trigger point to extract the QRS complex locations in the ECG signal. (*Note:* The QRS template matches with the P and T waves with cross-correlation values of about 0.5; wider QRS complexes may yield higher cross-correlation values with taller P and T waves. The threshold has to be chosen so as to detect only the QRS complexes.) A threshold

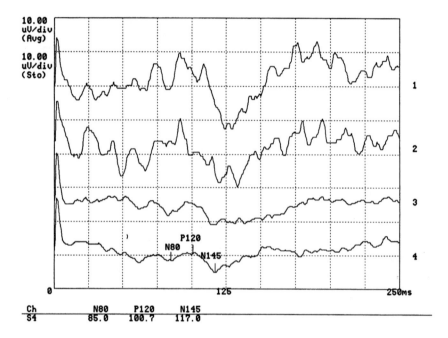

Figure 3.12   Traces 1 and 2: Two sample acquisitions of individual flash visual ERPs from the occipital midline (oz) position of a normal adult male. The ear lobes were used to form the reference lead (a1a2), and the left forehead was used as the reference (see Figure 1.20). Trace 3: Average of 10 ERPs. Trace 4: Average of 20 ERPs. The latencies of interest have been labeled on Trace 4 by an EEG technologist. See also Figure 3.2. Data courtesy of L. Alfaro and H. Darwish, Alberta Children's Hospital, Calgary.

of 0.9 was applied to $\gamma_{xy}(k)$, and the QRS positions of all of the 12 beats in the signal were detected.

Figure 3.14 illustrates two ECG cycles extracted using the trigger points obtained by thresholding the cross-correlation function, as well as the result of averaging the first 11 cycles in the signal. It is seen that the noise has been effectively suppressed by synchronized averaging. The low-level base-line variation and power-line interference present in the signal have caused minor artifacts in the result, which are negligible in this illustration.

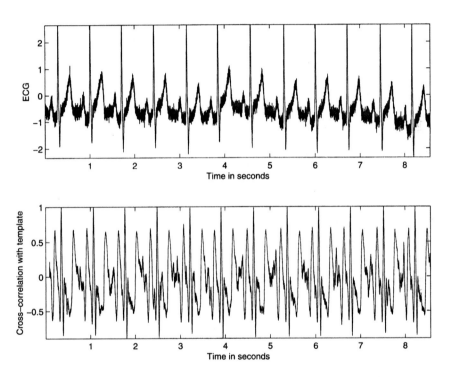

Figure 3.13 An ECG signal with noise (upper trace) and the result of cross-correlation (lower trace) with the QRS template selected from the first cycle. The cross-correlation coefficient is normalized to the range $(-1, 1)$.

The most important requirement in synchronized averaging is indicated by the first word in the name of the process: the realizations of the signal that are added for averaging *must be aligned* such that the repetitive part of the signal appears at exactly the same instant in each realization of the signal. If this condition is not met, the waveform of the event in the signal will be blurred or smudged along the time axis.

A major advantage of synchronized averaging is that no frequency-domain filtering is performed — either explicitly or implicitly. No spectral content of the signal is lost as is the case with frequency-domain (lowpass) filters or other time-domain filters such as moving-window averaging filters.

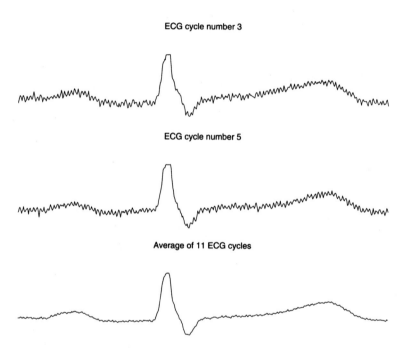

Figure 3.14   Upper two traces: two cycles of the ECG extracted from the signal in Figure 3.13. Bottom trace: the result of synchronized averaging of 11 cycles from the same ECG signal.

Structured noise such as power-line interference may be suppressed by synchronized averaging if the phase of the interference in each realization is different. To facilitate this feature, the repetition rate of the stimulus should be set so that it is not directly related to the power-line frequency (for example, the flashes used to acquire the averaged ERPs in Figure 3.12 were delivered at 2.1 *pps*). Physiological interference such as background EEG in ERPs and SEPs may also be suppressed by synchronized averaging, as such activity may bear no inter-relationship from one epoch of the desired signal to another.

3.3.2 Moving-average filters

Problem: *Propose a time-domain technique to remove random noise given only one realization of the signal or event of interest.*

Solution: When an ensemble of several realizations of an event is not available, synchronized averaging will not be possible. We are then forced to consider temporal averaging for noise removal, with the assumption that the processes involved are ergodic, that is, temporal statistics may be used instead of ensemble statistics. As temporal statistics are computed using a few samples of the signal along the time axis and the temporal window of samples is moved to obtain the output at various points of time, such a filtering procedure is called a moving-window averaging filter in general; the term moving-average (MA) filter is commonly used.

The general form of an MA filter is

$$y(n) = \sum_{k=0}^{N} b_k\, x(n-k), \tag{3.19}$$

where x and y are the input and output of the filter, respectively. The b_k values are the filter coefficients or tap weights, $k = 0, 1, 2, \ldots, N$, where N is the order of the filter. The effect of division by the number of samples used $(N+1)$ is included in the values of the filter coefficients. The signal-flow diagram of a generic MA filter is shown in Figure 3.15.

Applying the z-transform, we get the transfer function $H(z)$ of the filter as

$$H(z) = \frac{Y(z)}{X(z)} = \sum_{k=0}^{N} b_k\, z^{-k} = b_0 + b_1 z^{-1} + b_2 z^{-2} + \cdots + b_N z^{-N}, \tag{3.20}$$

where $X(z)$ and $Y(z)$ are the z-transforms of $x(n)$ and $y(n)$, respectively. (See Lathi [1], Oppenheim et al. [2], or Oppenheim and Schafer [14] for background details on system analysis using the z-transform and the Fourier transform.)

A simple MA filter for filtering noise is the von Hann or Hanning filter [27], given by

$$y(n) = \frac{1}{4}[x(n) + 2x(n-1) + x(n-2)]. \tag{3.21}$$

The signal-flow diagram of the Hanning filter is shown in Figure 3.16. The impulse response of the filter is obtained by letting $x(n) = \delta(n)$, resulting in $h(n) = \frac{1}{4}[\delta(n) + 2\delta(n-1) + \delta(n-2)]$.

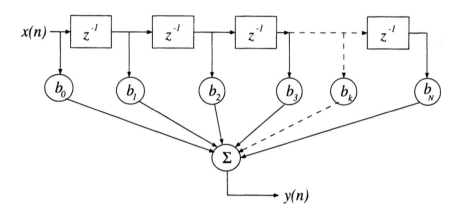

Figure 3.15   Signal-flow diagram of a moving-average filter of order N. Each block with the symbol z^{-1} represents a delay of one sample, and serves as a memory unit for the corresponding signal sample value.

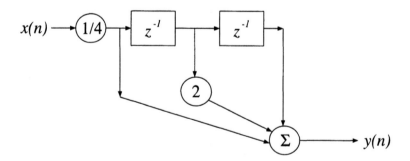

Figure 3.16   Signal-flow diagram of the Hanning filter.

The transfer function of the Hanning filter is

$$H(z) = \frac{1}{4}[1 + 2z^{-1} + z^{-2}]. \tag{3.22}$$

The transfer function has a double-zero at $z = -1$.

An MA filter is a finite impulse response (FIR) filter with the following attributes and advantages:

- The impulse response $h(k)$ has a finite number of terms: $h(k) = b_k,\ k = 0, 1, 2, \ldots, N$.

- An FIR filter may be realized non-recursively with no feedback.

- The output depends only on the present input sample and a few past input samples.

- The filter is merely a set of tap weights of the delay stages, as illustrated in Figure 3.15.

- The filter transfer function has no poles except at $z = 0$: the filter is inherently stable.

- The filter has linear phase if the series of tap weights is symmetric or antisymmetric.

The frequency response of a filter is obtained by substituting $z = e^{j\omega T}$ in the expression for $H(z)$, where T is the sampling interval in seconds and ω is the radian frequency ($\omega = 2\pi f$, where f is the frequency in Hz). Note that we may set $T = 1$ and deal with normalized frequency in the range $0 \leq \omega \leq 2\pi$ or $0 \leq f \leq 1$; then $f = 1$ or $\omega = 2\pi$ represents the sampling frequency, with lower frequency values being represented as a normalized fraction of the sampling frequency.

The frequency response of the Hanning filter is given as

$$H(\omega) = \frac{1}{4}[1 + 2e^{-j\omega} + e^{-j2\omega}]. \tag{3.23}$$

Letting $e^{-j\omega} = \cos(\omega) - j\sin(\omega)$, we obtain

$$H(\omega) = \frac{1}{4}[\{2 + 2\cos(\omega)\}e^{-j\omega}]. \tag{3.24}$$

The magnitude and phase responses are given as

$$|H(\omega)| = \left|\frac{1}{2}\{1 + \cos(\omega)\}\right| \tag{3.25}$$

and

$$\angle H(\omega) = -\omega. \tag{3.26}$$

The magnitude and phase responses of the Hanning filter are plotted in Figure 3.17. It is clear that the filter is a lowpass filter with linear phase.

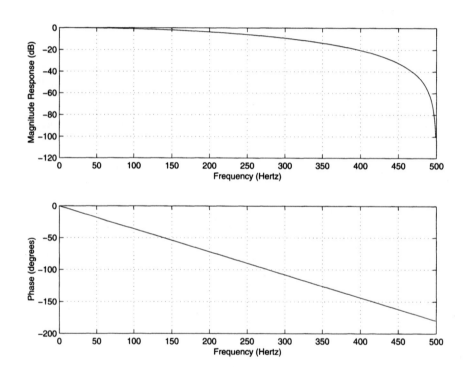

Figure 3.17   Magnitude and phase responses of the Hanning (smoothing) filter.

Note that, although we started with a description of the Hanning filter in the time domain, subsequent analysis of the filter was performed in the frequency domain using the z-transform and the frequency response. System analysis is easier to perform in the z domain in terms of the poles and zeros of the transfer function and in the frequency domain in terms of the magnitude and phase responses. The magnitude and phase responses assist in understanding the effect of the filter on the frequency components of the signal (and noise).

It is seen from the magnitude response of the Hanning filter (Figure 3.17) that components beyond about 20% of the sampling frequency of $1,000\ Hz$ are reduced in amplitude by more than 3 dB, that is, to less than half of their levels in the input. High-frequency components beyond 40% of the sampling frequency are suppressed to less than 20 dB below their input levels. The filter will perform adequate filtering of ECG signals sampled at $200\ Hz$, with the gain being lower than $-20\ dB$ beyond 80 Hz. However, if the signal is sampled at $1,000\ Hz$ (as in the present example), the gain remains above $-20\ dB$ for frequencies up to $400\ Hz$; such a lowpass filter may not be adequate for filtering ECG signals, but may be appropriate for other signals such as the PCG and the EMG.

Increased smoothing may be achieved by averaging signal samples over longer time windows, at the expense of increased filter delay. If the signal samples over a window of eight samples are averaged, we get the output as

$$y(n) = \frac{1}{8} \sum_{k=0}^{7} x(n - k). \tag{3.27}$$

The impulse response of the filter is $h(n) = \frac{1}{8}[\delta(n) + \delta(n-1) + \delta(n-2) + \delta(n-3) + \delta(n-4) + \delta(n-5) + \delta(n-6) + \delta(n-7)]$. The transfer function of the filter is

$$H(z) = \frac{1}{8} \sum_{k=0}^{7} z^{-k}, \tag{3.28}$$

and the frequency response is given by

$$
\begin{aligned}
H(\omega) &= \frac{1}{8} \sum_{k=0}^{7} \exp(-j\omega k) \\
&= \frac{1}{8}[1 + \exp(-j4\omega)] \\
&\times \ \{1 + 2\cos(\omega) + 2\cos(2\omega) + 2\cos(3\omega)\}].
\end{aligned} \tag{3.29}
$$

The frequency response of the 8-point MA filter is shown in Figure 3.18; the pole-zero plot of the filter is depicted in Figure 3.19. It is seen that the filter has zeros at $\frac{f_s}{8} = 125\ Hz$, $\frac{f_s}{4} = 250\ Hz$, $\frac{3f_s}{8} = 375\ Hz$, and $\frac{f_s}{2} = 500\ Hz$. Comparing the frequency response of the 8-point MA filter with that of the Hanning filter in Figure 3.17, we see that the former provides increased attenuation in the range $90 - 400\ Hz$ over the latter. Note that the attenuation provided by the filter after about 100 Hz is nonuniform, which may not be desirable in certain applications. Furthermore, the phase response of the filter is not linear, although it is piece-wise linear.

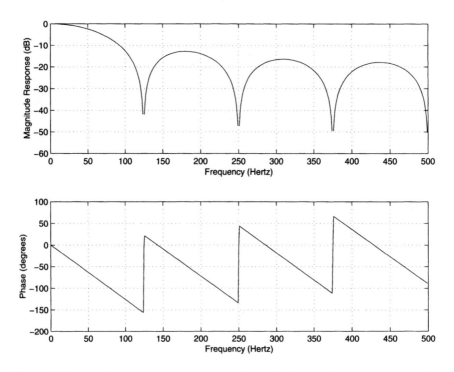

Figure 3.18   Magnitude and phase responses of the 8-point moving-average (smoothing) filter.

Relationship of moving-average filtering to integration: Disregarding the $\frac{1}{8}$ scale factor for a moment, the operation in Equation 3.27 may be interpreted as the summation or integration of the signal over the duration $n - 7$ to n. A comparable integration of a continuous-time signal $x(t)$ over the interval t_1 to t_2 is expressed as

$$y(t) = \int_{t_1}^{t_2} x(t) \, dt. \tag{3.30}$$

The general definition of the integral of a signal is

$$y(t) = \int_{-\infty}^{t} x(t) \, dt, \tag{3.31}$$

or, if the signal is causal,

$$y(t) = \int_{0}^{t} x(t) \, dt. \tag{3.32}$$

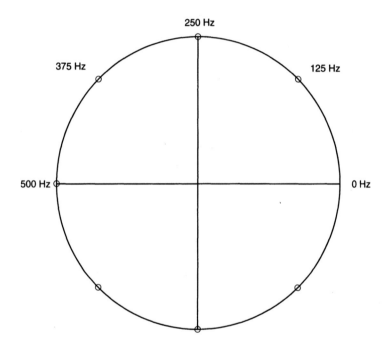

Figure 3.19   Pole-zero plot of the 8-point moving-average (smoothing) filter.

The Fourier transforms of the signals in the relationship above are related as [1, 2]

$$Y(\omega) = \frac{1}{j\omega} X(\omega) + \pi X(0)\delta(\omega). \tag{3.33}$$

The frequency response of the integration operator is

$$H(\omega) = \frac{1}{j\omega}, \tag{3.34}$$

with the magnitude response

$$|H(\omega)| = \left|\frac{1}{\omega}\right| \tag{3.35}$$

and phase response

$$\angle H(\omega) = -\frac{\pi}{2}. \tag{3.36}$$

It is seen from the frequency response that the gain of the filter reduces (nonlinearly) as the frequency is increased; therefore, the corresponding filter has lowpass characteristics.

Integration or accumulation of a discrete-time signal for all samples up to the present sample results in the transfer function $H(z) = \frac{1}{1-z^{-1}}$ [1, 2]. Such an operation is seldom used in practice. Instead, a moving-window sum is computed as in Equation 3.27. The 8-point MA filter may be rewritten as

$$y(n) = y(n-1) + \frac{1}{8}x(n) - \frac{1}{8}x(n-8). \tag{3.37}$$

The recursive form as above clearly depicts the integration aspect of the filter. The transfer function of this expression is easily derived to be

$$H(z) = \frac{1}{8}\left[\frac{1-z^{-8}}{1-z^{-1}}\right]. \tag{3.38}$$

The frequency response of the filter is given by

$$H(\omega) = \frac{1}{8}\left[\frac{1-e^{-j8\omega}}{1-e^{-j\omega}}\right] = \frac{1}{8}e^{-j\frac{7}{2}\omega}\left[\frac{\sin(4\omega)}{\sin(\frac{\omega}{2})}\right], \tag{3.39}$$

which is equivalent to that in Equation 3.29. Summation over a limited discrete-time window results in a frequency response having sinc-type characteristics, as illustrated in Figure 3.18. See Tompkins [27] for a discussion on other types of integrators.

Illustration of application: Figure 3.20 shows a segment of an ECG signal with high-frequency noise. Figure 3.21 shows the result of filtering the signal with the 8-point MA filter described above. Although the noise level has been reduced, some noise is still present in the result. This is due to the fact that the attenuation of the simple 8-point MA filter is not more than $-20\ dB$ at most frequencies (except near the zeros of the filter).

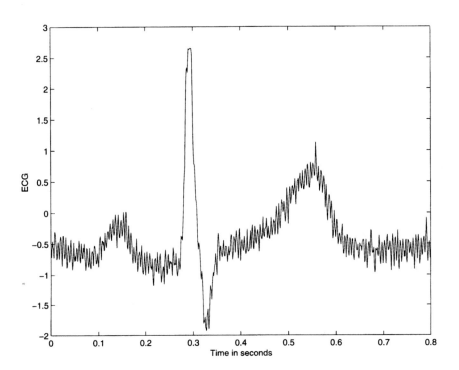

Figure 3.20   ECG signal with high-frequency noise; $f_s = 1,000\ Hz$.

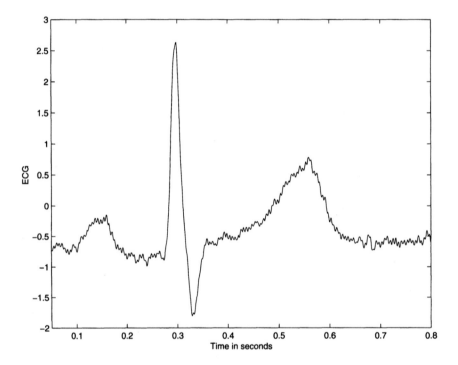

Figure 3.21   The ECG signal with high-frequency noise in Figure 3.20 after filtering by the 8-point MA filter shown in Figure 3.18.

3.3.3   Derivative-based operators to remove low-frequency artifacts

Problem: *Develop a time-domain technique to remove base-line drift in the ECG signal.*

Solution: The derivative operator in the time domain removes the parts of the input that are constant (the output is zero). Large changes in the input lead to high values in the output of the derivative operator. Improved understanding of the derivative operation may be obtained by studying its transform in the frequency domain.

The ideal $\frac{d}{dt}$ operator in the time domain results in multiplication of the Fourier transform of the original signal by $j\,\omega = j\,2\pi f$ in the frequency domain. If $X(f)$ represents the Fourier transform of the signal $x(t)$, then the Fourier transform of $\frac{dx}{dt}$ is $j\,2\pi f X(f)$ or $j\,\omega X(\omega)$. The frequency response of the operation is $H(\omega) = j\,\omega$. It is seen that the gain of the frequency response increases linearly with frequency, starting with $H(\omega) = 0$ at $\omega = 0$. Thus the DC component is removed by the derivative operator, and higher frequencies receive linearly increasing gain: the operation represents a highpass filter. The derivative operator may be used to remove DC and suppress low-frequency components (and boost high-frequency components).

It follows readily that the second-order derivative operator $\frac{d^2}{dt^2}$ has the frequency response $H(\omega) = -\omega^2$, with a quadratic increase in gain for higher frequencies. The second-order derivative operator may be used to obtain even higher gain for higher frequencies than the first-order derivative operator; the former may be realized as a cascade of two of the latter.

In digital signal processing, the basic derivative is given by the first-order difference operator [27]

$$y(n) = \frac{1}{T} \left[x(n) - x(n-1) \right]. \tag{3.40}$$

The scale factor including the sampling interval T is required in order to obtain the rate of change of the signal with respect to the true time. The transfer function of the operator is

$$H(z) = \frac{1}{T} \left(1 - z^{-1} \right). \tag{3.41}$$

The filter has a zero at $z = 1$, the DC point.

The frequency response of the operator is

$$H(\omega) = \frac{1}{T} \left[1 - \exp(-j\omega) \right] = \frac{1}{T} \exp\left(-j\frac{\omega}{2} \right) \left[2j \sin\left(\frac{\omega}{2} \right) \right], \tag{3.42}$$

which leads to

$$|H(\omega)| = \frac{2}{T} \left| \sin\left(\frac{\omega}{2} \right) \right| \tag{3.43}$$

and

$$\angle H(\omega) = \frac{\pi}{2} - \frac{\omega}{2}. \tag{3.44}$$

The magnitude and phase responses of the first-order difference operator are plotted in Figure 3.22. The gain of the filter increases for higher frequencies up to the folding frequency $f_s/2$ (half the sampling frequency f_s). The gain may be taken to approximate that of the ideal derivative operator, that is, $|\omega|$, for low values of ω. Any high-frequency noise present in the signal will be amplified significantly: the result could thus be noisy.

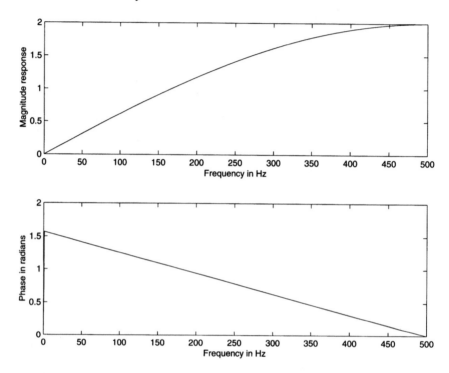

Figure 3.22 Magnitude and phase responses of the first-order difference operator. The magnitude response is shown on a linear scale in order to illustrate better its proportionality to frequency.

The noise-amplification problem with the first-order difference operator in Equation 3.40 may be controlled by taking the average of two successive output values:

$$
\begin{aligned}
y_3(n) &= \frac{1}{2}\left[y(n) + y(n-1)\right] \\
&= \frac{1}{2T}\left[\{x(n) - x(n-1)\} + \{x(n-1) - x(n-2)\}\right] \\
&= \frac{1}{2T}\left[x(n) - x(n-2)\right].
\end{aligned}
\tag{3.45}
$$

The transfer function of the operator above, known as the three-point central difference [27], is

$$
H(z) = \frac{1}{2T}(1 - z^{-2}) = \left[\frac{1}{T}(1 - z^{-1})\right]\left[\frac{1}{2}(1 + z^{-1})\right].
\tag{3.46}
$$

Observe that the transfer function of the three-point central-difference operator is the product of the transfer functions of the simple first-order difference operator and a two-point MA filter. The three-point central-difference operation may therefore be performed by the simple first-order difference operator and a two-point MA filter in series (cascade).

The magnitude and phase responses of the three-point central-difference operator are plotted in Figure 3.23. The transfer function has zeros at $z = 1$ and $z = -1$, with the latter pulling the gain at the folding frequency to zero: the operator is a bandpass filter. Although the operator does not have the noise-amplification problem of the first-order difference operator, the approximation of the $\frac{d}{dt}$ operation is poor after about $f_s/10$ [27].

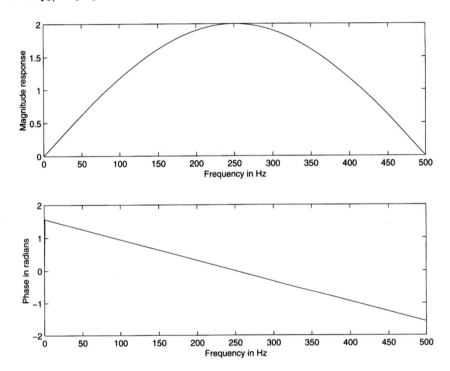

Figure 3.23 Magnitude and phase responses of the three-point central-difference operator. The magnitude response is shown on a linear scale.

Illustration of application: Figures 3.24 and 3.25 show the results of filtering the ECG signal with low-frequency noise shown in Figure 3.6, using the first-order difference and three-point central-difference operators, respectively. It is seen that the base-line drift has been removed, with the latter being less noisy than the former. However, it is obvious that the highpass and high-frequency emphasis effects inherent in both operators have removed the slow P and T waves, and altered the QRS complexes to such an extent as to make the resulting waveforms look unlike ECG signals. (We shall see in Section 4.3 that, although the derivative operators are not useful in the present application, they are indeed useful in detecting the QRS complex and the dicrotic notch.)

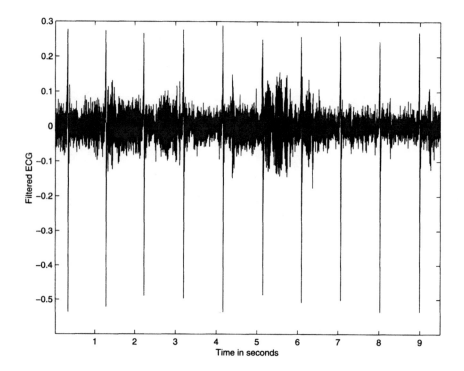

Figure 3.24   Result of filtering the ECG signal with low-frequency noise shown in Figure 3.6, using the first-order difference operator.

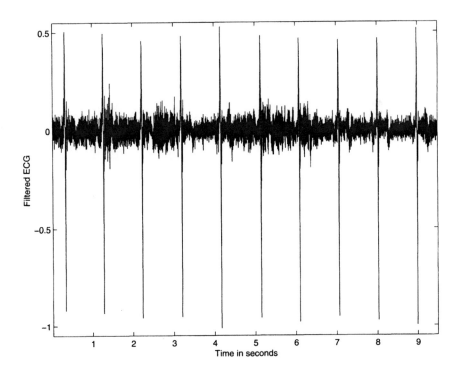

Figure 3.25 Result of filtering the ECG signal with low-frequency noise shown in Figure 3.6, using the three-point central-difference operator.

Problem: *How could we improve the performance of the basic first-order differ-ence operator as a filter to remove low-frequency noise or base-line wander without distorting the QRS complex?*

Solution: The drawback of the first-order difference and the three-point central-difference operators lies in the fact that their magnitude responses remain low for a significant range of frequencies well beyond the band related to base-line wander. The zero of the first-order difference operator at $z = 1$ is desired in order to reject the DC component and very low frequencies. However, we would like to maintain the levels of the components present in the signal beyond about $0.5 - 1\ Hz$, that is, we would like the gain of the filter to be close to unity after about $0.5\ Hz$.

The gain of a filter at specific frequencies may be boosted by placing poles at related locations around the unit circle in the z-plane. For the sake of stability of the filter, the poles should be placed within the unit circle. Since we are interested in maintaining a high gain at very low frequencies, we could place a pole on the real axis (zero frequency), at say $z = 0.995$ [80]. The transfer function of the modified first-order difference filter is then

$$H(z) = \frac{1}{T}\left[\frac{1 - z^{-1}}{1 - 0.995\ z^{-1}}\right], \tag{3.47}$$

or equivalently,

$$H(z) = \frac{1}{T}\left[\frac{z - 1}{z - 0.995}\right]. \tag{3.48}$$

The time-domain input – output relationship is given as

$$y(n) = \frac{1}{T}\ [x(n) - x(n - 1)] + 0.995\ y(n - 1). \tag{3.49}$$

Two equivalent signal-flow diagrams of the filter are shown in Figure 3.26. (*Note: The filter is no longer an FIR filter; details on infinite impulse response or IIR filters will be presented later in Section 3.4.1.*)

The form of $H(z)$ in Equation 3.48 in terms of z helps in understanding a graphical method for the evaluation of the frequency response of discrete-time filters [1, 2, 27]. The frequency response of a system is obtained by evaluating its transfer function at various points on the unit circle in the z-plane, that is, by letting $z = \exp(j\omega)$ and evaluating $H(z)$ for various values of the frequency variable ω of interest. The numerator in Equation 3.48 expresses the vector distance between a specified point in the z-plane and the zero at $z = 1$; the denominator gives the distance to the pole at $z = 0.995$. In general, the magnitude transfer function of a system for a particular value of z is given by the product of the distances from the corresponding point in the z-plane to all the zeros of the system's transfer function, divided by the product of the distances to its poles. The phase response is given by the sum of the angles of the vectors joining the point to all the zeros, minus the sum of the angles to the poles [1, 2, 27]. It is obvious that the magnitude response of the filter in Equations 3.47 and 3.48 is zero at $z = 1$, due to the presence of a zero at that point. Furthermore, for values of z away from $z = 1$, the distances to the zero at $z = 1$

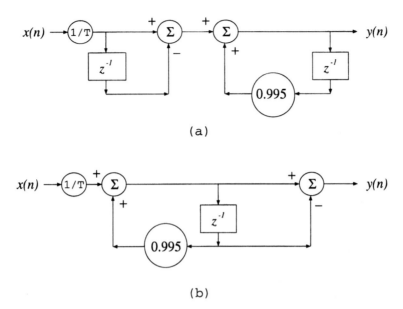

(a)

(b)

Figure 3.26 Two equivalent signal-flow diagrams of the filter to remove low-frequency noise or base-line wander. The form in (a) uses two delays, whereas that in (b) uses only one delay.

and the pole at $z = 0.995$ will be almost equal; therefore, the gain of the filter will be close to unity for frequencies greater than about 1 Hz. The magnitude and phase responses of the filter shown in Figure 3.26 confirm these observations: the filter is a highpass filter with nonlinear phase.

The result of application of the filter to the ECG signal with low-frequency noise shown in Figure 3.6 is displayed in Figure 3.28. It is evident that the low-frequency base-line artifact has been removed without any significant distortion of the ECG waveforms, as compared with the results of differentiation in Figures 3.24 and 3.25. Close inspection, however, reveals that the S wave has been enhanced (made deeper) and that a negative undershoot has been introduced after the T wave. Removal of the low-frequency base-line artifact has been achieved at the cost of a slight distortion of the ECG waves due to the use of a derivative-based filter and its nonlinear phase response.

3.4 FREQUENCY-DOMAIN FILTERS

The filters described in the previous section performed relatively simple operations in the time domain; although their frequency-domain characteristics were explored, the operators were not specifically designed to possess any particular frequency response at the outset. The frequency response of the MA filter, in particular, was seen to be

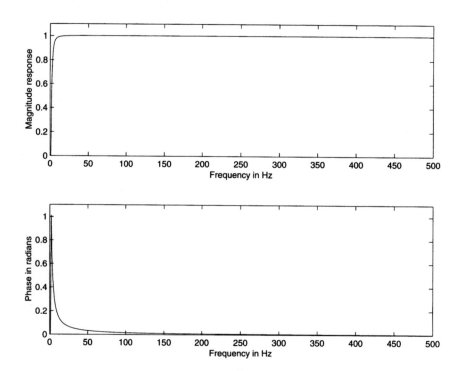

Figure 3.27 Normalized magnitude and phase responses of the filter to remove base-line wander as in Equation 3.47. The magnitude response is shown on a linear scale.

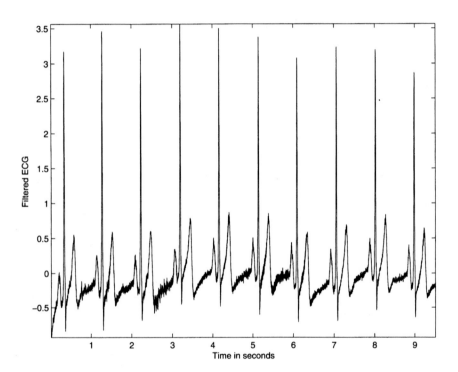

Figure 3.28   Result of processing the ECG signal with low-frequency noise shown in Figure 3.6, using the filter to remove base-line wander as in Equation 3.47. (Compare with the results in Figures 3.24 and 3.25.)

not attractive: the attenuation in the stop-band was not high and was not uniform, with the gain falling below -20 dB only around the zeros of the transfer function.

Filters may be designed in the frequency domain to provide specific lowpass, highpass, bandpass, or band-reject (notch) characteristics. Frequency-domain filters may be implemented in software after obtaining the Fourier transform of the input signal, or converted into equivalent time-domain filters and applied directly upon the signal samples.

Many design procedures are available in the literature to design various types of filters: the most-commonly used designs are the Butterworth, Chebyshev, elliptic, and Bessel filters [14, 81, 82, 83, 84, 85, 26]. Since these filters have been well-established in the analog-filter domain, it is common to commence with an analog design and apply the bilinear transformation to obtain a digital filter in the z-domain. It is also common to design a lowpass filter with the desired pass-band, transition, and stop-band characteristics on a normalized-frequency axis, and then transform it to the desired lowpass, highpass, bandpass, or band-reject characteristics [14, 81]. Frequency-domain filters may also be specified directly in terms of the values of the desired frequency response at certain frequency samples only, and then transformed into the equivalent time-domain filter coefficients via the inverse Fourier transform.

3.4.1 Removal of high-frequency noise: Butterworth lowpass filters

Problem: *Design a frequency-domain filter to remove high-frequency noise with minimal loss of signal components in the specified pass-band.*

Solution: The Butterworth filter is perhaps the most commonly used frequency-domain filter due to its simplicity and the property of a maximally flat magnitude response in the pass-band. For a Butterworth lowpass filter of order N, the first $2N - 1$ derivatives of the squared magnitude response are zero at $\Omega = 0$, where Ω represents the analog radian frequency. The Butterworth filter response is monotonic in the pass-band as well as in the stop-band.

The basic Butterworth lowpass filter function is given as [14, 86]

$$|H_a(j\Omega)|^2 = \frac{1}{1 + \left(\frac{j\Omega}{j\Omega_c}\right)^{2N}}, \tag{3.50}$$

where H_a is the frequency response of the analog filter and Ω_c is the cutoff frequency (in $radians/s$). A Butterworth filter is completely specified by its cutoff frequency Ω_c and order N. As the order N increases, the filter response becomes more flat in the pass-band, and the transition to the stop-band becomes faster or sharper. $|H_a(j\Omega_c)|^2 = \frac{1}{2}$ for all N.

Changing to the Laplace variable s, we get

$$H_a(s)H_a(-s) = \frac{1}{1 + \left(\frac{s}{j\Omega_c}\right)^{2N}}. \tag{3.51}$$

The poles of the squared transfer function are located with equal spacing around a circle of radius Ω_c in the s-plane, distributed symmetrically on either side of the imaginary axis $s = j\Omega$. No pole will lie on the imaginary axis itself; poles will appear on the real axis for odd N. The angular spacing between the poles is $\frac{\pi}{N}$. If $H_a(s)H_a(-s)$ has a pole at $s = s_p$, it will have a pole at $s = -s_p$ as well. Furthermore, for the filter coefficients to be real, complex poles must appear in conjugate pairs. In order to obtain a stable and causal filter, we need to form $H_a(s)$ with only the N poles on the left-hand side of the s-plane. The pole positions in the s-plane are given by

$$ s_k = \Omega_c \exp\left[j\pi\left(\frac{1}{2} + \frac{(2k-1)}{2N}\right)\right], \tag{3.52} $$

$k = 1, 2, \ldots, 2N$ [81].

Once the pole positions are obtained in the s-plane, they may be combined to obtain the transfer function in the analog Laplace domain as

$$ H_a(s) = \frac{G}{(s - p_1)(s - p_2)(s - p_3)\cdots(s - p_N)}, \tag{3.53} $$

where p_k, $k = 1, 2, \ldots, N$, are the N poles of the transfer function in the left-half of the s-plane, and G is a gain factor specified as needed or calculated to normalize the gain at DC ($s = 0$) to be unity.

If we use the bilinear transformation

$$ s = \frac{2}{T}\left[\frac{1 - z^{-1}}{1 + z^{-1}}\right], \tag{3.54} $$

the Butterworth circle in the s-plane maps to a circle in the z-plane with its real-axis intercepts at $z = \frac{2 - \Omega_c T}{2 + \Omega_c T}$ and $z = \frac{2 + \Omega_c T}{2 - \Omega_c T}$. The poles at $s = s_p$ and $s = -s_p$ in the s-plane map to the locations $z = z_p$ and $z = 1/z_p$, respectively. The poles in the z-plane are not uniformly spaced around the transformed Butterworth circle. For stability, all poles of $H(z)$ must lie within the unit circle in the z-plane.

Consider the unit circle in the z-plane given by $z = e^{j\omega}$. For points on the unit circle, we have

$$ s = \sigma + j\Omega = \frac{2}{T}\left(\frac{1 - e^{-j\omega}}{1 + e^{-j\omega}}\right) = \frac{2j}{T}\tan\left(\frac{\omega}{2}\right). \tag{3.55} $$

For the unit circle, $\sigma = 0$; therefore, we get the relationships between the continuous-time (s-domain) frequency variable Ω and the discrete-time (z-domain) frequency variable ω as

$$ \Omega = \frac{2}{T}\tan\left(\frac{\omega}{2}\right) \tag{3.56} $$

and

$$ \omega = 2\tan^{-1}\left(\frac{\Omega T}{2}\right). \tag{3.57} $$

This is a nonlinear relationship that warps the frequency values as they are mapped from the imaginary (vertical) axis in the s-plane to the unit circle in the z-plane (or vice-versa), and should be taken into account in specifying cutoff frequencies.

The transfer function $H_a(s)$ may be mapped to the z-domain by applying the bilinear transformation, that is, by substituting $s = \frac{2}{T} \frac{1-z^{-1}}{1+z^{-1}}$. The transfer function $H(z)$ may then be simplified to the form

$$H(z) = \frac{G' (1 + z^{-1})^N}{\sum_{k=0}^{N} a_k z^{-k}}, \tag{3.58}$$

where a_k, $k = 0, 1, 2, \ldots, N$, are the filter coefficients or tap weights (with $a_0 = 1$), and G' is the gain factor (usually calculated so as to obtain $|H(z)| = 1$ at DC, that is, at $z = 1$. Observe that the filter has N zeros at $z = -1$ due to the use of the bilinear transformation. The filter is now in the familiar form of an IIR filter. Two forms of realization of a generic IIR filter are illustrated as signal-flow diagrams in Figures 3.29 and 3.30: the former represents a direct realization using $2N$ delays and $2N - 1$ multipliers (with $a_0 = 1$), whereas the latter uses only N delays and $2N - 1$ multipliers.

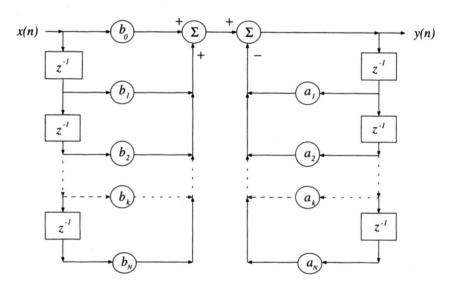

Figure 3.29   Signal-flow diagram of a direct realization of a generic infinite impulse response (IIR) filter. This form uses $2N$ delays and $2N - 1$ multipliers for a filter of order N.

A time-domain representation of the filter will be required if the filter is to be applied to data samples directly in the time domain. From the filter transfer function $H(z)$ in Equation 3.58, it becomes easy to represent the filter in the time domain with the difference equation

$$y(n) = \sum_{k=0}^{N} b_k\, x(n - k) - \sum_{k=1}^{N} a_k\, y(n - k). \tag{3.59}$$

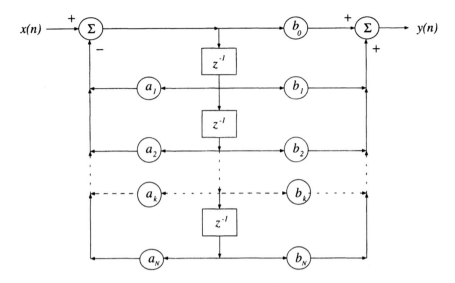

Figure 3.30 Signal-flow diagram of a realization of an IIR filter that uses only N delays and $(2N - 1)$ multipliers for a filter of order N.

The coefficients b_k are given by the coefficients of the expansion of $G'(1 + z^{-1})^N$. The MATLAB [87] command *butter* and its variants provide Butterworth filters obtained using the procedure described above.

It is also possible to directly specify the Butterworth filter as

$$|H(\omega)|^2 = \frac{1}{1 + \left(\frac{\omega}{\omega_c}\right)^{2N}}, \tag{3.60}$$

with ω normalized to the range $(0, 2\pi)$ for sampled or discrete-time signals; in such a case, the equation is valid only for the range $(0, \pi)$, with the function in the range $(\pi, 2\pi)$ being a reflection of that over $(0, \pi)$. The cutoff frequency ω_c should be specified in the range $(0, \pi)$.

If the discrete Fourier transform (DFT) is used to compute the Fourier transforms of the signals being filtered, Equation 3.60 may be modified to

$$|H(k)|^2 = \frac{1}{1 + \left(\frac{k}{k_c}\right)^{2N}}, \tag{3.61}$$

where k is the index of the DFT array standing for discretized frequency. With K being the number of points in the DFT array, k_c is the array index corresponding to the cutoff frequency ω_c (that is, $k_c = K\frac{\omega_c}{\omega_s}$). The equation above is valid for $k = 0, 1, 2, \ldots, \frac{K}{2}$, with the second half over $(\frac{K}{2} + 1, K - 1)$ being a reflection of the first half (that is, $H(k) = H(K - k)$, $k = \frac{K}{2} + 1, \ldots, K - 1$). Note that the DFT includes two unique values: the DC component in $H(0)$ and the folding-frequency component in $H(\frac{K}{2})$. The variable k in the filter equation could also be used to represent normalized frequency in the range $(0, 1)$, with unity standing for the sampling frequency, 0.5 standing for the maximum frequency present in the sampled signal (that is, the folding frequency), and k_c being specified in the range $(0, 0.5)$. (*Note:* MATLAB normalizes half the sampling frequency to unity; the maximum normalized frequency present in the sampled signal is then unity. MATLAB and a few other programming languages do not allow an array index to be zero: in such a case, the indices mentioned above must be incremented by one.)

One could compute the DFT of the given signal, multiply the result by $|H(k)|$, and compute the inverse DFT to obtain the filtered signal. The advantage of this procedure is that no phase change is involved: the filter is a strictly magnitude-only transfer function. The time-domain implementation described earlier will include a phase response which may not be desired. However, time-domain implementation will be required in on-line signal processing applications.

Butterworth lowpass filter design example: In order to design a Butterworth lowpass filter, we need to specify two parameters: ω_c and N. The two parameters may be specified based on a knowledge of the characteristics of the filter as well as those of the signal and noise. It is also possible to specify the required minimum gain at a certain frequency in the pass-band and the required minimum attenuation at another frequency in the stop-band. The two values may then be used with Equation 3.50 to obtain two equations in the two unknowns ω_c and N, which may be solved to derive the filter parameters [86].

Given the 3 dB cutoff frequency f_c and order N, the procedure to design a Butterworth lowpass filter is as follows:

1. Convert the specified 3 dB cutoff frequency f_c to radians in the normalized range $(0, 2\pi)$ as $\omega_c = \frac{f_c}{f_s} 2\pi$. Then, $T = 1$. Prewarp the cutoff frequency ω_c by using Equation 3.56 and obtain Ω_c.

2. Derive the positions of the poles of the filter in the s-plane as given by Equation 3.52.

3. Form the transfer function $H_a(s)$ of the Butterworth lowpass filter in the Laplace domain by using the poles in the left-half plane only as given by Equation 3.53.

4. Apply the bilinear transformation as per Equation 3.54 and obtain the transfer function of the filter $H(z)$ in the z-domain as in Equation 3.58.

5. Convert the filter to the series of coefficients b_k and a_k as in Equation 3.59.

Let us now design a Butterworth lowpass filter with $f_c = 40 \ Hz$, $f_s = 200 \ Hz$, and $N = 4$. We have $\omega_c = \frac{40}{200} 2\pi = 0.4\pi \ radians/s$. The prewarped s-domain cutoff frequency is $\Omega_c = \frac{2}{T} \tan\left(\frac{\omega_c}{2}\right) = 1.453085 \ radians/s$.

The poles of $H_a(s)H_a(-s)$ are placed around a circle of radius 1.453085 with an angular separation of $\frac{\pi}{N} = \frac{\pi}{4} \ radians$. The poles of interest are located at angles $\frac{5}{8}\pi$

and $\frac{7}{8}\pi$ and the corresponding conjugate positions. Figure 3.31 shows the positions of the poles of $H_a(s)H_a(-s)$ in the Laplace plane. The coordinates of the poles of interest are $(-0.556072 \pm j\,1.342475)$ and $(-1.342475 \pm j\,0.556072)$. The transfer function of the filter is

$$H_a(s) = \frac{4.458247}{(s^2 + 1.112143s + 2.111456)(s^2 + 2.684951s + 2.111456)}. \quad (3.62)$$

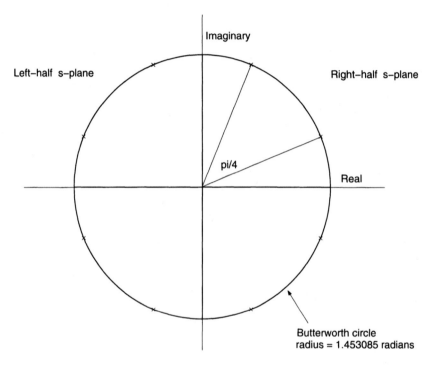

Figure 3.31 Pole positions in the s-plane of the squared magnitude response of the Butterworth lowpass filter with $f_c = 40\ Hz$, $f_s = 200\ Hz$, and $N = 4$.

Applying the bilinear transformation, we get

$$H(z) = \frac{0.046583(1 + z^{-1})^4}{(1 - 0.447765z^{-1} + 0.460815z^{-2})(1 - 0.328976z^{-1} + 0.064588z^{-2})}. \quad (3.63)$$

The filter has four poles at $(0.223882 \pm j\,0.640852)$ and $(0.164488 \pm j\,0.193730)$, and four zeros at $-1 + j\,0$. The b_k coefficients of the filter as in Equation 3.59 are $\{0.0465829, 0.186332, 0.279497, 0.186332, 0.046583\}$, and the a_k coefficients are $\{1, -0.776740, 0.672706, -0.180517, 0.029763\}$. The pole-zero plot and the frequency response of the filter are given in Figures 3.32 and 3.33, respectively. The frequency response displays the expected monotonic decrease in gain and $-3\ dB$ power point or 0.707 gain at $40\ Hz$.

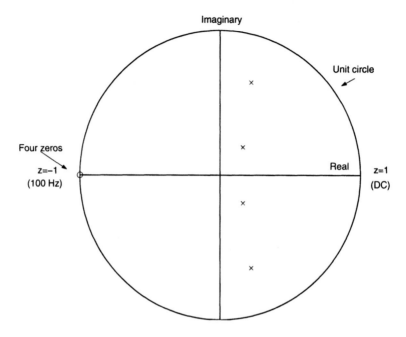

Figure 3.32   Positions of the poles and zeros in the z-plane of the Butterworth lowpass filter with $f_c = 40\ Hz$, $f_s = 200\ Hz$, and $N = 4$.

Figure 3.34 compares the magnitude responses of three Butterworth lowpass filters with $f_c = 40\ Hz$ and $f_s = 200\ Hz$, with the order increasing from $N = 4$ (dotted) to $N = 8$ (dashed) to $N = 12$ (solid). All three filters have their half-power points (gain = 0.707) at 40 Hz, but the transition band becomes sharper as the order N is increased.

The Butterworth design is popular because of its simplicity, a monotonically decreasing magnitude response, and a maximally flat magnitude response in the pass-band. Its main disadvantages are a slow (or wide) transition from the pass-band to the stop-band, and a nonlinear phase response. The nonlinear phase may be corrected for by passing the filter output again through the same filter but after a reversal in time [82]. This process, however, leads to a magnitude response that is the square of that provided by the initial filter design. The squaring effect may be compensated for in the initial design; however, the approach cannot be applied in real time. The elliptic filter design provides a sharp transition band at the expense of ripples in the pass-band and the stop-band. The Bessel design provides a group delay that is maximally flat at DC, and a phase response that approximates a linear response. Details on the design of Bessel, Chebyshev, elliptic, and other filters may be found in other sources on filter design [14, 81, 82, 83, 84, 85, 26].

Illustration of application: The upper trace in Figure 3.35 illustrates a carotid pulse signal with high-frequency noise and effects of clipping. The lower trace in the same figure shows the result of processing in the time domain with the MATLAB *filter*

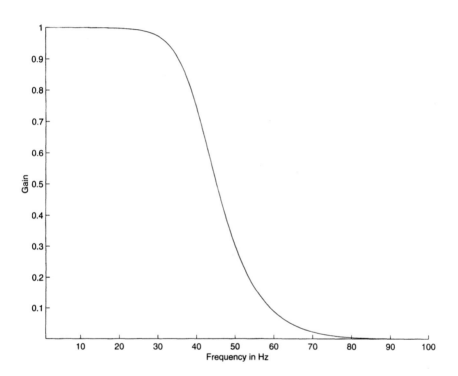

Figure 3.33 Magnitude response of the Butterworth lowpass filter with $f_c = 40\ Hz$, $f_s = 200\ Hz$, and $N = 4$.

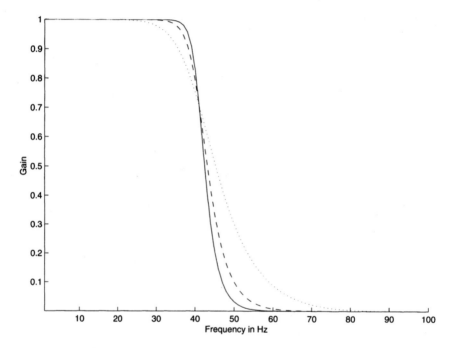

Figure 3.34   Magnitude responses of three Butterworth lowpass filters with $f_c = 40$ Hz, $f_s = 200$ Hz, and variable order: $N = 4$ (dotted), $N = 8$ (dashed), and $N = 12$ (solid).

command; the Butterworth lowpass filter coefficients as designed in the preceding paragraphs and indicated in Equation 3.63 were used ($f_c = 40\ Hz$, $f_s = 200\ Hz$, and $N = 4$). The high-frequency noise has been effectively removed; furthermore, the effects of clipping have been smoothed. However, the low-frequency artifacts in the signal remain (for example, around the 14 s time mark).

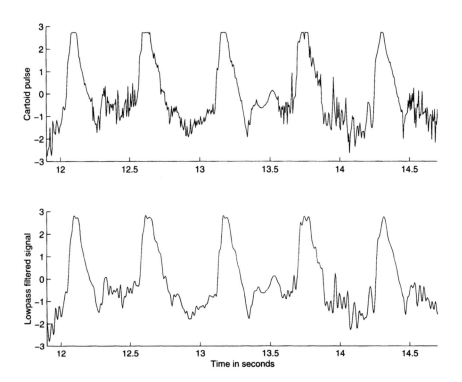

Figure 3.35 Upper trace: a carotid pulse signal with high-frequency noise and effects of clipping. Lower trace: result of filtering with a Butterworth lowpass filter with $f_c = 40\ Hz$, $f_s = 200\ Hz$, and $N = 4$. The filtering operation was performed in the time domain using the MATLAB *filter* command.

Figure 3.36 shows the result of filtering the noisy ECG signal shown in Figure 3.20 with an eighth-order Butterworth lowpass filter as in Equations 3.60 and 3.61 and a cutoff frequency of 70 Hz. The frequency response $|H(\omega)|$ of the filter is shown in Figure 3.37. It is evident that the high-frequency noise has been suppressed by the filter.

3.4.2 Removal of low-frequency noise: Butterworth highpass filters

Problem: *Design a frequency-domain filter to remove low-frequency noise with minimal loss of signal components in the pass-band.*

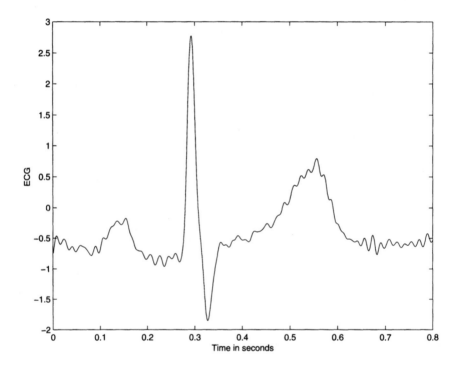

Figure 3.36 Result of frequency-domain filtering of the noisy ECG signal in Figure 3.20 with an eighth-order Butterworth lowpass filter with cutoff frequency = 70 Hz.

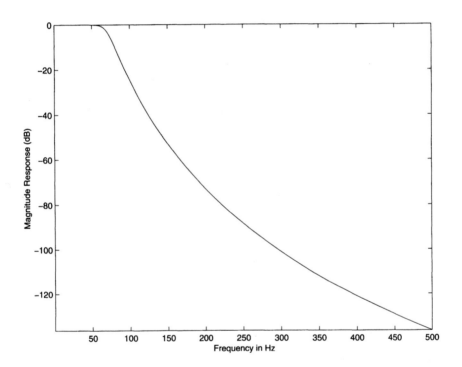

Figure 3.37 Frequency response of the eighth-order Butterworth lowpass filter with cutoff frequency = $f_c = 70\,Hz$ and $f_s = 1,000\,Hz$.

Solution: Highpass filters may be designed on their own, or obtained by transforming a normalized prototype lowpass filter [86, 81]. The latter approach is easier since lowpass filter prototypes with various characteristics are readily available, as are the transformations required to derive highpass, bandpass, and bandstop filters [86, 81]. MATLAB provides highpass filters with the simple command *butter(N, f_c, 'high')*.

As in the case of the Butterworth lowpass filter in Equation 3.61, the Butterworth highpass filter may be specified directly in the discrete-frequency domain as

$$|H(k)|^2 = \frac{1}{1 + \left(\frac{k_c}{k}\right)^{2N}}. \tag{3.64}$$

Illustration of application: Figure 3.6 shows a segment of an ECG signal with low-frequency noise appearing in the form of a wandering base-line (base-line drift). Figure 3.38 shows the result of filtering the signal with an eighth-order Butterworth highpass filter as in Equation 3.64 and a cutoff frequency of 2 Hz. The frequency response of the filter is shown in Figure 3.39. While the low-frequency artifact has been removed by the filter, it should be noted that the high-frequency noise present in the signal has not been affected.

Observe that the filtered result retains the characteristics of the QRS complex, unlike the case with the derivative-based time-domain filters (compare Figure 3.38 with Figures 3.24 and 3.25.) This advantage is due to the fact that the Butterworth highpass filter that was used has a gain of almost unity over the frequency range of $3 - 100$ Hz; the derivative-based filters severely attenuate these components and hence distort the QRS complex. However, it should be observed that the filter has distorted the P and T waves to some extent. The result in Figure 3.38 compares well with that in Figure 3.28, obtained using the much simpler IIR filter in Equation 3.47. (Compare the frequency responses in Figures 3.39, 3.22, 3.23, and 3.27.)

3.4.3   Removal of periodic artifacts: Notch and comb filters

Problem: *Design a frequency-domain filter to remove periodic artifacts such as power-line interference.*

Solution: The simplest method to remove periodic artifacts is to compute the Fourier transform of the signal, delete the undesired component(s) from the spectrum, and then compute the inverse Fourier transform. The undesired components could be set to zero, or better, to the average level of the signal components over a few frequency samples around the component that is to be removed; the former method will remove the noise components as well as the signal components at the frequencies of concern, whereas the latter assumes that the signal spectrum is smooth in the affected regions.

Periodic interference may also be removed by notch filters with zeros on the unit circle in the z-domain at the specific frequencies to be rejected. If f_o is the interference frequency, the angles of the (complex conjugate) zeros required will be $\pm\frac{f_o}{f_s}(2\pi)$; the radius of the zeros will be unity. If harmonics are also present, multiple zeros will be required at $\pm\frac{nf_o}{f_s}(2\pi)$, n representing the orders of all of the harmonics present. The zero angles are limited to the range $(-\pi, \pi)$. The filter is then called

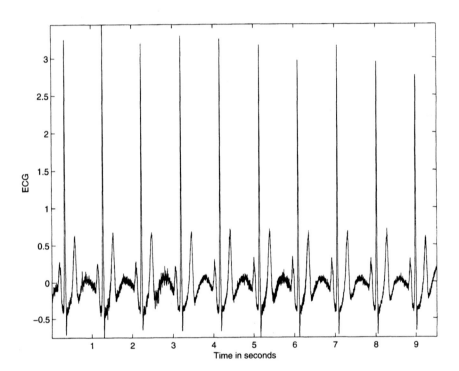

Figure 3.38 Result of frequency-domain filtering of the ECG signal with low-frequency noise in Figure 3.6 with an eighth-order Butterworth highpass filter with cutoff frequency = 2 Hz. (Compare with the results in Figures 3.24, 3.25, and 3.28.)

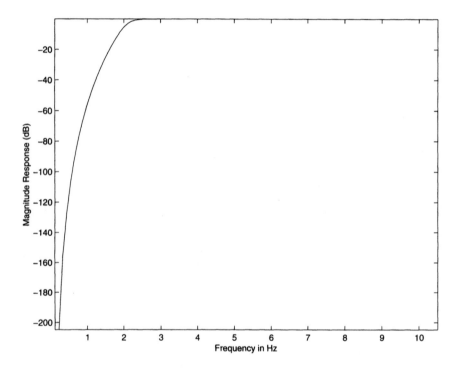

Figure 3.39 Frequency response of an eighth-order Butterworth highpass filter with cutoff frequency = 2 Hz. $f_s = 1,000\ Hz$. The frequency response is shown on an expanded scale for the range $0 - 10\ Hz$ only.

a "comb" filter. In some situations, higher-order harmonics beyond $\frac{f_s}{2}$ may appear at aliased locations (see Figures 3.8 and 3.57); zeros may then be placed at such frequencies as well.

Notch filter design example: Consider a signal with power-line interference at $f_o = 60\ Hz$ and sampling rate of $f_s = 1,000\ Hz$ (see Figures 3.7 and 3.8). The notch filter is then required to have zeros at $w_o = \pm\frac{f_o}{f_s}(2\pi) = \pm0.377\ radians = \pm21.6°$. The zero locations are then given by $\cos(w_o) \pm j\sin(w_o)$ or $z_1 = 0.92977 + j0.36812$ and $z_2 = 0.92977 - j0.36812$. The transfer function is

$$H(z) = (1 - z^{-1}z_1)(1 - z^{-1}z_2) = 1 - 1.85955z^{-1} + z^{-2}. \qquad (3.65)$$

If the gain at DC ($z = 1$) is required to be unity, $H(z)$ should be divided by 0.14045.

Figure 3.40 shows a plot of the zeros of the notch filter in the z-plane. Figure 3.41 shows the magnitude and phase responses of the notch filter obtained using MATLAB. Observe that the filter attenuates not only the 60 Hz component but also a band of frequencies around 60 Hz. The sharpness of the notch may be improved by placing a few poles near or symmetrically around the zeros and inside the unit circle [1, 80]. Note also that the gain of the filter is at its maximum at $f_s/2$; additional lowpass filtering in the case of application to ECG signals could be used to reduce the gain at frequencies beyond about 80 Hz.

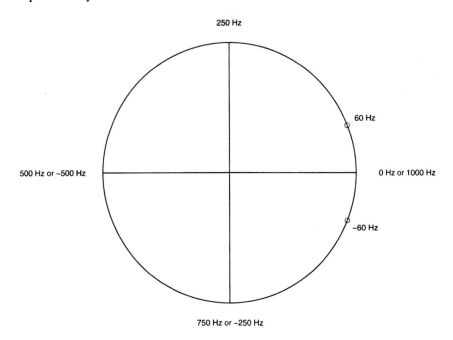

Figure 3.40   Zeros of the notch filter to remove 60 Hz interference, the sampling frequency being 1,000 Hz.

Comb filter design example: Let us consider the presence of a periodic artifact with the fundamental frequency of 60 Hz and odd harmonics at 180 Hz, 300 Hz,

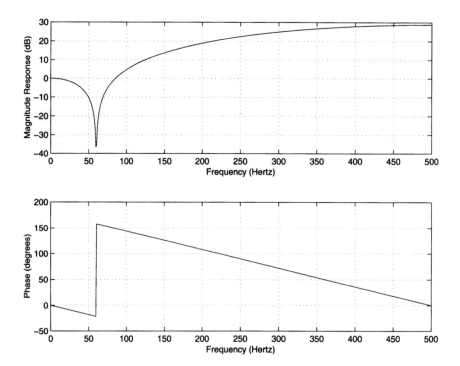

Figure 3.41 Magnitude and phase responses of the 60 Hz notch filter with zeros as shown in Figure 3.40. $f_s = 1,000\ Hz$.

and 420 Hz. Let $f_s = 1,000\ Hz$, and assume the absence of any aliasing error. Zeros are then desired at 60 Hz, 180 Hz, 300 Hz, and 420 Hz, which translate to $\pm 21.6°$, $\pm 64.8°$, $\pm 108°$, and $\pm 151.2°$, with $360°$ corresponding to $1,000\ Hz$. The coordinates of the zeros are $0.92977 \pm j0.36812$, $0.42578 \pm j0.90483$, $-0.30902 \pm j0.95106$, and $-0.87631 \pm j0.48175$. The transfer function of the filter is

$$
\begin{aligned}
H(z) &= G\left(1 - 1.85955z^{-1} + z^{-2}\right)\left(1 - 0.85156z^{-1} + z^{-2}\right) \\
&\times \left(1 + 0.61803z^{-1} + z^{-2}\right)\left(1 - 1.75261z^{-1} + z^{-2}\right), \quad (3.66)
\end{aligned}
$$

where G is the desired gain or scaling factor. With G computed so as to set the gain at DC to be unity, the filter transfer function becomes

$$
\begin{aligned}
H(z) &= 0.6310 - 0.2149z^{-1} + 0.1512z^{-2} - 0.1288z^{-3} + 0.1227z^{-4} \\
&- 0.1288z^{-5} + 0.1512z^{-6} - 0.2149z^{-7} + 0.6310z^{-8}. \quad (3.67)
\end{aligned}
$$

A plot of the locations of the zeros in the z-plane is shown in Figure 3.42. The frequency response of the comb filter is shown in Figure 3.43. Observe the low gain at not only the notch frequencies but also in the adjacent regions.

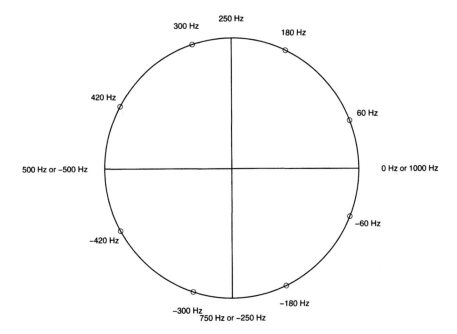

Figure 3.42 Zeros of the comb filter to remove 60 Hz interference with odd harmonics; the sampling frequency is $1,000\ Hz$.

Illustration of application: Figure 3.44 shows an ECG signal with power-line interference at $f_o = 60\ Hz$. Figure 3.45 shows the result of applying the notch filter in Equation 3.65 to the signal. The 60 Hz interference has been effectively removed, with no perceptible distortion of the ECG waveform.

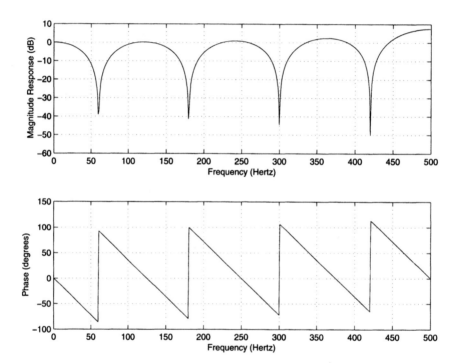

Figure 3.43   Magnitude and phase responses of the comb filter with zeros as shown in Figure 3.42.

An illustration of the application of the comb filter will be provided at the end of the chapter, in Section 3.8.

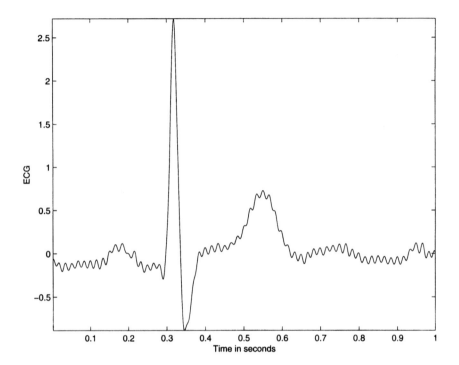

Figure 3.44   ECG signal with 60 Hz interference.

3.5   OPTIMAL FILTERING: THE WIENER FILTER

The filters described in the preceding sections can take into account only limited information about the temporal or spectral characteristics of the signal and noise processes. They are often labeled as *ad hoc* filters: one may have to try several filter parameters and settle upon the filter that appears to provide a usable result. The output is not guaranteed to be the best achievable result: it is not optimized in any sense.

Problem: *Design an optimal filter to remove noise from a signal, given that the signal and noise processes are independent, stationary, random processes. You may assume the "desired" or ideal characteristics of the uncorrupted signal to be known. The noise characteristics may also be assumed to be known.*

Solution: Wiener filter theory provides for *optimal* filtering by taking into account the statistical characteristics of the signal and noise processes. The filter parameters are *optimized* with reference to a *performance criterion*. The output is guaranteed to be the best achievable result under the conditions imposed and the information

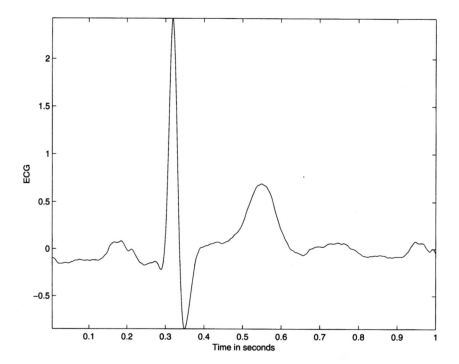

Figure 3.45   The ECG signal in Figure 3.44 after filtering with the 60 *Hz* notch filter shown in Figures 3.40 and 3.41.

provided. The Wiener filter is a powerful conceptual tool that changed traditional approaches to signal processing.

Considering the application of filtering a biomedical signal to remove noise, let us limit ourselves to a single-input, single-output, FIR filter with real input signal values and real coefficients. Figure 3.46 shows the general signal-flow diagram of a transversal filter with coefficients or tap weights $w_i, i = 0, 1, 2, \ldots, M - 1$, input $x(n)$, and output $\tilde{d}(n)$ [77]. The output is usually considered to be an estimate of some "desired" signal $d(n)$ that represents the ideal, uncorrupted signal, and is, therefore, indicated as $\tilde{d}(n)$. If we assume for the moment that the desired signal is available, we could compute the *estimation error* between the output and the desired signal as

$$e(n) = d(n) - \tilde{d}(n). \tag{3.68}$$

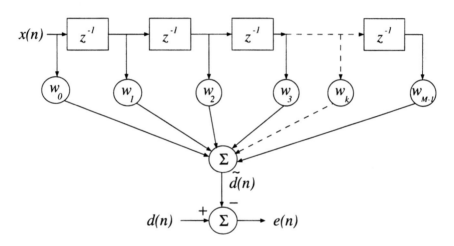

Figure 3.46 Block diagram of the Wiener filter.

Since $\tilde{d}(n)$ is the output of a linear FIR filter, it can be expressed as the convolution of the input $x(n)$ with the tap-weight sequence w_i (which is also the impulse response of the filter) as

$$\tilde{d}(n) = \sum_{k=0}^{M-1} w_k \, x(n - k). \tag{3.69}$$

For easier handling of the optimization procedures, the tap-weight sequence may be written as an $M \times 1$ *tap-weight vector*

$$\mathbf{w} = [w_0, w_1, w_2, \ldots, w_{M-1}]^T, \tag{3.70}$$

where the bold-faced character $\mathbf{w}$ represents a vector and the superscript T indicates vector transposition. As the tap weights are combined with M values of the input in the convolution expression, we could also write the M input values as an $M \times 1$ vector:

$$\mathbf{x}(n) = [x(n), x(n - 1), \ldots, x(n - M + 1)]^T. \tag{3.71}$$

Note that the vector $\mathbf{x}(n)$ varies with time: at a given instant n the vector contains the current input sample $x(n)$ and the preceding $(M-1)$ input samples from $x(n-1)$ to $x(n-M+1)$. The convolution expression in Equation 3.69 may now be written in a simpler form as the inner or dot product of the vectors $\mathbf{w}$ and $\mathbf{x}(n)$:

$$\tilde{d}(n) = \mathbf{w}^T\mathbf{x}(n) = \mathbf{x}^T(n)\mathbf{w} = \langle \mathbf{x}, \mathbf{w} \rangle. \tag{3.72}$$

The estimation error is then given by

$$e(n) = d(n) - \mathbf{w}^T\mathbf{x}(n). \tag{3.73}$$

Wiener filter theory estimates the tap-weight sequence that minimizes the MS value of the estimation error; the output could then be called the *minimum mean-squared error* (MMSE) estimate of the desired response, the filter being then an *optimal filter*. The mean-squared error (MSE) is defined as

$$\begin{aligned}
J(\mathbf{w}) &= E[e^2(n)] \\
&= E[\{d(n) - \mathbf{w}^T\mathbf{x}(n)\}\{d(n) - \mathbf{x}^T(n)\mathbf{w}\}] \\
&= E[d^2(n)] - \mathbf{w}^T E[\mathbf{x}(n)d(n)] - E[d(n)\mathbf{x}^T(n)]\mathbf{w} \\
&\quad + \mathbf{w}^T E[\mathbf{x}(n)\mathbf{x}^T(n)]\mathbf{w}.
\end{aligned} \tag{3.74}$$

Note that the expectation operator is not applicable to $\mathbf{w}$ as it is not a random variable.

Under the assumption that the input vector $\mathbf{x}(n)$ and the desired response $d(n)$ are jointly stationary, the expectation expressions in the equation above have the following interpretations [77]:

- $E[d^2(n)]$ is the variance of $d(n)$, written as σ_d^2, with the further assumption that the mean of $d(n)$ is zero.

- $E[\mathbf{x}(n)d(n)]$ is the cross-correlation between the input vector $\mathbf{x}(n)$ and the desired response $d(n)$, which is an $M \times 1$ vector:

$$\Theta = E[\mathbf{x}(n)d(n)]. \tag{3.75}$$

Note that $\Theta = [\theta(0), \theta(-1), \ldots, \theta(1-M)]^T$, where

$$\theta(-k) = E[x(n-k)d(n)], \;\; k = 0, 1, 2, \ldots, M-1. \tag{3.76}$$

- $E[d(n)\mathbf{x}^T(n)]$ is simply the transpose of $E[\mathbf{x}(n)d(n)]$; therefore

$$\Theta^T = E[d(n)\mathbf{x}^T(n)]. \tag{3.77}$$

- $E[\mathbf{x}(n)\mathbf{x}^T(n)]$ represents the autocorrelation of the input vector $\mathbf{x}(n)$ computed as the outer product of the vector with itself, written as

$$\Phi = E[\mathbf{x}(n)\mathbf{x}^T(n)] \tag{3.78}$$

or in its full $M \times M$ matrix form as

$$
\mathbf{\Phi} =
\begin{bmatrix}
\phi(0) & \phi(1) & \cdots & \phi(M-1) \\
\phi(-1) & \phi(0) & \cdots & \phi(M-2) \\
\vdots & \vdots & \ddots & \vdots \\
\phi(-M+1) & \phi(-M+2) & \cdots & \phi(0)
\end{bmatrix}
\tag{3.79}
$$

with the element in row k and column i given by

$$
\phi(i-k) = E[x(n-k)x(n-i)],
\tag{3.80}
$$

with the property that $\phi(i-k) = \phi(k-i)$. (*Note:* $\phi = \phi_{xx}$.) With the assumption of wide-sense stationarity, the $M \times M$ matrix $\mathbf{\Phi}$ is completely specified by M values of the autocorrelation $\phi(0), \phi(1), \ldots, \phi(M-1)$ for lags $0, 1, \ldots, M-1$.

With the interpretations as listed above, the MSE expression in Equation 3.74 is simplified to

$$
J(\mathbf{w}) = \sigma_d^2 - \mathbf{w}^T\mathbf{\Theta} - \mathbf{\Theta}^T\mathbf{w} + \mathbf{w}^T\mathbf{\Phi}\mathbf{w}.
\tag{3.81}
$$

This expression indicates that the MSE is a second-order function of the tap-weight vector $\mathbf{w}$. To determine the optimal tap-weight vector, denoted by $\mathbf{w}_o$, we could differentiate $J(\mathbf{w})$ with respect to $\mathbf{w}$, set it to zero, and solve the resulting equation. To perform this differentiation, we should note the following derivatives:

$$
\frac{d}{d\mathbf{w}}(\mathbf{\Theta}^T\mathbf{w}) = \mathbf{\Theta},
$$

$$
\frac{d}{d\mathbf{w}}(\mathbf{w}^T\mathbf{\Theta}) = \mathbf{\Theta},
$$

$$
\frac{d}{d\mathbf{w}}(\mathbf{w}^T\mathbf{\Phi}\mathbf{w}) = 2\mathbf{\Phi}\mathbf{w}.
$$

Now, we obtain the derivative of $J(\mathbf{w})$ with respect to $\mathbf{w}$ as

$$
\frac{dJ(\mathbf{w})}{d\mathbf{w}} = -2\mathbf{\Theta} + 2\mathbf{\Phi}\mathbf{w}.
\tag{3.82}
$$

Setting this expression to zero, we obtain the condition for the optimal filter as

$$
\mathbf{\Phi}\mathbf{w}_o = \mathbf{\Theta}.
\tag{3.83}
$$

This equation is known as the *Wiener-Hopf* equation. It is also known as the *normal equation* as it can be shown that [77], for the optimal filter, each element of the input vector $\mathbf{x}(n)$ and the estimation error $e(n)$ are mutually orthogonal, and furthermore, that the filter output $\tilde{d}(n)$ and the error $e(n)$ are mutually orthogonal (that is, the expectation of their products is zero). The optimal filter is obtained as

$$
\mathbf{w}_o = \mathbf{\Phi}^{-1}\mathbf{\Theta}.
\tag{3.84}
$$

In expanded form, we have the Wiener-Hopf equation as

$$
\begin{bmatrix}
\phi(0) & \phi(1) & \cdots & \phi(M-1) \\
\phi(-1) & \phi(0) & \cdots & \phi(M-2) \\
\vdots & \vdots & \ddots & \vdots \\
\phi(-M+1) & \phi(-M+2) & \cdots & \phi(0)
\end{bmatrix}
\begin{bmatrix}
w_{o0} \\
w_{o1} \\
\vdots \\
w_{o(M-1)}
\end{bmatrix}
=
$$

$$
\begin{bmatrix}
\theta(0) \\
\theta(-1) \\
\vdots \\
\theta(1-M)
\end{bmatrix}
\tag{3.85}
$$

or as

$$
\sum_{i=0}^{M-1} w_{oi}\,\phi(i-k) = \theta(-k), \quad k = 0, 1, 2, \ldots, M-1. \tag{3.86}
$$

The minimum MSE is given by

$$
J_{\min} = \sigma_d^2 - \boldsymbol{\Theta}^T \boldsymbol{\Phi}^{-1} \boldsymbol{\Theta}. \tag{3.87}
$$

Given the condition that the signals involved are stationary, we have $\phi(i-k) = \phi(k-i)$ and $\theta(-k) = \theta(k)$. Then, we may write Equation 3.86 as

$$
\sum_{i=0}^{M-1} w_{oi}\,\phi(k-i) = \theta(k), \quad k = 0, 1, 2, \ldots, M-1. \tag{3.88}
$$

Thus we have the convolution relationship

$$
w_{ok} * \phi(k) = \theta(k). \tag{3.89}
$$

Applying the Fourier transform to the equation above, we get

$$
W(\omega)S_{xx}(\omega) = S_{xd}(\omega), \tag{3.90}
$$

which may be modified to obtain the Wiener filter frequency response $W(\omega)$ as

$$
W(\omega) = \frac{S_{xd}(\omega)}{S_{xx}(\omega)}, \tag{3.91}
$$

where $S_{xx}(\omega)$ is the PSD of the input signal and $S_{xd}(\omega)$ is the cross-spectral density (CSD) between the input signal and the desired signal.

Note that derivation of the optimal filter requires rather specific knowledge about the input and the desired response in the form of the autocorrelation $\boldsymbol{\Phi}$ of the input $x(n)$ and the cross-correlation $\boldsymbol{\Theta}$ between the input $x(n)$ and the desired response $d(n)$. In practice, although the desired response $d(n)$ may not be known, it should be possible to obtain an estimate of its temporal or spectral statistics, which may be used to estimate Θ. Proper estimation of the statistical entities mentioned above requires a large number of samples of the corresponding signals.

(*Note:* Haykin [77] allows all the entities involved to be complex. Vector transposition ^{T} is then Hermitian or complex-conjugate transposition ^{H}. Products of two entities require one to be conjugated: for example, $e^2(n)$ is obtained as $e(n)e^*(n)$; Equation 3.69 will have w_k^* in place of w_k, and so on. Furthermore, $\frac{d}{dw}(\Theta^H w) = 0$ and $\frac{d}{dw}(w^H \Theta) = 2\Theta$. The final Wiener-Hopf equation, however, simplifies to the same as above in Equation 3.86.)

Let us now consider the problem of removing noise from a corrupted input signal. For this case, let the input $x(n)$ contain a mixture of the desired (original) signal $d(n)$ and noise $\eta(n)$, that is,

$$x(n) = d(n) + \eta(n). \tag{3.92}$$

Using the vector notation as before, we have

$$\mathbf{x}(n) = \mathbf{d}(n) + \boldsymbol{\eta}(n), \tag{3.93}$$

where $\boldsymbol{\eta}(n)$ is the vector representation of the noise function $\eta(n)$. The autocorrelation matrix of the input is given by

$$\boldsymbol{\Phi} = E[\mathbf{x}(n)\mathbf{x}^T(n)] = E[\{\mathbf{d}(n) + \boldsymbol{\eta}(n)\}\{\mathbf{d}(n) + \boldsymbol{\eta}(n)\}^T]. \tag{3.94}$$

If we now assume that the noise process is statistically independent of the signal process, we have

$$E[\mathbf{d}(n)\boldsymbol{\eta}^T(n)] = E[\boldsymbol{\eta}^T(n)\mathbf{d}(n)] = \mathbf{0}. \tag{3.95}$$

Then,

$$\boldsymbol{\Phi} = E[\mathbf{d}(n)\mathbf{d}^T(n)] + E[\boldsymbol{\eta}(n)\boldsymbol{\eta}^T(n)] = \boldsymbol{\Phi}_d + \boldsymbol{\Phi}_\eta, \tag{3.96}$$

where $\boldsymbol{\Phi}_d$ and $\boldsymbol{\Phi}_\eta$ are the $M \times M$ autocorrelation matrices of the signal and noise, respectively. Furthermore,

$$\boldsymbol{\Theta} = E[\mathbf{x}(n)d(n)] = E[\{\mathbf{d}(n) + \boldsymbol{\eta}(n)\}d(n)] = E[\mathbf{d}(n)d(n)] = \boldsymbol{\Phi}_{1d}, \tag{3.97}$$

where $\boldsymbol{\Phi}_{1d}$ is an $M \times 1$ autocorrelation vector of the desired signal. The optimal Wiener filter is then given by

$$\mathbf{w}_o = (\boldsymbol{\Phi}_d + \boldsymbol{\Phi}_\eta)^{-1}\boldsymbol{\Phi}_{1d}. \tag{3.98}$$

The frequency response of the Wiener filter may be obtained by modifying Equation 3.91 by taking into account the spectral relationships

$$S_{xx}(\omega) = S_d(\omega) + S_\eta(\omega) \tag{3.99}$$

and

$$S_{xd}(\omega) = S_d(\omega), \tag{3.100}$$

which leads to

$$W(\omega) = \frac{S_d(\omega)}{S_d(\omega) + S_\eta(\omega)} = \frac{1}{1 + \frac{S_\eta(\omega)}{S_d(\omega)}}, \tag{3.101}$$

where $S_d(\omega)$ and $S_\eta(\omega)$ are the PSDs of the desired signal and the noise process, respectively. Note that designing the optimal filter requires knowledge of the PSDs of the desired signal and the noise process (or models thereof).

Illustration of application: The upper trace in Figure 3.47 shows one ECG cycle extracted from the signal with noise in Figure 3.5. A piece-wise linear model of the desired version of the signal was created by concatenating linear segments to provide P, QRS, and T waves with amplitudes, durations, and intervals similar to those in the given noisy signal. The base-line of the model signal was set to zero. The noise-free model signal is shown in the middle trace of Figure 3.47. The log PSDs of the given noisy signal and the noise-free model, the latter being $S_d(\omega)$ in Equation 3.101, are shown in the upper two plots of Figure 3.48.

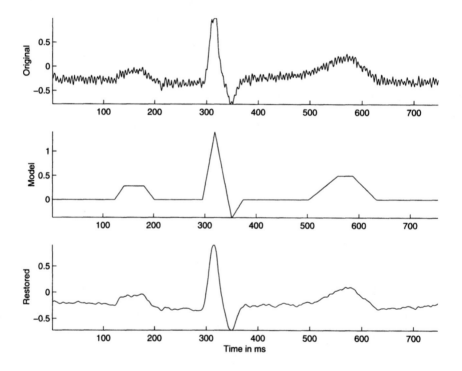

Figure 3.47 From top to bottom: one cycle of the noisy ECG signal in Figure 3.5 (labeled as Original); a piece-wise linear model of the desired noise-free signal (Model); and the output of the Wiener filter (Restored).

The T – P intervals between successive cardiac cycles in an ECG (the inter-beat intervals) may be taken to represent the iso-electric base-line. Then, any activity present in these intervals constitutes noise. Four T – P intervals were selected from the noisy signal in Figure 3.5 and their Fourier power spectra were averaged to derive the noise PSD $S_\eta(\omega)$ required in the Wiener filter (Equation 3.101). The estimated log PSD of the noise is shown in the third trace of Figure 3.48. Observe the relatively high levels of energy in the noise PSD above 100 Hz compared to the PSDs of

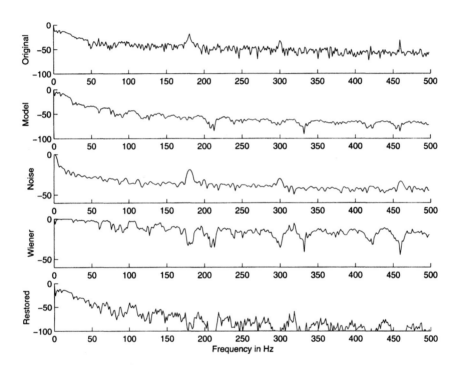

Figure 3.48 From top to bottom: log PSD (in *dB*) of the given noisy signal (labeled as Original); log PSD of the noise-free model (Model); estimated log PSD of the noise process (Noise); log frequency response of the Wiener filter (Wiener); and log PSD of the filter output (Restored).

the original noisy signal and the model. Observe also the peaks in the original and noise PSDs near 180 Hz, 300 Hz, and 420 Hz, representing the third, fifth, and seventh harmonics of 60 Hz, respectively; the peak at 460 Hz is an aliased version of the ninth harmonic at 540 Hz. The 60 Hz component itself appears to have been suppressed by a notch filter in the signal acquisition system. (See Sections 3.2.4 and 3.4.3 for more details.)

The Wiener filter frequency response was derived as in Equation 3.101, and is shown in the fourth plot in Figure 3.48. Observe the low gain of the filter near 180 Hz, 300 Hz, 420 Hz, and 460 Hz corresponding to the peaks in the noise spectrum. As indicated by Equation 3.101, the Wiener filter gain is inversely related to the noise PSD and directly related to the signal PSD. The result of application of the Wiener filter to the given signal is shown in the third trace of Figure 3.47. It is evident that almost all of the noise has been effectively removed by the filter.

The most important point to observe here is that the filter was derived with models of the noise and signal processes (PSDs), which were obtained from the given signal itself in the present application. No cutoff frequency was required to be specified in designing the Wiener filter, whereas the Butterworth filter requires the specification of a cutoff frequency and filter order.

Most signal acquisition systems should permit the measurement of at least the variance or power level of the noise present. A uniform (white) PSD model may then be easily derived. Models of the ideal signal and the noise processes may also be created using parametric Gaussian or Laplacian models either in the time domain (ACF) or directly in the frequency domain (PSD).

3.6 ADAPTIVE FILTERS FOR REMOVAL OF INTERFERENCE

Filters with fixed characteristics (tap weights or coefficients), as seen in the preceding sections, are suitable when the characteristics of the signal and noise (random or structured) are stationary and known. Design of frequency-domain filters requires detailed knowledge of the spectral contents of the signal and noise. Such filters are not applicable when the characteristics of the signal and/or noise vary with time, that is, when they are nonstationary. They are also not suitable when the spectral contents of the signal and the interference overlap significantly.

Consider the situation when two ECG signals such as those of a fetus and the mother, or two vibration signals such as the VAG and the VMG, arrive at the recording site and get added in some proportion. The spectra of the signals in the mixture span the same or similar frequency ranges, and hence fixed filtering cannot separate them. In the case of the VAG/VMG mixture, it is also possible for the spectra of the signals to vary from one point in time to another, due to changes in the characteristics of the cartilage surfaces causing the VAG signal, and due to the effect of variations in the recruitment of muscle fibers on the VMG signal. Such a situation calls for the use of a filter that can learn and adapt to the characteristics of the interference, estimate the interfering signal, and remove it from the mixture to obtain the desired signal.

This requires the filter to automatically adjust its impulse response (and hence its frequency response) as the characteristics of the signal and/or noise vary.

Problem: *Design an optimal filter to remove a nonstationary interference from a nonstationary signal. An additional channel of information related to the interference is available for use. The filter should continuously adapt to the changing characteristics of the signal and interference.*

Solution: We need to address two different concerns in this problem:

1. The filter should be *adaptive*; the tap-weight vector of the filter will then vary with time. The principles of the adaptive filter, also known as the adaptive noise canceler (ANC), will be explained in Section 3.6.1.

2. The filter should be *optimal*. Two well-established methods for optimization of the adaptive filter will be presented in Sections 3.6.2 and 3.6.3.

Illustrations of the application of the methods will be presented at the end of Sections 3.6.2 and 3.6.3, as well as the end of the chapter, in Sections 3.9 and 3.10.

3.6.1 The adaptive noise canceler

Figure 3.49 shows a generic block diagram of an adaptive filter or ANC [62, 88]. The "primary input" to the filter $x(n)$ is a mixture of the signal of interest $v(n)$ and the "primary noise" $m(n)$:

$$x(n) = v(n) + m(n). \tag{3.102}$$

$x(n)$ is the primary observed signal; it is desired that the interference or noise $m(n)$ be estimated and removed from $x(n)$ in order to obtain the signal of interest $v(n)$. It is assumed that $v(n)$ and $m(n)$ are uncorrelated. Adaptive filtering requires a second input, known as the "reference input" $r(n)$, that is uncorrelated with the signal of interest $v(n)$ but closely related to or correlated with the interference or noise $m(n)$ in some manner that need not be known. The ANC filters or modifies the reference input $r(n)$ to obtain a signal $y(n)$ that is as close to the noise $m(n)$ as possible. $y(n)$ is then subtracted from the primary input to estimate the desired signal:

$$\tilde{v}(n) = e(n) = x(n) - y(n). \tag{3.103}$$

Let us now analyze the function of the filter. Let us assume that the signal of interest $v(n)$, the primary noise $m(n)$, the reference input $r(n)$, and the primary noise estimate $y(n)$ are statistically stationary and have zero means. (*Note:* The requirement of stationarity will be removed later when the expectations are computed in moving windows.) We have already stated that $v(n)$ is uncorrelated with $m(n)$ and $r(n)$, and that $r(n)$ is correlated with $m(n)$. The output of the ANC is

$$
\begin{aligned}
e(n) &= x(n) - y(n) \\
&= v(n) + m(n) - y(n), \tag{3.104}
\end{aligned}
$$

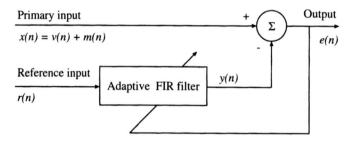

Figure 3.49   Block diagram of a generic adaptive noise canceler (ANC) or adaptive filter.

where $y(n) = \tilde{m}(n)$ is the estimate of the primary noise obtained at the output of the adaptive filter. By taking the square and expectation (statistical average) of both sides of Equation 3.104, we obtain

$$E[e^2(n)] = E[v^2(n)] + E[\{m(n) - y(n)\}^2] + 2E[v(n)\{m(n) - y(n)\}]. \quad (3.105)$$

Since $v(n)$ is uncorrelated with $m(n)$ and $y(n)$ and all of them have zero means, we have

$$E[v(n)\{m(n) - y(n)\}] = E[v(n)]E[m(n) - y(n)] = 0. \quad (3.106)$$

Equation 3.105 can be rewritten as

$$E[e^2(n)] = E[v^2(n)] + E[\{m(n) - y(n)\}^2]. \quad (3.107)$$

Note from Figure 3.49 that the output $e(n)$ is used (fed back) to control the adaptive filter. In ANC applications, the objective is to obtain an output $e(n)$ that is a least-squares fit to the desired signal $v(n)$. This is achieved by feeding the output back to the adaptive filter and adjusting the filter to minimize the total system output power. The system output serves as the error signal for the adaptive process.

The signal power $E[v^2(n)]$ will be unaffected as the filter is adjusted to minimize $E[e^2(n)]$; accordingly, the minimum output power is

$$\min\ E[e^2(n)] = E[v^2(n)] + \min\ E[\{m(n) - y(n)\}^2]. \quad (3.108)$$

As the filter is adjusted so that $E[e^2(n)]$ is minimized, $E[\{m(n) - y(n)\}^2]$ is also minimized. Thus the filter output $y(n)$ is the MMSE estimate of the primary noise $m(n)$. Moreover, when $E[\{m(n) - y(n)\}^2]$ is minimized, $E[\{e(n) - v(n)\}^2]$ is also minimized, since from Equation 3.104

$$e(n) - v(n) = m(n) - y(n). \quad (3.109)$$

Adjusting or adapting the filter to minimize the total output power is, therefore, equivalent to causing the *output $e(n)$ to be the MMSE estimate of the signal of interest $v(n)$* for the given structure and adjustability of the adaptive filter and for the given reference input.

The output $e(n)$ will contain the signal of interest $v(n)$ and some noise. From Equation 3.109, the output noise is given by $e(n) - v(n) = \tilde{v}(n) - v(n) = m(n) - y(n)$. Since minimizing $E[e^2(n)]$ minimizes $E[\{m(n) - y(n)\}^2]$, *minimizing the total output power minimizes the output noise power.* Since the signal component $v(n)$ in the output remains unaffected, *minimizing the total output power maximizes the output SNR.*

Note from Equation 3.107 that the output power is minimum when $E[e^2(n)] = E[v^2(n)]$. When this condition is achieved, $E[\{m(n) - y(n)\}^2] = 0$. We then have $y(n) = m(n)$ and $e(n) = v(n)$; that is, the output is a perfect and noise-free estimate of the desired signal.

Optimization of the filter may be performed by expressing the error in terms of the tap-weight vector and applying the procedure of choice. The output $y(n)$ of the adaptive filter (see Figure 3.49) in response to its input $r(n)$ is given by

$$y(n) = \sum_{k=0}^{M-1} w_k \, r(n - k), \tag{3.110}$$

where w_k, $k = 0, 1, 2, \ldots, M - 1$, are the tap weights, and M is the order of the filter. The estimation error $e(n)$ or the output of the ANC system is

$$e(n) = x(n) - y(n). \tag{3.111}$$

For the sake of notational simplicity, let us define the tap-weight vector at time n as

$$\mathbf{w}(n) = [w_0(n), w_1(n), \ldots, w_{M-1}(n)]^T. \tag{3.112}$$

Similarly, the tap-input vector at each time instant n may be defined as the M-dimensional vector

$$\mathbf{r}(n) = [r(n), r(n - 1), \ldots, r(n - M + 1)]^T. \tag{3.113}$$

Then, the estimation error $e(n)$ given in Equation 3.111 may be rewritten as

$$e(n) = x(n) - \mathbf{w}^T(n)\mathbf{r}(n). \tag{3.114}$$

It is worth noting that the derivations made above required no knowledge about the processes behind $v(n)$, $m(n)$, and $r(n)$ or their inter-relationships, other than the assumptions of statistical independence between $v(n)$ and $m(n)$ and some form of correlation between $m(n)$ and $r(n)$. The arguments can be extended to situations where the primary and reference inputs contain additive random noise processes that are mutually uncorrelated and also uncorrelated with $v(n)$, $m(n)$, and $r(n)$. The procedures may also be extended to cases where $m(n)$ and $r(n)$ are deterministic or structured rather than stochastic, such as power-line interference or an ECG or a VMG signal [62].

Several methods are available to maximize the output SNR; two such methods based on the least-mean-squares (LMS) and the recursive least-squares (RLS) approaches are described in the following sections.

3.6.2   The least-mean-squares adaptive filter

The purpose of adaptive filtering algorithms is to adjust the tap-weight vector to minimize the MSE. By squaring the expression for the estimation error $e(n)$ given in Equation 3.114, we get

$$e^2(n) = x^2(n) - 2x(n)\mathbf{r}^T(n)\mathbf{w}(n) + \mathbf{w}^T(n)\mathbf{r}(n)\mathbf{r}^T(n)\mathbf{w}(n). \qquad (3.115)$$

The squared error is a second-order (quadratic) function of the tap-weight vector (and the inputs), and may be depicted as a concave hyper-paraboloidal (bowl-like) surface that is never negative. The aim of the filter optimization procedure would be to reach the bottom of the bowl-like function. Gradient-based methods may be used for this purpose.

By taking the expected values of the entities in Equation 3.115 and taking the derivative with respect to the tap-weight vector, we may derive the Wiener-Hopf equation for the present application. The LMS algorithm takes a simpler approach by assuming the square of the instantaneous error as in Equation 3.115 to stand for an estimate of the MSE [62]. The LMS algorithm is based on the method of steepest descent, where the new tap-weight vector $\mathbf{w}(n+1)$ is given by the present tap-weight vector $\mathbf{w}(n)$ plus a correction proportional to the negative of the gradient $\nabla(n)$ of the squared error:

$$\mathbf{w}(n + 1) = \mathbf{w}(n) - \mu\nabla(n). \qquad (3.116)$$

The parameter μ controls the stability and rate of convergence of the algorithm: the larger the value of μ, the larger is the gradient of the noise that is introduced and the faster is the convergence of the algorithm, and vice-versa.

The LMS algorithm approximates $\nabla(n)$ by the derivative of the squared error in Equation 3.115 with respect to the tap-weight vector as

$$\tilde{\nabla}(n) = -2x(n)\mathbf{r}(n) + 2\{\mathbf{w}^T(n)\mathbf{r}(n)\}\mathbf{r}(n) = -2e(n)\mathbf{r}(n). \qquad (3.117)$$

Using this estimate of the gradient in Equation 3.116, we get

$$\mathbf{w}(n + 1) = \mathbf{w}(n) + 2\mu\, e(n)\, \mathbf{r}(n). \qquad (3.118)$$

This expression is known as the Widrow-Hoff LMS algorithm.

The advantages of the LMS algorithm lie in its simplicity and ease of implementation: although the method is based on the MSE and gradient-based optimization, the filter expression itself is free of differentiation, squaring, or averaging. It has been shown that the expected value of the tap-weight vector provided by the LMS algorithm converges to the optimal Wiener solution when the input vectors are uncorrelated over time [89, 62]. The procedure may be started with an arbitrary tap-weight vector; it will converge in the mean and remain stable as long as μ is greater than zero but less than the reciprocal of the largest eigenvalue of the autocorrelation matrix of the reference input [62].

Illustration of application: Zhang et al. [63] used a two-stage adaptive LMS filter to cancel muscle-contraction interference from VAG signals. The first stage was used to remove the measurement noise in the accelerometers and associated amplifiers, and the second stage was designed to cancel the muscle signal.

Zhang et al. [63] also proposed a procedure for optimization of the step size μ by using an RMS-error-based misadjustment factor and a time-varying estimate of the input signal power, among other entities. The LMS algorithm was implemented as

$$\mathbf{w}(n+1) = \mathbf{w}(n) + 2\mu(n)\, e(n)\, \mathbf{r}(n). \tag{3.119}$$

The step size μ was treated as a variable, its value being determined dynamically as

$$\mu(n) = \frac{\mu}{(M+1)\, \bar{x}^2(n)\, (\alpha, r(n), \bar{x}^2(n-1))}, \tag{3.120}$$

with $0 < \mu < 1$. A forgetting factor α was introduced in the adaptation process, with $0 \le \alpha \ll 1$; this feature was expected to overcome problems caused by high levels of nonstationarity in the signal. $\bar{x}^2(n)$ is a time-varying estimate of the input signal power, computed as $\bar{x}^2(n) = \alpha r^2(n) + (1-\alpha)\bar{x}^2(n-1)$.

The filtered versions of the VAG signals recorded from the mid-patella and the tibial tuberosity positions, as shown in Figure 3.11 (traces (b) and (c), right-hand column), are shown in Figure 3.50. The muscle-contraction signal recorded at the distal rectus femoris position was used as the reference input (Figure 3.11, right-hand column, trace (a)). It is seen that the low-frequency muscle-contraction artifact has been successfully removed from the VAG signals (compare the second half of each signal in Figure 3.50 with the corresponding part in Figure 3.11).

3.6.3 The recursive least-squares adaptive filter

When the input process of an adaptive system is (quasi-) stationary, the best steady-state performance results from slow adaptation. However, when the input statistics are time-variant (nonstationary), the best performance is obtained by a compromise between fast adaptation (necessary to track variations in the input process) and slow adaptation (necessary to limit the noise in the adaptive process). The LMS adaptation algorithm is a simple and efficient approach for ANC; however, it is not appropriate for fast-varying signals due to its slow convergence, and due to the difficulty in selecting the correct value for the step size μ. An alternative approach based on the exact minimization of the least-squares criterion is the RLS method [77, 90]. The RLS algorithm has been widely used in real-time system identification and noise cancellation because of its fast convergence, which is about an order of magnitude higher than that of the LMS method. (The derivation of the RLS filter in this section has been adapted from Sesay [90] and Krishnan [88] with permission.)

An important feature of the RLS algorithm is that it utilizes information contained in the input data, and extends it back to the instant of time when the algorithm was initiated [77]. Given the least-squares estimate of the tap-weight vector of the filter at time $n-1$, the updated estimate of the vector at time n is computed upon the arrival of new data.

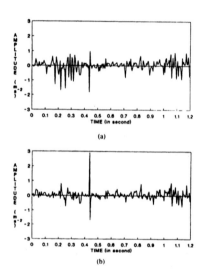

Figure 3.50 LMS-filtered versions of the VAG signals recorded from the mid-patella and the tibial tuberosity positions, as shown in Figure 3.11 (traces (b) and (c), right-hand column). The muscle-contraction signal recorded at the distal rectus femoris position was used as the reference input (Figure 3.11, right-hand column, trace (a)). The recording setup is shown in Figure 3.10. Reproduced with permission from Y.T Zhang, R.M. Rangayyan, C.B. Frank, and G.D. Bell, Adaptive cancellation of muscle-contraction interference from knee joint vibration signals, *IEEE Transactions on Biomedical Engineering,* 41(2):181–191, 1994. ©IEEE.

In the derivation of the RLS algorithm, the *performance index* or *objective function* $\xi(n)$ to be minimized in the sense of least squares is defined as

$$\xi(n) = \sum_{i=1}^{n} \lambda^{n-i} |e(i)|^2, \tag{3.121}$$

where $1 \leq i \leq n$ is the observation interval, $e(i)$ is the estimation error as defined in Equation 3.114, and λ is a weighting factor (also known as the *forgetting factor*) with $0 < \lambda \leq 1$. The values of $\lambda^{n-i} < 1$ give more "weight" to the more recent error values. Such weighting is desired in the case of nonstationary data, where changes in the signal statistics make the inclusion of past data less appropriate. The inverse of $(1 - \lambda)$ is a measure of the memory of the algorithm.

The Wiener-Hopf equation is the necessary and sufficient condition [77] for minimizing the performance index in the least-squares sense and for obtaining the optimal values of the tap weights, and may be derived in a manner similar to that presented in Section 3.5 for the Wiener filter. The normal equation to be solved in the RLS procedure is

$$\mathbf{\Phi}(n)\tilde{\mathbf{w}}(n) = \mathbf{\Theta}(n), \tag{3.122}$$

where $\tilde{\mathbf{w}}(n)$ is the optimal tap-weight vector for which the performance index is at its minimum, $\mathbf{\Phi}(n)$ is an $M \times M$ time-averaged (and weighted) autocorrelation matrix of the reference input $\mathbf{r}(i)$ defined as

$$\mathbf{\Phi}(n) = \sum_{i=1}^{n} \lambda^{n-i} \, \mathbf{r}(i) \, \mathbf{r}^T(i), \tag{3.123}$$

and $\mathbf{\Theta}(n)$ is an $M \times 1$ time-averaged (and weighted) cross-correlation matrix between the reference input $\mathbf{r}(i)$ and the primary input $x(i)$, defined as

$$\mathbf{\Theta}(n) = \sum_{i=1}^{n} \lambda^{n-i} \, \mathbf{r}(i) \, x(i). \tag{3.124}$$

The general scheme of the RLS filter is illustrated in Figure 3.51.

Because of the difficulty in solving the normal equation for the optimal tap-weight vector, recursive techniques need to be considered. In order to obtain a recursive solution, we could isolate the term corresponding to $i = n$ from the rest of the summation on the right-hand side of Equation 3.123, and obtain

$$\mathbf{\Phi}(n) = \lambda \left[\sum_{i=1}^{n-1} \lambda^{n-i-1} \, \mathbf{r}(i) \, \mathbf{r}^T(i) \right] + \mathbf{r}(n) \, \mathbf{r}^T(n). \tag{3.125}$$

According to the definition in Equation 3.123, the expression inside the square brackets on the right-hand side of Equation 3.125 equals the time-averaged and weighted autocorrelation matrix $\mathbf{\Phi}(n - 1)$. Hence, Equation 3.125 can be rewritten as a recursive expression, given by

$$\mathbf{\Phi}(n) = \lambda \mathbf{\Phi}(n - 1) + \mathbf{r}(n)\mathbf{r}^T(n). \tag{3.126}$$

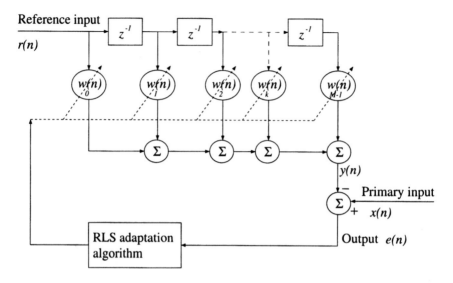

Figure 3.51   General structure of the adaptive RLS filter.

Similarly, Equation 3.124 can be written as the recursive equation

$$\Theta(n) = \lambda\Theta(n - 1) + \mathbf{r}(n)x(n). \tag{3.127}$$

To compute the least-squares estimate $\tilde{\mathbf{w}}(n)$ for the tap-weight vector in accordance with Equation 3.122, we have to determine the inverse of the correlation matrix $\mathbf{\Phi}(n)$. In practice, such an operation is time-consuming (particularly if M is large). To reduce the computational requirements, a matrix inversion lemma known as the "ABCD lemma" could be used (a similar form of the lemma can be found in Haykin [77]). According to the ABCD lemma, given matrices $\mathbf{A}, \mathbf{B}, \mathbf{C},$ and $\mathbf{D}$,

$$(\mathbf{A} + \mathbf{BCD})^{-1} = \mathbf{A}^{-1} - \mathbf{A}^{-1}\mathbf{B}(\mathbf{DA}^{-1}\mathbf{B} + \mathbf{C}^{-1})^{-1}\mathbf{DA}^{-1}. \tag{3.128}$$

The matrices $\mathbf{A}, \mathbf{C}, (\mathbf{A} + \mathbf{BCD})$, and $(\mathbf{DA}^{-1}\mathbf{B}+\mathbf{C}^{-1})$ are assumed to be invertible. With the correlation matrix $\mathbf{\Phi}(n)$ assumed to be positive definite and therefore nonsingular, we may apply the matrix inversion lemma to Equation 3.126 by assigning

$$\begin{aligned}
\mathbf{A} &= \lambda\mathbf{\Phi}(n - 1), \\
\mathbf{B} &= \mathbf{r}(n), \\
\mathbf{C} &= 1, \\
\mathbf{D} &= \mathbf{r}^T(n).
\end{aligned}$$

We then have

$$\begin{aligned}
\mathbf{\Phi}^{-1}(n) &= \lambda^{-1}\mathbf{\Phi}^{-1}(n - 1) \\
&- \lambda^{-1}\mathbf{\Phi}^{-1}(n - 1)\mathbf{r}(n)\left[\lambda^{-1}\mathbf{r}^T(n)\mathbf{\Phi}^{-1}(n - 1)\mathbf{r}(n) + 1\right]^{-1} \\
&\times \lambda^{-1}\mathbf{r}^T(n)\mathbf{\Phi}^{-1}(n - 1).
\end{aligned} \tag{3.129}$$

Since the expression inside the brackets of the above equation is a scalar, the equation can be rewritten as

$$\Phi^{-1}(n) = \lambda^{-1}\Phi^{-1}(n-1) - \frac{\lambda^{-2}\Phi^{-1}(n-1)\mathbf{r}(n)\mathbf{r}^T(n)\Phi^{-1}(n-1)}{1 + \lambda^{-1}\mathbf{r}^T(n)\Phi^{-1}(n-1)\mathbf{r}(n)}. \quad (3.130)$$

For convenience of notation, let

$$\mathbf{P}(n) = \Phi^{-1}(n), \quad (3.131)$$

with $\mathbf{P}(0) = \delta^{-1}\mathbf{I}$, where δ is a small constant and $\mathbf{I}$ is the identity matrix. Furthermore, let

$$\mathbf{k}(n) = \frac{\lambda^{-1}\mathbf{P}(n-1)\mathbf{r}(n)}{1 + \lambda^{-1}\mathbf{r}^T(n)\mathbf{P}(n-1)\mathbf{r}(n)}. \quad (3.132)$$

$\mathbf{k}(n)$ is analogous to the *Kalman gain vector* in Kalman filter theory [77]. Equation 3.130 may then be rewritten in a simpler form as

$$\mathbf{P}(n) = \lambda^{-1}\mathbf{P}(n-1) - \lambda^{-1}\mathbf{k}(n)\mathbf{r}^T(n)\mathbf{P}(n-1). \quad (3.133)$$

By multiplying both sides of Equation 3.132 by the denominator on its right-hand side, we get

$$\mathbf{k}(n)\left[1 + \lambda^{-1}\mathbf{r}^T(n)\mathbf{P}(n-1)\mathbf{r}(n)\right] = \lambda^{-1}\mathbf{P}(n-1)\mathbf{r}(n), \quad (3.134)$$

or,

$$\mathbf{k}(n) = \left[\lambda^{-1}\mathbf{P}(n-1) - \lambda^{-1}\mathbf{k}(n)\mathbf{r}^T(n)\mathbf{P}(n-1)\right]\mathbf{r}(n). \quad (3.135)$$

Comparing the expression inside the brackets on the right-hand side of the above equation with Equation 3.133, we have

$$\mathbf{k}(n) = \mathbf{P}(n)\mathbf{r}(n). \quad (3.136)$$

$\mathbf{P}(n)$ and $\mathbf{k}(n)$ have the dimensions $M \times M$ and $M \times 1$, respectively.

By using Equations 3.122, 3.127, and 3.131, a recursive equation for updating the least-squares estimate $\tilde{\mathbf{w}}(n)$ of the tap-weight vector can be obtained as

$$
\begin{aligned}
\tilde{\mathbf{w}}(n) &= \Phi^{-1}(n)\Theta(n) \\
&= \mathbf{P}(n)\Theta(n) \\
&= \lambda\mathbf{P}(n)\Theta(n-1) + \mathbf{P}(n)\mathbf{r}(n)x(n). \quad (3.137)
\end{aligned}
$$

Substituting Equation 3.133 for $\mathbf{P}(n)$ in the first term of Equation 3.137, we get

$$
\begin{aligned}
\tilde{\mathbf{w}}(n) &= \mathbf{P}(n-1)\Theta(n-1) - \mathbf{k}(n)\mathbf{r}^T(n)\mathbf{P}(n-1)\Theta(n-1) \\
&\quad + \mathbf{P}(n)\mathbf{r}(n)x(n) \\
&= \Phi^{-1}(n-1)\Theta(n-1) - \mathbf{k}(n)\mathbf{r}^T(n)\Phi^{-1}(n-1)\Theta(n-1) \\
&\quad + \mathbf{P}(n)\mathbf{r}(n)x(n) \\
&= \tilde{\mathbf{w}}(n-1) - \mathbf{k}(n)\mathbf{r}^T(n)\tilde{\mathbf{w}}(n-1) + \mathbf{P}(n)\mathbf{r}(n)x(n). \quad (3.138)
\end{aligned}
$$

Finally, from Equation 3.136, using the fact that $\mathbf{P}(n)\mathbf{r}(n)$ equals the gain vector $\mathbf{k}(n)$, the above equation can be rewritten as

$$
\begin{aligned}
\tilde{\mathbf{w}}(n) &= \tilde{\mathbf{w}}(n-1) - \mathbf{k}(n)\left[x(n) - \mathbf{r}^T(n)\tilde{\mathbf{w}}(n-1)\right] \\
&= \tilde{\mathbf{w}}(n-1) + \mathbf{k}(n)\alpha(n),
\end{aligned}
\tag{3.139}
$$

where $\tilde{\mathbf{w}}(0) = \mathbf{0}$, and

$$
\begin{aligned}
\alpha(n) &= x(n) - \mathbf{r}^T(n)\tilde{\mathbf{w}}(n-1) \\
&= x(n) - \tilde{\mathbf{w}}^T(n-1)\mathbf{r}(n).
\end{aligned}
\tag{3.140}
$$

The quantity $\alpha(n)$ is often referred to as the *a priori error*, reflecting the fact that it is the error obtained using the "old" filter (that is, the filter before being updated with the new data at the nth time instant). It is evident that in the case of ANC applications, $\alpha(n)$ will be the estimated signal of interest $\tilde{v}(n)$ after the filter has converged, that is,

$$
\alpha(n) = \tilde{v}(n) = x(n) - \tilde{\mathbf{w}}^T(n-1)\mathbf{r}(n).
\tag{3.141}
$$

Furthermore, after convergence, the primary noise estimate, that is, the output of the adaptive filter $y(n)$, can be written as

$$
y(n) = \tilde{m}(n) = \tilde{\mathbf{w}}^T(n-1)\mathbf{r}(n).
\tag{3.142}
$$

By substituting Equations 3.104 and 3.142 in Equation 3.141, we get

$$
\begin{aligned}
\tilde{v}(n) &= v(n) + m(n) - \tilde{m}(n) \\
&= v(n) + m(n) - \tilde{\mathbf{w}}^T(n-1)\mathbf{r}(n) \\
&= x(n) - \tilde{\mathbf{w}}^T(n-1)\mathbf{r}(n).
\end{aligned}
\tag{3.143}
$$

Equation 3.139 gives a recursive relationship for obtaining the optimal values of the tap weights, which, in turn, provide the least-squares estimate $\tilde{v}(n)$ of the signal of interest $v(n)$ as in Equation 3.143.

Illustration of application: Figure 3.52 shows plots of the VAG signal of a normal subject (trace (a)) and a simultaneously recorded channel of muscle-contraction interference (labeled as MCI, trace (b)). The characteristics of the vibration signals in this example are different from those of the signals in Figure 3.11, due to a different recording protocol in terms of speed and range of swinging motion of the leg [88]. The results of adaptive filtering of the VAG signal with the muscle-contraction interference channel as the reference are also shown in Figure 3.52: trace (c) shows the result of LMS filtering, and trace (d) shows that of RLS filtering. A single-stage LMS filter with variable step size $\mu(n)$ as in Equation 3.120 was used; no attempt was made to remove instrumentation noise. The LMS filter used $M = 7$, $\mu = 0.05$, and a forgetting factor $\alpha = 0.98$; other values resulted in poor results. The RLS filter used $M = 7$ and $\lambda = 0.98$.

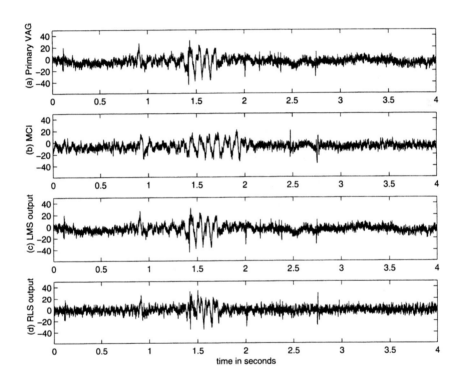

Figure 3.52 (a) VAG signal of a normal subject. (b) Muscle-contraction interference (MCI). (c) Result of LMS filtering. (d) Result of RLS filtering. The recording setup is shown in Figure 3.10.

The relatively low-frequency muscle-contraction interference has been removed better by the RLS filter than by the LMS filter; the latter failed to track the nonstationarities in the interference, and has caused additional artifacts in the result. The spectrograms of the primary, reference, and RLS-filtered signals are shown in Figures 3.53, 3.54, and 3.55, respectively. (The logarithmic scale is used to display better the minor differences between the spectrograms.) It is seen that the predominantly low-frequency artifact, indicated by the high energy levels at low frequencies for the entire duration in the spectrograms in Figures 3.53 and 3.54, has been removed by the RLS filter.

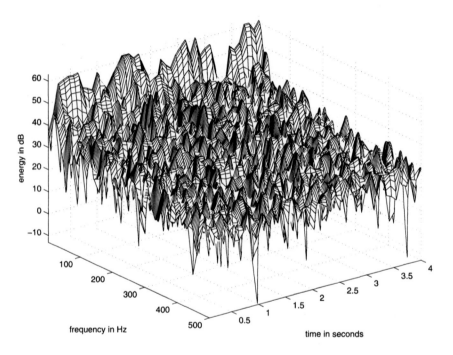

Figure 3.53 Spectrogram of the VAG signal in Figure 3.52 (a). A Hanning window of length 256 samples (128 *ms*) was used; an overlap of 32 samples (16 *ms*) was allowed between adjacent segments.

3.7 SELECTING AN APPROPRIATE FILTER

We have so far examined five approaches to remove noise and interference: (1) synchronized or ensemble averaging of multiple realizations or copies of a signal, (2) MA filtering, (3) frequency-domain filtering, (4) optimal (Wiener) filtering, and (5) adaptive filtering. The first two approaches work directly with the signal in the time domain. Frequency-domain (fixed) filtering is performed on the spectrum of the

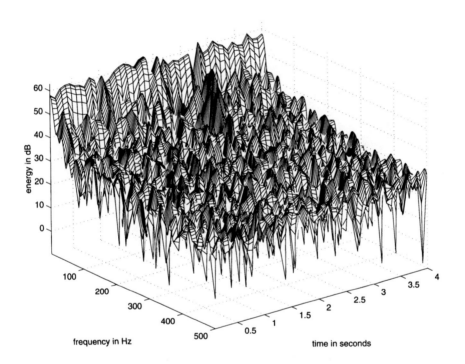

Figure 3.54 Spectrogram of the muscle-contraction interference signal in Figure 3.52 (b). A Hanning window of length 256 samples (128 ms) was used; an overlap of 32 samples (16 ms) was allowed between adjacent segments.

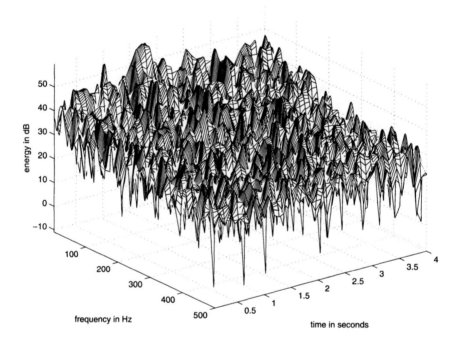

Figure 3.55 Spectrogram of the RLS-filtered VAG signal in Figure 3.52 (d). A Hanning window of length 256 samples (**128** *ms*) was used; an overlap of 32 samples (**16** *ms*) was allowed between adjacent segments.

signal. Note that the impulse response of a filter designed in the frequency domain could be used to implement the filter in the time domain as an IIR or FIR filter. Furthermore, time-domain filters may be analyzed in the frequency domain via their transfer function or frequency response to understand better their characteristics and effects on the input signal. The Wiener filter may be implemented either in the time domain as a transversal filter or in the frequency domain. Adaptive filters work directly on the signal in the time domain, but dynamically alter their characteristics in response to changes in the interference; their frequency response thus varies from one point in time to another.

What are the guiding principles to determine which of these filters is the best for a given application? The following points should assist in making this decision.

Synchronized or ensemble averaging is possible when:

- The signal is statistically stationary, (quasi-)periodic, or cyclo-stationary.

- Multiple realizations or copies of the signal of interest are available.

- A trigger point or time marker is available, or can be derived to extract and align the copies of the signal.

- The noise is a stationary random process that is uncorrelated with the signal and has a zero mean (or a known mean).

Temporal MA filtering is suitable when:

- The signal is statistically stationary at least over the duration of the moving window.

- The noise is a zero-mean random process that is stationary at least over the duration of the moving window and is independent of the signal.

- The signal is a relatively slow (low-frequency) phenomenon.

- Fast, on-line, real-time filtering is desired.

Frequency-domain fixed filtering is applicable when:

- The signal is statistically stationary.

- The noise is a stationary random process that is statistically independent of the signal.

- The signal spectrum is limited in bandwidth compared to that of the noise (or vice-versa).

- Loss of information in the spectral band removed by the filter does not seriously affect the signal.

- On-line, real-time filtering is not required (if implemented in the spectral domain via the Fourier transform).

The optimal Wiener filter can be designed if:

- The signal is statistically stationary.

- The noise is a stationary random process that is statistically independent of the signal.

- Specific details (or models) are available regarding the ACFs or the PSDs of the signal and noise.

Adaptive filtering is called for and possible when:

- The noise or interference is not stationary and not necessarily a random process.

- The noise is uncorrelated with the signal.

- No information is available about the spectral characteristics of the signal and noise, which may also overlap significantly.

- A second source or recording site is available to obtain a reference signal that is strongly correlated with the noise but uncorrelated with the signal.

It is worth noting that an adaptive filter acts as a fixed filter when the signal and noise are stationary. An adaptive filter can also act as a notch or a comb filter when the interference is periodic. It should be noted that all of the filters mentioned above are applicable only when the noise is additive. Techniques such as homomorphic filtering (see Section 4.8) may be used as preprocessing steps if signals combined with operations other than addition need to be separated.

3.8   APPLICATION: REMOVAL OF ARTIFACTS IN THE ECG

Problem: *Figure 3.56 (top trace) shows an ECG signal with a combination of baseline drift, high-frequency noise, and power-line interference. Design filters to remove the artifacts.*

Solution: The power spectrum of the given signal is shown in the top-most plot in Figure 3.57. Observe the relatively high amount of spectral energy present near DC, from $100\ Hz$ to $500\ Hz$, and at the power-line frequency and its harmonics located at $60\ Hz$, $180\ Hz$, $300\ Hz$, and $420\ Hz$. The fundamental component at $60\ Hz$ is lower than the third, fifth, and seventh harmonics due perhaps to a notch filter included in the signal acquisition system, which has not been effective.

A Butterworth lowpass filter with order $N = 8$ and $f_c = 70\ Hz$ (see Section 3.4.1 and Equation 3.61), a Butterworth highpass filter of order $N = 8$ and $f_c = 2\ Hz$ (see Section 3.4.2 and Equation 3.64), and a comb filter with zeros at $60\ Hz$, $180\ Hz$, $300\ Hz$, and $420\ Hz$ (see Section 3.4.3 and Equation 3.67) were applied in series to the signal. The signal spectrum displays the presence of further harmonics (ninth and eleventh) of the power-line interference at $540\ Hz$ and $660\ Hz$ that have been aliased to the peaks apparent at $460\ Hz$ and $340\ Hz$, respectively. However, the comb filter in the present example was not designed to remove these components. The lowpass and highpass filters were applied in the frequency domain to the Fourier transform of the signal using the form indicated by Equations 3.61 and 3.64. The comb filter was applied in the time domain using the MATLAB *filter* command and the coefficients in Equation 3.67.

The combined frequency response of the filters is shown in the middle plot in Figure 3.57. The spectrum of the ECG signal after the application of all three filters is shown in the bottom plot is Figure 3.57. The filtered signal spectrum has no appreciable energy beyond about 100 Hz, and displays significant attenuation at 60 Hz.

The outputs after the lowpass filter, the highpass filter, and the comb filter are shown in Figure 3.56. Observe that the base-line drift is present in the output of the lowpass filter, and that the power-line interference is present in the outputs of the lowpass and highpass filters. The final trace is free of all three types of interference. Note, however, that the highpass filter has introduced a noticeable distortion (undershoot) in the P and T waves.

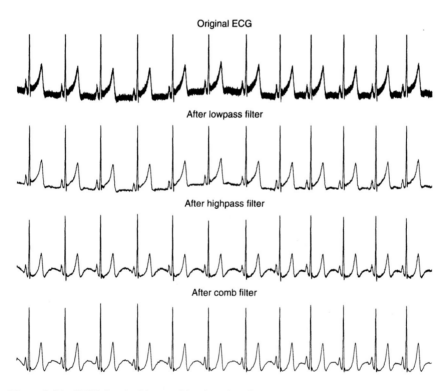

Figure 3.56   ECG signal with a combination of artifacts and its filtered versions. The duration of the signal is 10.7 s.

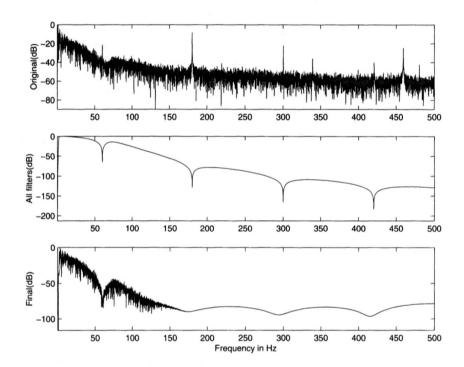

Figure 3.57   Top and bottom plots: Power spectra of the ECG signals in the top and bottom traces of Figure 3.56. Middle plot: Frequency response of the combination of lowpass, highpass, and comb filters. The cutoff frequency of the highpass filter is 2 Hz; the highpass portion of the frequency response is not clearly seen in the plot.

3.9   APPLICATION: ADAPTIVE CANCELLATION OF THE MATERNAL ECG TO OBTAIN THE FETAL ECG

Problem: *Propose an adaptive noise cancellation filter to remove the maternal ECG signal from the abdominal-lead ECG shown in Figure 3.9 to obtain the fetal ECG. Chest-lead ECG signals of the mother may be used for reference.*

Solution: Widrow et al. [62] describe a multiple-reference ANC for removal of the maternal ECG in order to obtain the fetal ECG. The combined ECG was obtained from a single abdominal lead, whereas the maternal ECG was obtained via four chest leads. The model was designed to permit the treatment of not only multiple sources of interference, but also of components of the desired signal present in the reference inputs, and further to consider the presence of uncorrelated noise components in the reference inputs. It should be noted that the maternal cardiac vector is projected onto the axes of different ECG leads in different ways, and hence the characteristics of the maternal ECG in the abdominal lead would be different from those of the chest-lead ECG signals used as reference inputs.

Each filter channel used by Widrow et al. [62] had 32 taps and a delay of 129 ms. The signals were pre-filtered to the bandwidth $3 - 35$ Hz and a sampling rate of 256 Hz was used. The optimal Wiener filter (see Section 3.5) included transfer functions and cross-spectral vectors between the input source and each reference input. Further extension of the method to more general multiple-source, multiple-reference noise cancelling problems was also discussed by Widrow et al.

The result of cancellation of the maternal ECG from the abdominal lead ECG signal in Figure 3.9 is shown in Figure 3.58. Comparing the two figures, it is seen that the filter output has successfully extracted the fetal ECG and suppressed the maternal ECG. See Widrow et al. [62] for details; see also Ferrara and Widrow [91].

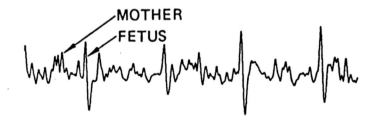

Figure 3.58   Result of adaptive cancellation of the maternal chest ECG from the abdominal ECG in Figure 3.9. The QRS complexes extracted correspond to the fetal ECG. Reproduced with permission from B. Widrow, J.R. Glover, Jr., J.M. McCool, J. Kaunitz, C.S. Williams, R.H. Hearn, J.R. Zeidler, E. Dong, Jr., R.C. Goodlin, Adaptive noise cancelling: Principles and applications, *Proceedings of the IEEE*, 63(12):1692–1716, 1975. ©IEEE.

3.10   APPLICATION: ADAPTIVE CANCELLATION OF MUSCLE-CONTRACTION INTERFERENCE IN KNEE-JOINT VIBRATION SIGNALS

Problem: *Study the applicability of adaptive noise cancellation filters to remove the muscle-contraction interference caused by the rectus femoris in the VAG signal recorded at the patella.*

Solution: Rangayyan et al. [92] conducted a study on the impact of muscle-contraction interference cancellation on modeling and classification of VAG signals and further classification of the filtered signals as normal or abnormal. Both the LMS (see Section 3.6.2) and the RLS (see Section 3.6.3) methods were investigated, and the RLS method was chosen for its more efficient tracking of nonstationarities in the input and reference signals.

Figure 3.59 shows plots of the VAG signal of a subject with chondromalacia patella of grade II (trace (a)) and a simultaneously recorded channel of muscle-contraction interference (labeled as MCI, trace (b)). The results of adaptive filtering of the VAG signal with the muscle-contraction interference channel as the reference are also shown in Figure 3.59: trace (c) shows the result of LMS filtering, and trace (d) shows that of RLS filtering. A single-stage LMS filter with variable step size $\mu(n)$ as in Equation 3.120 was used, with $M = 7$, $\mu = 0.05$, and $\alpha = 0.98$. The RLS filter used $M = 7$ and $\lambda = 0.98$.

As in the earlier example in Figure 3.52, it is seen that the muscle-contraction interference has been removed by the RLS filter; however, the LMS filter failed to perform well, due to its limited capabilities in tracking the nonstationarities in the interference. The spectrograms of the primary, reference, and RLS-filtered signals are shown in Figures 3.60, 3.61, and 3.62, respectively. (The logarithmic scale is used to display better the minor differences between the spectrograms.) It is seen that the frequency components of the muscle-contraction interference have been suppressed by the RLS filter.

The primary (original) and filtered VAG signals of 53 subjects were adaptively segmented and modeled in the study of Rangayyan et al. [92] (see Chapter 8). The segment boundaries were observed to be markedly different for the primary and the filtered VAG signals. Parameters extracted from the filtered VAG signals were expected to provide higher discriminant power in pattern classification when compared to the same parameters of the unfiltered or primary VAG signals. However, classification experiments indicated otherwise: the filtered signals gave a lower classification accuracy by almost 10%. It was reasoned that after removal of the predominantly low-frequency muscle-contraction interference, the transient VAG signals of clinical interest were not modeled well by the prediction-based methods. It was concluded that the adaptive filtering procedure used was not an appropriate preprocessing step before signal modeling for pattern classification. However, it was noted that cancellation of muscle-contraction interference may be a desirable step before spectral analysis of VAG signals.

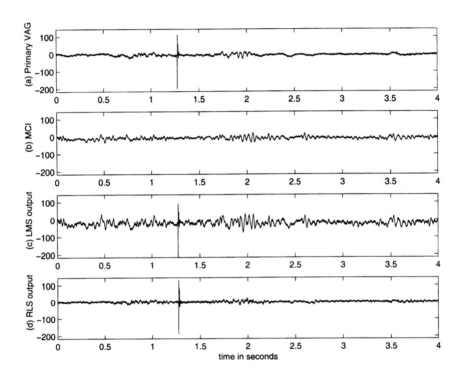

Figure 3.59 Top to bottom: (a) VAG signal of a subject with chondromalacia patella of grade II; (b) Muscle-contraction interference (MCI); (c) Result of LMS filtering; and (d) Result of RLS filtering. The recording setup in shown in Figure 3.10.

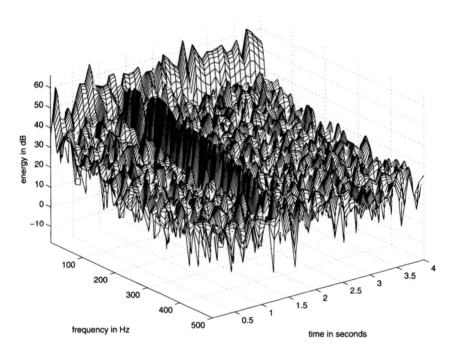

Figure 3.60 Spectrogram of the original VAG signal in Figure 3.59 (a). A Hanning window of length 256 samples (128 ms) was used; an overlap of 32 samples (16 ms) was allowed between adjacent segments.

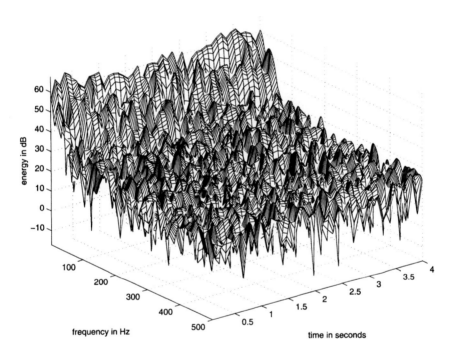

Figure 3.61   Spectrogram of the muscle-contraction interference signal in Figure 3.59 (b). A Hanning window of length 256 samples (128 ms) was used; an overlap of 32 samples (16 ms) was allowed between adjacent segments.

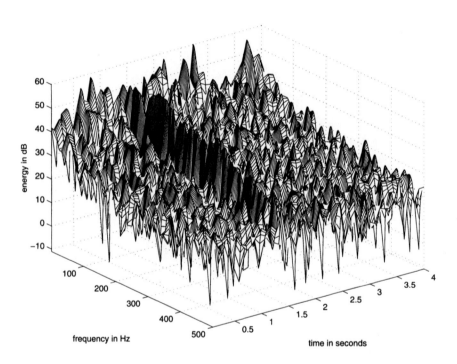

Figure 3.62   Spectrogram of the RLS-filtered VAG signal in Figure 3.59 (d). A Hanning window of length 256 samples (128 *ms*) was used; an overlap of 32 samples (16 *ms*) was allowed between adjacent segments.

3.11 REMARKS

We have investigated problems posed by artifact, noise, and interference of various forms in the acquisition and analysis of several biomedical signals. Random noise, structured interference, and physiological interference have been identified and analyzed separately. Attention has been drawn to the different characteristics of various types of noise, such as frequency content and nonstationarity. Fixed, optimal, and adaptive filters were developed in the time and frequency domains for several applications, and guidelines were drawn to assist in choosing the appropriate filter for various types of artifacts. Advanced methods for adaptive denoising based on wavelet and time-frequency decomposition methods have not been discussed in this chapter, but are described by Krishnan and Rangayyan [93] for filtering VAG signals. Another category of filters that has not been considered in this chapter is that of morphological filters [94, 95], which include nonlinear statistics-based operations and could be formulated under certain conditions to include linear filter operations as well.

It is important to observe that each practical problem needs to be studied carefully to determine the type and characteristics of the artifact present; the nature of the signal and its relationship to, or interaction with, the artifact; and the effect of the filter being considered on the desired signal or features computed from the filtered result. Different filters may be suitable for different subsequent steps of signal analysis. It is unlikely that a single filter will address all of the problems and the requirements in a wide variety of practical situations and applications. Regardless of one's expertise in filters, it should be remembered that *prevention is better than cure*: most filters, while removing an artifact, may introduce another. Attempts should be made at the outset to acquire artifact-free signals to the extent possible.

3.12 STUDY QUESTIONS AND PROBLEMS

(*Note:* Some of the questions deal with the fundamentals of signals and systems, and may require background preparation with other sources such as Lathi [1] or Oppenheim et al. [2]. Such problems are included for the sake of recollection of the related concepts.)

1. What are the potential sources of instrumentation and physiological artifacts in recording the PCG signal? Propose non-electronic methods to prevent or suppress the latter type of artifacts.

2. List four potential sources of instrumentation and physiological artifacts in recording the ECG signal. Describe methods to prevent or remove each artifact. Identify the possible undesired effects of your procedures on the ECG signal.

3. Identify at least three potential sources of physiological artifacts in recording the EEG signal.

4. In recording the EEG in a clinical laboratory, some channels were found to contain the ECG as an artifact. Will simple lowpass or bandpass filtering help in removing the artifact? Why (not)? Propose a scheme to remove the artifact.

5. A biomedical signal is bandpass filtered to the range $0 - 150\ Hz$. Assume the filter to be ideal, and assume any distribution of spectral energy over the bandwidth of the signal.

    (a) What is the minimum frequency at which the signal should be sampled in order to avoid aliasing errors?

    (b) A researcher samples the signal at $500\ Hz$. Draw a schematic representation of the spectrum of the sampled signal.

    (c) Another researcher samples the signal at $200\ Hz$. Draw a schematic representation of the spectrum of the sampled signal. Explain the differences between case (b) and case (c).

6. Distinguish between ensemble averages and temporal (time) averages. Identify applications of first-order and second-order averages of both types in EEG analysis.

7. Explain how one may apply ensemble averaging and temporal (time) averaging procedures to process ECG signals. Identify applications of first-order and second-order averages of both types in ECG analysis.

8. Explain how you would apply synchronized averaging to remove noise in

    (a) ECG signals,

    (b) event-related (or evoked) potentials,

    (c) heart sound (PCG) signals,

    (d) EMG signals.

    In each case, explain

    (i) how you will obtain the information required for synchronization of the signals epochs or episodes;

    (ii) sources of artifacts and how you will deal with them;

    (iii) limitations and practical difficulties; and

    (iv) potential for success of the method.

9. Draw a typical ECG waveform over one cardiac cycle indicating the important component waves. How is the waveform affected by passage through

    (a) a lowpass filter with a cutoff frequency of $40\ Hz$?

    (b) a highpass filter with a cutoff frequency of $5\ Hz$?

    Draw schematic representations of the expected outputs and explain their characteristics.

10. What is the z-transform of a signal whose samples are given in the series $\{4, 3, 2, 1, 0, -1, 0, 1, 0\}$?
    (The first sample represents zero time in all the signal sample arrays given in the problems, unless stated otherwise.)

11. A digital filter is used to process a signal at a sampling rate of $2,000\ Hz$.

    (a) Draw the unit circle in the complex z-plane and identify the frequencies corresponding to the points $z = (1 + j0)$, $z = (0 + j1)$, $z = (-1 + j0)$, $z = (0 - j1)$, and the point $z = (1 + j0)$ again as approached in the counter-clockwise direction.

    (b) What are the frequencies corresponding to these same points if the sampling rate is 500 Hz?

12. What is the transfer function of a linear shift-invariant system whose impulse response is given by the series $\{2, 1, 0, 0, -1, 0, 1, 0\}$ for $n = 0, 1, 2, \ldots, 7.?$

13. The impulse response of a digital filter is $\{1, -2, 1\}$. What will be the response of the filter to the unit step?

14. The impulse response of a filter is $\{3, -2, 2\}$. What will be the response of the filter to the input $\{6, 4, 2, 1\}$?

15. The transfer function of a filter is $H(z) = z^{-1} - 3z^{-2} + 2z^{-4} - z^{-6}$. What is the difference equation relating the output to the input? What is the impulse response of the filter?

16. The impulse response of a filter is given by the series of values $\{3, 2, 1, 0, -1, 0, 0, 1\}$. What is its transfer function?

17. The impulse response of a filter is specified by the series of sample values $\{3, 1, -1\}$.

    (a) What will be the response of the filter to the input whose sample values are $\{4, 4, 2, 1\}$?

    (b) Is the filter response obtained by linear convolution or circular convolution of the input with the impulse response?

    (c) What will be the response with the type of convolution other than the one you indicated as the answer to the question above?

    (d) How would you implement convolution of the two signals listed above using the FFT? Which type of convolution will this procedure provide? How would you get the other type of convolution for the signals in this problem via the FFT-based procedure?

18. A biomedical signal is expected to be band-limited to $100\ Hz$, with significant components of interest up to $80\ Hz$. However, the signal is contaminated with a periodic artifact with a fundamental frequency of $60\ Hz$ and significant third and fifth harmonics. A researcher samples the signal at $200\ Hz$ without pre-filtering the signal.

    Draw a schematic representation of the spectrum of the signal and indicate the artifact components. Label the frequency axis clearly in Hz.

    What kind of a filter would you recommend to remove the artifact?

19. A biomedical signal sampled at $500\ Hz$ was found to have a significant amount of $60\ Hz$ interference.

    (a) Design a notch filter with two zeros to remove the interference.

    (b) What is the effect of the filter if a signal sampled at $100\ Hz$ is applied as the input?

20. Two filters with transfer functions $H_1(z) = \frac{1}{3}(1 + z^{-1} + z^{-2})$ and $H_2(z) = 1 - z^{-1}$ are cascaded.

    (a) What is the transfer function of the complete system?

    (b) What is its impulse response?

    (c) What is its gain at DC and at the folding frequency (that is, $f_s/2$)?

21. A filter has the transfer function $H(z) = (1 + 2z^{-1} + z^{-2})/(1 - z^{-2})$.

    (a) Write the difference equation relating the output to the input.

    (b) Draw the signal-flow diagram of a realization of the filter.

    (c) Draw its pole-zero diagram.

22. A digital filter has zeros at $0.5 \pm j0.5$ and poles at $-0.6 \pm j0.3$.

    (a) Derive the transfer function of the filter.

    (b) Derive the time-domain difference equation (input – output relationship) of the filter.

(c) If the filter is used at a sampling frequency of $1,000\ Hz$, what are the frequencies at which the gain of the filter is maximum and minimum?

23. Two filters with transfer functions $H_1(z) = \frac{1}{2T}(1 - z^{-2})$ and $H_2(z) = \frac{1}{1 - \frac{1}{2}z^{-1}}$ are cascaded.

(a) What is the transfer function of the complete system?

(b) Draw its pole-zero diagram.

(c) Write the difference equation relating the output to the input.

(d) Draw the signal-flow diagram of a realization of the filter.

(e) Compute the first six values of the impulse response of the filter.

(f) The filter is used to process a signal sampled at $1,000\ Hz$. What is its gain at $0, 250$, and $500\ Hz$?

24. A filter is described by the difference equation $y(n) = y(n-1) + \frac{1}{4}x(n) - \frac{1}{4}x(n-4)$.

(a) What is its transfer function?

(b) Draw the signal-flow diagram of a realization of the filter.

(c) Draw its pole-zero diagram.

25. Under what conditions will synchronized averaging fail to reduce noise?

26. A signal sampled at the rate of 100 Hz has the samples $\{0, 10, 0, -5, 0\}$ in mV. The signal is passed through a filter described by the transfer function $H(z) = \frac{1}{T}(1 - z^{-1})$. What will be the output sequence? Plot the output and indicate the amplitude and time scales in detail with appropriate units.

27. A signal sampled at the rate of 100 Hz has the samples $\{0, 10, 0, -5, 0\}$ in mV. It is supposed to be processed by a differentiator with the difference equation $y(n) = \frac{1}{T}[x(n) - x(n-1)]$ and then squared. By mistake the squaring operation is performed before the differentiation. What will be the output sequence? Plot the outputs for both cases and indicate the amplitude and time scales in detail with appropriate units. Explain the differences between the two results.

28. A certain signal analysis technique requires the following operations in order: (a) differentiation, (b) squaring, and (c) lowpass filtering with a filter $H(\omega)$. Considering a generic signal $x(t)$ as the input, write the time-domain and frequency-domain expressions for the output of each stage.

Will changing the order of the operations change the final result? Why (not)?

29. A signal sampled at the rate of 100 Hz has the samples $\{0, 10, 0, -5, 0\}$ in mV. The signal is processed by a differentiator with the difference equation $y(n) = \frac{1}{T}[x(n) - x(n-1)]$, and then filtered with a 4-point moving-average filter.

(a) Derive the transfer function and frequency response of each filter and the combined system.

(b) Derive the values of the signal samples at each stage.

(c) Does it matter which filter is placed first? Why (not)?

(d) Plot the output and indicate the amplitude and time scales in detail with appropriate units.

30. Distinguish between ensemble averages and temporal (time) averages.

Identify potential applications of first-order and second-order averages of both types in heart sound (PCG) analysis. Explain how you would obtain a trigger for synchronization.

31. Is the heart sound signal (PCG) a stationary signal or not? Provide your answer in the context of one full cardiac cycle and give reasons.

    If you say that the PCG signal is nonstationary, identify parts (segments) that could possibly be stationary, considering the possibility of murmurs in both systole and diastole.

32. A signal $x(t)$ is transmitted through a channel. The received signal $y(t)$ is a scaled, shifted, and noisy version of $x(t)$ given as $y(t) = \alpha x(t - t_0) + \eta(t)$ where α is a scale factor, t_0 is the time delay, and $\eta(t)$ is noise. Assume that the noise process has zero mean and is statistically independent of the signal process, and that all processes are stationary.

    Derive expressions for the PSD of $y(t)$ in terms of the PSDs of x and η.

33. A signal $x(n)$ that is observed in an experiment is modeled as a noisy version of a desired signal $d(n)$ as $x(n) = d(n) + \eta(n)$. The noise process η is a zero-mean, unit-variance random process with uncorrelated samples ("white" noise, with ACF $\phi_\eta(\tau) = \delta(\tau)$) that is statistically independent of the signal process d. The ACF $\phi_d(\tau)$ of d is given by the sequence $\{1.0, 0.6, 0.2\}$, for $\tau = 0, 1, 2$, respectively.

    Prepare the Wiener-Hopf equation and derive the coefficients of the optimal Wiener filter.

3.13 LABORATORY EXERCISES AND PROJECTS

Note: Data files related to the exercises are available at the site
ftp://ftp.ieee.org/uploads/press/rangayyan/

1. The data file ecg2x60.dat contains an ECG signal, sampled at 200 Hz, with a significant amount of 60 Hz power-line artifact. (See also the file ecg2x60.m.)

    (a) Design a notch filter with two zeros to remove the artifact and implement it in MATLAB.

    (b) Add two poles at the same frequencies as those of the zeros, but with a radius that is less than unity. Study the effect of the poles on the output of the filter as their radius is varied between 0.8 and 0.99.

2. A noisy ECG signal is provided in the file ecg_hfn.dat. (See also the file ecg_hfn.m.) The sampling rate of this signal is 1,000 Hz.

    Develop a MATLAB program to perform synchronized averaging as described in Section 3.3.1. Select a QRS complex from the signal for use as the template and use a suitable threshold on the cross-correlation function in Equation 3.18 for beat detection. Plot the resulting averaged QRS complex. Ensure that the averaged result covers one full cardiac cycle. Plot a sample ECG cycle from the noisy signal for comparison.

    Select the QRS complex from a different beat for use as the template and repeat the experiment. Observe the results when the threshold on the cross-correlation function is low (say, 0.4) or high (say, 0.95) and comment.

3. Filter the noisy ECG signal in the file ecg_hfn.dat (See also the file ecg_hfn.m; $f_s = 1,000\ Hz$.) using four different Butterworth lowpass filters (individually) realized through MATLAB with the following characteristics:

    (a) Order 2, cutoff frequency 10 Hz;

(b) Order 8, cutoff frequency 20 Hz;

(c) Order 8, cutoff frequency 40 Hz.

(d) Order 8, cutoff frequency 70 Hz.

Use "help butter" and "help filter" in MATLAB to get details about the Butterworth filter.

Compare the results obtained using each of the four Butterworth filters (individually) with those obtained by synchronized averaging, and comment upon the improvements or distortions in the outputs. Relate your discussions to specific characteristics observed in plots of the signals.

4. The ECG signal in the file ecg_lfn.dat has a wandering base-line (low-frequency artifact). (See also the file ecg_lfn.m.) Filter the signal with the derivative-based filters described in Section 3.3.3 and study the results. Study the effect of variation of the position of the pole in the filter in Equation 3.47 on the signal.

5. Filter the signal in the file ecg_lfn.dat using Butterworth highpass filters with orders $2 - 8$ and cutoff frequencies $0.5 - 5$ Hz. (See also the file ecg_lfn.m.) Study the efficacy of the filters in removing the base-line artifact and the effect on the ECG waveform itself. Determine the best compromise acceptable.

6. Design a Wiener filter to remove the artifacts in the ECG signal in the file ecg_hfn.dat. (See also the file ecg_hfn.m.) The equation of the desired filter is given in Equation 3.101. The required model PSDs may be obtained as follows:

Create a piece-wise linear model of the desired version of the signal by concatenating linear segments to provide P, QRS, and T waves with amplitudes, durations, and intervals similar to those in the given noisy ECG signal. Compute the PSD of the model signal.

Select a few segments from the given ECG signal that are expected to be iso-electric (for example, the T – P intervals). Compute their PSDs and obtain their average. The selected noise segments should have zero mean or have the mean subtracted out.

Compare the results of the Wiener filter with those obtained by synchronized averaging and lowpass filtering.

4

Event Detection

Biomedical signals carry signatures of physiological events. The part of a signal related to a specific event of interest is often referred to as an *epoch*. Analysis of a signal for monitoring or diagnosis requires the identification of epochs and investigation of the corresponding events. Once an event has been identified, the corresponding waveform may be segmented and analyzed in terms of its amplitude, waveshape (morphology), time duration, intervals between events, energy distribution, frequency content, and so on. Event detection is thus an important step in biomedical signal analysis.

4.1 PROBLEM STATEMENT

A generic problem statement applicable to the theme of this chapter may be formulated as follows:

> *Given a biomedical signal, identify discrete signal epochs and correlate them with events in the related physiological process.*

In the sections to follow, we shall first study a few examples of epochs in different biomedical signals, with the aim of understanding the nature of the related physiological events. Such an understanding will help in the subsequent development of signal processing techniques to emphasize, detect, and analyze epochs.

4.2 ILLUSTRATION OF THE PROBLEM WITH CASE-STUDIES

The following sections provide illustrations of several events in biomedical signals. The aim of the illustrations is to develop an appreciation of the nature of signal events. A good understanding of signal events will help in designing appropriate signal processing techniques for their detection.

4.2.1 The P, QRS, and T waves in the ECG

As we have already observed in Section 1.2.4, a cardiac cycle is reflected in a period of the repetitive ECG signal as the series of waves labeled as P, QRS, and T. If we view the cardiac cycle as a series of events, we have the following epochs in an ECG waveform:

- **The P wave:** Contraction of the atria is triggered by the SA-node impulse. The atria do not possess any specialized conduction nerves as the ventricles do; as such, contraction of the atrial muscles takes place in a slow squeezing manner, with the excitation stimulus being propagated by the muscle cells themselves. For this reason, the P wave is a slow waveform, with a duration of about 80 ms. The P wave amplitude is much smaller (about $0.1 - 0.2\ mV$) than that of the QRS because the atria are smaller than the ventricles. The P wave is the epoch related to the event of atrial contraction. (Atrial relaxation does not produce any distinct waveform in the ECG as it is overshadowed by the following QRS wave.)

- **The PQ segment:** The AV node provides a delay to facilitate completion of atrial contraction and transfer of blood to the ventricles before ventricular contraction is initiated. The resulting PQ segment, of about 80 ms duration, is thus a "non-event"; however, it is important in recognizing the base-line as the interval is almost always iso-electric.

- **The QRS wave:** The specialized system of Purkinje fibers stimulate contraction of ventricular muscles in a rapid sequence from the apex upwards. The almost-simultaneous contraction of the entire ventricular musculature results in a sharp and tall QRS complex of about 1 mV amplitude and $80 - 100\ ms$ duration. The event of ventricular contraction is represented by the QRS epoch.

- **The ST segment:** The normally flat (iso-electric) ST segment is related to the plateau in the action potential of the left ventricular muscle cells (see Figure 1.3). The duration of the plateau in the action potential is about 200 ms; the ST segment duration is usually about $100 - 120\ ms$. As in the case of the PQ segment, the ST segment may also be termed as a non-event. However, myocardial ischemia or infarction could change the action potentials of a portion of the left ventricular musculature, and cause the ST segment to be depressed (see Figure 1.28) or elevated. The PQ segment serves as a useful reference when the iso-electric nature of the ST segment needs to be verified.

- **The T wave:** The T wave appears in a normal ECG signal as a discrete wave separated from the QRS by an iso-electric ST segment. However, it relates to the last phase of the action potential of ventricular muscle cells, when the potential returns from the plateau of the depolarized state to the resting potential through the process of repolarization [23]. The T wave is commonly referred to as the wave corresponding to ventricular relaxation. While this is indeed correct, it should be noted that relaxation through repolarization is but the final phase of contraction: contraction and relaxation are indicated by the upstroke and downstroke of the same action potential. For this reason, the T wave may be said to relate to a nonspecific event.

  The T wave is elusive, being low in amplitude ($0.1 - 0.3 \ mV$) and being a slow wave extending over $120 - 160 \ ms$. It is almost absent in many ECG recordings. Rather than attempt to detect the often obscure T wave, one may extract a segment of the ECG $80 - 360 \ ms$ from the beginning of the QRS and use it to represent the ST segment and the T wave.

4.2.2 The first and second heart sounds

We observed in Section 1.2.8 that the normal cardiac cycle manifests as a series of the first and second heart sounds — S1 and S2. Murmurs and additional sounds may appear in the presence of cardiovascular diseases or defects. We shall concentrate on S1, S2, and murmurs only.

- **The first heart sound S1:** S1 reflects a sequence of events related to ventricular contraction — closure of the atrio-ventricular valves, isovolumic contraction, opening of the semilunar valves, and ejection of the blood from the ventricles [23]. The epoch of S1 is directly related to the event of ventricular contraction.

- **The second heart sound S2:** S2 is related to the end of ventricular contraction, signified by closure of the aortic and pulmonary valves. As we observed in the case of the T wave, the end of ventricular contraction cannot be referred to as a specific event per se. However, in the case of S2, we do have the specific events of closure of the aortic and pulmonary valves to relate to, as indicated by the corresponding A2 and P2 components of S2. Unfortunately, separate identification of A2 and P2 is confounded by the fact that they usually overlap in normal signals. If A2 and P2 are separated due to a cardiovascular disorder, simultaneous multi-site PCG recordings will be required to identify each component definitively as they may be reversed in order (see Tavel [41] and Rushmer [23]).

- **Murmurs:** Murmurs, if present, could be viewed as specific events. For example, the systolic murmur of aortic stenosis relates to the event of turbulent ejection of blood from the left ventricle through a restricted aortic valve opening. The diastolic murmur in the case of aortic insufficiency corresponds to the event of regurgitation of blood from the aorta back into the left ventricle through a leaky aortic valve.

4.2.3 The dicrotic notch in the carotid pulse

As we saw in Sections 1.2.9 and 1.2.10, closure of the aortic valve causes a sudden drop in aortic pressure that is already on a downward slope at the end of ventricular systole. The dicrotic notch inscribed in the carotid pulse is a delayed, upstream manifestation of the incisura in the aortic pressure wave. The dicrotic notch is a specific signature on the relatively nondescript carotid pulse signal, and may be taken as an epoch related to the event of aortic valve closure (albeit with a time delay); the same event also signifies the end of ventricular systole and ejection as well as the beginning of S2 and diastole.

4.2.4 EEG rhythms, waves, and transients

We have already studied a few basic characteristics of the EEG in Section 1.2.5, and noted the nature of the α, β, δ, and θ waves. We shall now consider a few events and transients that occur in EEG signals [32, 33, 34, 96, 97, 98]. Figure 4.1 shows typical manifestations of the activities described below [32].

- **K-complex:** This is a transient complex waveform with slow waves, sometimes associated with sharp components, and often followed by 14 Hz waves. It occurs spontaneously or in response to a sudden stimulus during sleep, with an amplitude of about 200 μV.

- **Lambda waves:** These are monophasic, positive, sharp waves that occur in the occipital location with an amplitude of less than 50 μV. They are related to eye movement, and are associated with visual exploration.

- **Mu rhythm:** This rhythm appears as a group of waves in the frequency range of $7 - 11$ Hz with an arcade or comb shape in the central location. The mu rhythm usually has an amplitude of less than 50 μV, and is blocked or attenuated by contralateral movement, thought of movement, readiness to move, or tactile stimulation.

- **Spike:** A spike is defined as a transient with a pointed peak, having a duration in the range of $20 - 30$ ms.

- **Sharp wave:** A sharp wave is also a transient with a pointed peak, but with a longer duration than a spike, in the range of $70 - 200$ ms.

- **Spike-and-wave rhythm:** A sequence of surface-negative slow waves in the frequency range of $2.5 - 3.5$ Hz and having a spike associated with each wave is referred to as a spike-and-wave rhythm. There could be several spikes of amplitude up to $1,000$ μV in each complex, in which case the rhythm is called a polyspike-and-wave complex.

- **Sleep spindle:** This is an episodic rhythm at about 14 Hz and 50 μV, occurring maximally over the fronto-central regions during certain stages of

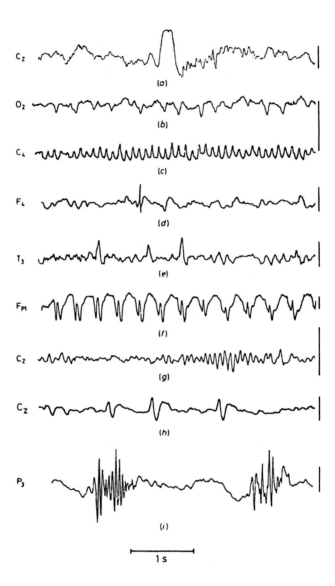

Figure 4.1 From top to bottom: (a) the K-complex; (b) the lambda wave; (c) the mu rhythm; (d) a spike; (e) sharp waves; (f) spike-and-wave complexes; (g) a sleep spindle; (h) vertex sharp waves; and (i) polyspike discharges. The horizontal bar at the bottom indicates a duration of 1 s; the vertical bars at the right indicate 100 μV. Reproduced with permission from R. Cooper, J.W. Osselton, and J.C. Shaw, *EEG Technology*, 3rd Edition, 1980. ©Butterworth Heinemann Publishers, a division of Reed Educational & Professional Publishing Ltd., Oxford, UK.

sleep. A spindle is defined, in general, as a short sequence of monomorphic waves having a fusiform appearance [33].

- **Vertex sharp transient** or **V-wave:** This wave is a sharp potential that is maximal at the vertex at about 300 μV and is negative in relation to the EEG in other areas. It occurs spontaneously during sleep or in response to a sensory stimulus during sleep or wakefulness.

In addition to the above, the term "burst" is used to indicate a phenomenon composed of two or more waves that are different from the principal (background) activity in terms of amplitude, frequency, or waveform. A burst is abrupt and has a relatively short duration [33].

An EEG record is described in terms of [32]

- the most persistent rhythm (for example, α);

- the presence of other rhythmic features, such as δ, θ, or β;

- discrete features of relatively long duration, such as an episode of spike-and-wave activity;

- discrete features of relatively short duration, such as isolated spikes or sharp waves;

- the activity remaining when all the previous features have been described, referred to as background activity; and

- artifacts, if any, giving rise to ambiguity in interpretation.

Each of the EEG waves or activities is described in chronological sequence in terms of amplitude; frequency, in the case of rhythmic features; waveform, in the case of both rhythmic and transient features; location or spatial distribution; incidence or temporal variability; right – left symmetry in location of activity; and responsiveness to stimuli, such as eye opening and closure. The EEG record at rest is first described as above; effects of evocative techniques are then specified in the same terms. Behavioral changes, such as the subject becoming drowsy or falling asleep, are also noted [32].

The EEG signals in Figure 1.22 demonstrate the presence of the α rhythm in all the channels. The EEG signals in Figure 1.23 depict spike-and-wave complexes in almost all the channels.

4.3   DETECTION OF EVENTS AND WAVES

We shall now see how the knowledge that we have gained so far of several biomedical signal events may be applied to develop signal processing techniques for their detection. Each of the following subsections will deal with the problem of detection of a specific type of event. The techniques described should find applications in the detection of other events of comparable characteristics.

4.3.1 Derivative-based methods for QRS detection

Problem: *Develop signal processing techniques to facilitate detection of the QRS complex, given that it is the sharpest wave in an ECG cycle.*

Solution 1: We noted in Section 1.2.4 that the QRS complex has the largest slope (rate of change of voltage) in a cardiac cycle by virtue of the rapid conduction and depolarization characteristics of the ventricles. As the rate of change is given by the derivative operator, the $\frac{d}{dt}$ operation would be the most logical starting point in an attempt to develop an algorithm to detect the QRS complex.

We saw in Section 3.3.3 that the derivative operator enhances the QRS, although the resulting wave does not bear any resemblance to a typical QRS complex. Observe in Figures 3.24 and 3.25 that the slow P and T waves have been suppressed by the derivative operators, while the output is the highest at the QRS. However, given the noisy nature of the results of the derivative-based operators, it is also evident that significant smoothing will be required before further processing can take place.

Balda et al. [99] proposed a derivative-based algorithm for QRS detection, which was further studied and evaluated by Ahlstrom and Tompkins [100], Friesen et al. [101], and Tompkins [27]. The algorithm progresses as follows. In a manner similar to Equation 3.45, the smoothed three-point first derivative $y_0(n)$ of the given signal $x(n)$ is approximated as

$$y_0(n) = |x(n) - x(n-2)|. \tag{4.1}$$

The second derivative is approximated as

$$y_1(n) = |x(n) - 2x(n-2) + x(n-4)|. \tag{4.2}$$

The two results are weighted and combined to obtain

$$y_2(n) = 1.3y_0(n) + 1.1y_1(n). \tag{4.3}$$

The result $y_2(n)$ is scanned with a threshold of 1.0. Whenever the threshold is crossed, the subsequent eight samples are also tested against the same threshold. If at least six of the eight points pass the threshold test, the segment of eight samples is taken to be a part of a QRS complex. The procedure results in a pulse with its width proportional to that of the QRS complex; however, the method is sensitive to noise.

Illustration of application: Figure 4.2 illustrates, in the top-most trace, two cycles of a filtered version of the ECG signal shown in Figure 3.5. The signal was filtered with an eighth-order Butterworth lowpass filter with $f_c = 90 \ Hz$, down-sampled by a factor of five, and filtered with a notch filter with $f_o = 60 \ Hz$. The effective sampling rate is 200 Hz. The signal was normalized by dividing by its maximum value.

The second and third plots in Figure 4.2 show the derivatives $y_0(n)$ and $y_1(n)$, respectively; the fourth plot illustrates the combined result $y_2(n)$. Observe the relatively high values in the derivative-based results at the QRS locations; the outputs are low or negligible at the P and T wave locations, in spite of the fact that the original signal possesses an unusually sharp and tall T wave. It is also seen that the results

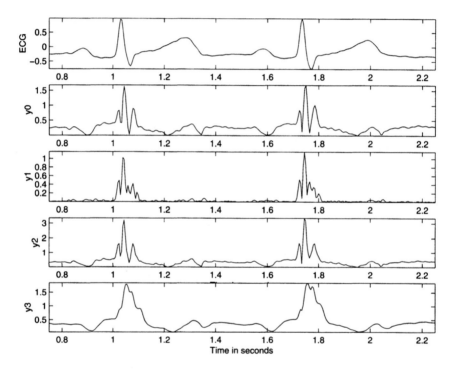

Figure 4.2 From top to bottom: two cycles of a filtered version of the ECG signal shown in Figure 3.5; output $y_0(n)$ of the first-derivative-based operator in Equation 4.1; output $y_1(n)$ of the second-derivative-based operator in Equation 4.2; the combined result $y_2(n)$ from Equation 4.3; and the result $y_3(n)$ of passing $y_2(n)$ through the 8-point MA filter in Equation 3.27.

have multiple peaks over the duration of the QRS wave, due to the fact that the QRS complex includes three major swings: Q – R, R – S, and S – ST base-line in the present example (an additional PQ base-line – Q swing may also be present in other ECG signals).

The last plot in Figure 4.2 shows the smoothed result $y_3(n)$ obtained by passing $y_2(n)$ through the 8-point MA filter in Equation 3.27. We now have a single pulse with amplitude greater than 1.0 over the duration of the corresponding QRS complex. A simple peak-searching algorithm may be used to detect each ECG beat. The net delay introduced by the filters should be subtracted from the detected peak location in order to obtain the corresponding QRS location.

Note that peak searching cannot be performed directly on an ECG signal: the QRS might not always be the highest wave in a cardiac cycle, and artifacts may easily upset the search procedure. Observe also that the ECG signal in the present illustration was filtered to a restricted bandwidth of 90 Hz before the derivatives were computed, and that it is free of base-line drift.

Solution 2: Murthy and Rangaraj [102] proposed a QRS detection algorithm based upon a weighted and squared first-derivative operator and an MA filter. In this method, a filtered-derivative operator was defined as

$$g_1(n) = \sum_{i=1}^{N} |x(n - i + 1) - x(n - i)|^2 (N - i + 1), \qquad (4.4)$$

where $x(n)$ is the ECG signal, and N is the width of a window within which first-order differences are computed, squared, and weighted by the factor $(N - i + 1)$. The weighting factor decreases linearly from the current difference to the difference N samples earlier in time, and provides a smoothing effect. Further smoothing of the result was performed by an MA filter over M points to obtain

$$g(n) = \frac{1}{M} \sum_{j=0}^{M-1} g_1(n - j). \qquad (4.5)$$

With a sampling rate of 100 Hz, the filter window widths were set as $M = N = 8$. The algorithm provides a single peak for each QRS complex and suppresses P and T waves.

Searching for the peak in a processed signal such as $g(n)$ may be accomplished by a simple peak-searching algorithm as follows:

1. Scan a portion of the signal $g(n)$ that may be expected to contain a peak and determine the maximum value $g_{\max}$. The maximum of $g(n)$ over its entire available duration may also be taken to be $g_{\max}$.

2. Define a threshold as a fraction of the maximum, for example, $Th = 0.5\, g_{\max}$.

3. For all $g(n) > Th$, select those samples for which the corresponding $g(n)$ values are greater than a certain predefined number M of preceding and suc-

ceeding samples of $g(n)$, that is,

$$
\begin{aligned}
\{p\} = \quad & [\quad n \mid g(n) > Th \ ] \text{ AND} \\
& [\quad g(n) > g(n-i), i = 1, 2, \ldots, M \ ] \text{ AND} \\
& [\quad g(n) > g(n+i), i = 1, 2, \ldots, M \ ].
\end{aligned}
\tag{4.6}
$$

The set $\{p\}$ defined as above contains the indices of the peaks in $g(n)$.

Additional conditions may be imposed to reject peaks due to artifacts, such as a minimum interval between two adjacent peaks. A more elaborate peak-searching algorithm will be described in Section 4.3.2.

Illustration of application: Figure 4.3 illustrates, in the top-most trace, two cycles of a filtered version of the ECG signal shown in Figure 3.5. The signal was filtered with an eighth-order Butterworth lowpass filter with $f_c = 40 \ Hz$, and down-sampled by a factor of ten. The effective sampling rate is $100 \ Hz$ to match the parameters used by Murthy and Rangaraj [102]. The signal was normalized by dividing by its maximum value.

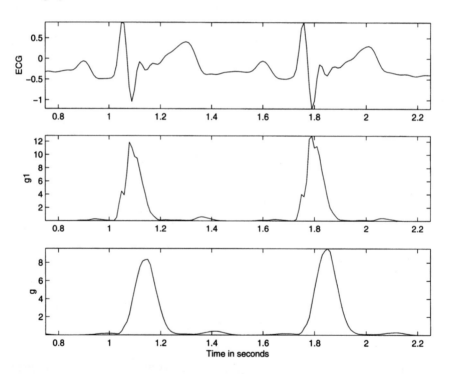

Figure 4.3 From top to bottom: two cycles of a filtered version of the ECG signal shown in Figure 3.5; output $g_1(n)$ of the weighted and squared first-derivative operator in Equation 4.4; output $g(n)$ of the smoothing filter in Equation 4.5.

The second and third plots in Figure 4.3 show the outputs of the derivative-based operator and the smoothing filter. Observe that the final output contains a single, smooth peak for each QRS, and that the P and T waves produce no significant output. A simple peak-searching algorithm may be used to detect and segment each beat [102].

4.3.2 The Pan-Tompkins algorithm for QRS detection

Problem: *Propose an algorithm to detect QRS complexes in an ongoing ECG signal.*
   Solution: Pan and Tompkins [103, 27] proposed a real-time QRS detection algorithm based on analysis of the slope, amplitude, and width of QRS complexes. The algorithm includes a series of filters and methods that perform lowpass, highpass, derivative, squaring, integration, adaptive thresholding, and search procedures. Figure 4.4 illustrates the steps of the algorithm in schematic form.

Figure 4.4   Block diagram of the Pan-Tompkins algorithm for QRS detection.

Lowpass filter: The recursive lowpass filter used in the Pan-Tompkins algorithm has integer coefficients to reduce computational complexity, with the transfer function defined as

$$H(z) = \frac{1}{32} \frac{(1 - z^{-6})^2}{(1 - z^{-1})^2}.$$ (4.7)

(See also Equations 3.37 and 3.38.) The output $y(n)$ is related to the input $x(n)$ as

$$y(n) = 2y(n-1) - y(n-2) + \frac{1}{32}[x(n) - 2x(n-6) + x(n-12)].$$ (4.8)

With the sampling rate being $200\ Hz$, the filter has a rather low cutoff frequency of $f_c = 11\ Hz$, and introduces a delay of 5 samples or $25\ ms$. The filter provides an attenuation greater than $35\ dB$ at $60\ Hz$, and effectively suppresses power-line interference, if present.
   Highpass filter: The highpass filter used in the algorithm is implemented as an allpass filter minus a lowpass filter. The lowpass component has the transfer function

$$H_{lp}(z) = \frac{(1 - z^{-32})}{(1 - z^{-1})};$$ (4.9)

the input – output relationship is

$$y(n) = y(n-1) + x(n) - x(n-32).$$ (4.10)

The transfer function $H_{hp}(z)$ of the highpass filter is specified as

$$H_{hp}(z) = z^{-16} - \frac{1}{32} H_{lp}(z).$$ (4.11)

Equivalently, the output $p(n)$ of the highpass filter is given by the difference equation

$$p(n) = x(n - 16) - \frac{1}{32} [y(n - 1) + x(n) - x(n - 32)], \qquad (4.12)$$

with $x(n)$ and $y(n)$ being related as in Equation 4.10. The highpass filter has a cutoff frequency of 5 Hz and introduces a delay of 80 ms.

Derivative operator: The derivative operation used by Pan and Tompkins is specified as

$$y(n) = \frac{1}{8} [2x(n) + x(n - 1) - x(n - 3) - 2x(n - 4)], \qquad (4.13)$$

and approximates the ideal $\frac{d}{dt}$ operator up to 30 Hz. The derivative procedure suppresses the low-frequency components of the P and T waves, and provides a large gain to the high-frequency components arising from the high slopes of the QRS complex. (See Section 3.3.3 for details on the properties of derivative-based filters.)

Squaring: The squaring operation makes the result positive and emphasizes large differences resulting from QRS complexes; the small differences arising from P and T waves are suppressed. The high-frequency components in the signal related to the QRS complex are further enhanced.

Integration: As observed in the previous subsection, the output of a derivative-based operation will exhibit multiple peaks within the duration of a single QRS complex. The Pan-Tompkins algorithm performs smoothing of the output of the preceding operations through a moving-window integration filter as

$$y(n) = \frac{1}{N} [x(n - (N - 1)) + x(n - (N - 2)) + \cdots + x(n)]. \qquad (4.14)$$

The choice of the window width N is to be made with the following considerations: too large a value will result in the outputs due to the QRS and T waves being merged, whereas too small a value could yield several peaks for a single QRS. A window width of $N = 30$ was found to be suitable for $f_s = 200$ Hz. Figure 4.5 illustrates the effect of the window width on the output of the integrator and its relationship to the QRS width. (See Section 3.3.2 for details on the properties of moving-average and integrating filters.)

Adaptive thresholding: The thresholding procedure in the Pan-Tompkins algorithm adapts to changes in the ECG signal by computing running estimates of signal and noise peaks. A peak is said to be detected whenever the final output changes direction within a specified interval. In the following discussion, $SPKI$ represents the peak level that the algorithm has learned to be that corresponding to QRS peaks, and $NPKI$ represents the peak level related to non-QRS events (noise, EMG, etc.). $THRESHOLD \ I1$ and $THRESHOLD \ I2$ are two thresholds used to categorize peaks detected as signal (QRS) or noise.

Every new peak detected is categorized as a signal peak or a noise peak. If a peak exceeds $THRESHOLD \ I1$ during the first step of analysis, it is classified as a QRS (signal) peak. If the searchback technique (described in the next paragraph) is used,

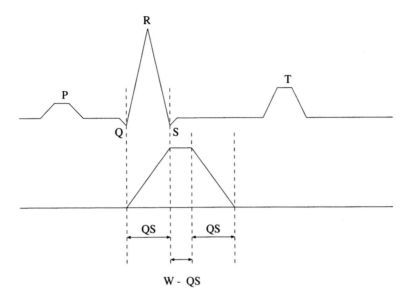

Figure 4.5 The relationship of a QRS complex to the moving-window integrator output. Upper plot: Schematic ECG signal. Lower plot: Output of the moving-window integrator. QS: QRS complex width. W: width of the integrator window, given as N/f_s s. Adapted from Tompkins [27].

the peak should be above $THRESHOLD\ I2$ to be called a QRS. The peak levels and thresholds are updated after each peak is detected and classified as

$$SPKI = 0.125\ PEAKI + 0.875\ SPKI \quad \text{if } PEAKI \text{ is a signal peak;} \quad (4.15)$$
$$NPKI = 0.125\ PEAKI + 0.875\ NPKI \quad \text{if } PEAKI \text{ is a noise peak;}$$

$$THRESHOLD\ I1 \quad = \quad NPKI + 0.25(SPKI - NPKI); \quad (4.16)$$
$$THRESHOLD\ I2 \quad = \quad 0.5\ THRESHOLD\ I1.$$

The updating formula for $SPKI$ is changed to

$$SPKI = 0.25\ PEAKI + 0.75\ SPKI \quad (4.17)$$

if a QRS is detected in the searchback procedure using $THRESHOLD\ I2$.

Searchback procedure: The Pan-Tompkins algorithm maintains two RR-interval averages: $RR\ AVERAGE1$ is the average of the eight most-recent beats, and $RR\ AVERAGE2$ is the average of the eight most-recent beats having RR intervals within the range specified by $RR\ LOW\ LIMIT = 0.92 \times RR\ AVERAGE2$ and $RR\ HIGH\ LIMIT = 1.16 \times RR\ AVERAGE2$. Whenever a QRS is not detected for a certain interval specified as $RR\ MISSED\ LIMIT = 1.66 \times RR\ AVERAGE2$, the QRS is taken to be the peak between the established thresholds applied in the searchback procedure.

The algorithm performed with a very low error rate of 0.68%, or 33 beats per hour on a database of about 116, 000 beats obtained from 24-hour records of the ECGs of 48 patients (see Tompkins [27] for details).

Illustration of application: Figure 4.6 illustrates, in the top-most trace, the same ECG signal as in Figure 4.2. The Pan-Tompkins algorithm as above was implemented in MATLAB. The outputs of the various stages of the algorithm are illustrated in sequence in the same figure. The observations to be made are similar to those in the preceding section on the derivative-based method. The derivative operator suppresses the P and T waves and provides a large output at the QRS locations. The squaring operation preferentially enhances large values, and boosts high-frequency components. The result still possesses multiple peaks for each QRS, and hence needs to be smoothed. The final output of the integrator is a single smooth pulse for each QRS. Observe the shift between the actual QRS location and the pulse output due to the cumulative delays of the various filters. The thresholding and search procedures and their results are not illustrated. More examples of QRS detection will be presented in Sections 4.9 and 4.10.

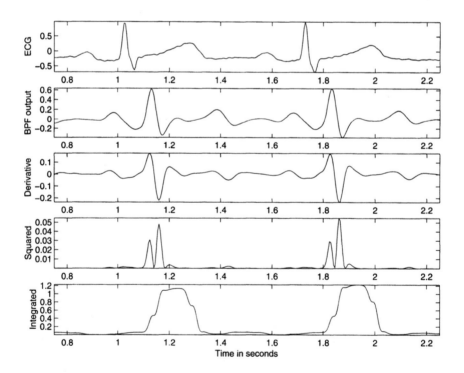

Figure 4.6   Results of the Pan-Tompkins algorithm. From top to bottom: two cycles of a filtered version of the ECG signal shown in Figure 3.5 (the same as that in Figure 4.2); output of the bandpass filter (BPF, a combination of lowpass and highpass filters); output of the derivative-based operator; the result of squaring; and $100\times$ the result of the final integrator.

4.3.3 Detection of the dicrotic notch

Problem: *Propose a method to detect the dicrotic notch in the carotid pulse signal.*

Solution: Lehner and Rangayyan [66] proposed a method for detection of the dicrotic notch that used the least-squares estimate of the second derivative $p(n)$ of the carotid pulse signal $y(n)$ defined as

$$p(n) = 2y(n-2) - y(n-1) - 2y(n) - y(n+1) + 2y(n+2). \qquad (4.18)$$

Observe that this expression is noncausal; it may be made causal by adding a delay of two samples.

The second derivative was used due to the fact that the dicrotic notch appears as a short wave riding on the downward slope of the carotid pulse signal (see also Starmer et al. [104]). A first-derivative operation would give an almost-constant output for the downward slope. The second-derivative operation removes the effect of the downward slope and enhances the notch itself. The result was squared and smoothed to obtain

$$s(n) = \sum_{k=1}^{M} p^2(n-k+1)w(k), \qquad (4.19)$$

where $w(k) = (M - k + 1)$ is a linear weighting function, and $M = 16$ for $f_s = 256\,Hz$.

The method yields two peaks for each period of the carotid pulse signal. The first peak in the result represents the onset of the carotid upstroke. The second peak that appears in the result within a cardiac cycle is due to the dicrotic notch. To locate the dicrotic notch, the local minimum in the carotid pulse within a $\pm 20\,ms$ interval of the second peak needs to be located.

Illustration of application: The upper plot in Figure 4.7 illustrates two cycles of a carotid pulse signal. The signal was lowpass filtered at $100\,Hz$ and sampled at $250\,Hz$. The result of application of the Lehner and Rangayyan method to the signal is shown in the lower plot. It is evident that the second derivative has successfully accentuated the dicrotic notch. A simple peak-searching algorithm may be used to detect the first and second peaks in the result. The dicrotic notch may then be located by searching for the minimum in the carotid pulse signal within a $\pm 20\,ms$ interval around the second peak location.

Observe that the result illustrated in Figure 4.7 may benefit from further smoothing by increasing the window width M in Equation 4.19. The window width needs to be chosen in accordance with the characteristics of the signal on hand as well as the lowpass filter and sampling rate used. Further illustration of the detection of the dicrotic notch will be provided in Section 4.10.

4.4 CORRELATION ANALYSIS OF EEG CHANNELS

EEG signals are usually acquired simultaneously over multiple channels. Event detection and epoch analysis of EEG signals becomes more complicated than the

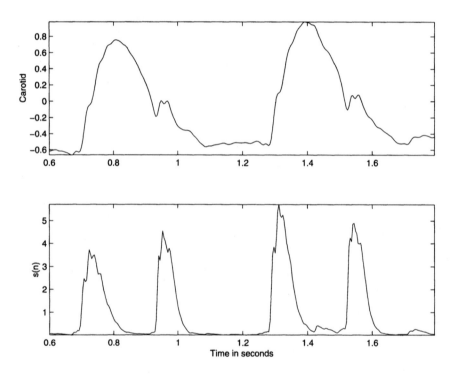

Figure 4.7   Two cycles of a carotid pulse signal and the result of the Lehner and Rangayyan method for detection of the dicrotic notch.

problems we have seen so far with the single-channel ECG and carotid pulse signals, due to the need to detect similar events across multiple channels. Autocorrelation and cross-correlation techniques in both the time and frequency domains serve such needs.

4.4.1 Detection of EEG rhythms

Problem: *Propose a method to detect the presence of the α rhythm in an EEG channel. How would you extend the method to detect the presence of the same rhythm simultaneously in two EEG channels?*

Solution: Two signals may be compared to detect common characteristics present in them via their dot product (also known as the inner or scalar product), defined as

$$x \cdot y = \langle x, y \rangle = \sum_{n=0}^{N-1} x(n)y(n), \tag{4.20}$$

where the signals $x(n)$ and $y(n)$ have N samples each. The dot product represents the projection of one signal onto the other, with each signal being viewed as an N-dimensional vector. The dot product may be normalized by the geometric mean of the energies of the two signals to obtain a correlation coefficient as [67]

$$\gamma_{xy} = \frac{\sum_{n=0}^{N-1} x(n)y(n)}{\left[\sum_{n=0}^{N-1} x^2(n) \sum_{n=0}^{N-1} y^2(n)\right]^{1/2}}. \tag{4.21}$$

The means of the signals may be subtracted out, if desired, as in Equation 3.18.

In the case of two continuous-time signals $x(t)$ and $y(t)$, the projection of one signal onto the other is defined as

$$\theta_{xy} = \int_{-\infty}^{\infty} x(t)y(t)dt. \tag{4.22}$$

When a shift or time delay may be present in the occurrence of the epoch of interest in the two signals being compared, it becomes necessary to introduce a time-shift parameter to compute the projection for every possible position of overlap. The shift parameter facilitates searching one signal for the occurrence of an event matching that in the other signal at any time instant within the available duration of the signals. The cross-correlation function (CCF) between two signals for a shift or delay of τ seconds or k samples may be obtained as

$$\theta_{xy}(\tau) = \int_{-\infty}^{\infty} x(t)y(t+\tau)dt, \quad \text{or} \tag{4.23}$$

$$\theta_{xy}(k) = \sum_{n} x(n)y(n+k). \tag{4.24}$$

The range of summation in the latter case needs to be limited to the range of the available overlapped data. A scale factor, depending upon the number of data samples used, needs to be introduced to obtain the true CCF, but will be neglected here (see Section 6.4). An extended version of the correlation coefficient γ_{xy} in Equation 4.21, to include time shift, is provided in Equation 3.18.

When the ACF or the CCF are computed for various shifts, a question arises about the data samples in one of the signal segments beyond the duration of the other. We may add zeros to one of the signals and increase its length by the maximum shift of interest, or we may use the true data samples from the original signal record, if available. The latter method was used wherever possible in the following illustrations.

In the case of random signals, we need to take the expectation or sample average of the outer product of the vectors formed by the available samples of the signals. Let $\mathbf{x}(n) = [x(n), x(n-1), \ldots, x(n-N+1)]^T$ and $\mathbf{y}(n) = [y(n), y(n-1), \ldots, y(n-N+1)]^T$ represent the N-dimensional vectorial form of the two signals $x(n)$ and $y(n)$ with the most-recent N samples being available in each signal at the time instant n. If $\mathbf{x}(n)$ and $\mathbf{y}(n)$ are sample observations of random processes, their CCF is defined as

$$\mathbf{\Theta}_{xy} = E[\mathbf{x}(n)\mathbf{y}^T(n)], \tag{4.25}$$

in a manner similar to what we saw in Equations 3.78 and 3.79. The outer product, which is an $N \times N$ matrix, provides the cross-terms that include all possible delays (shifts) within the duration of the signals.

All of the equations above may be modified to obtain the ACF by replacing the second signal y with the first signal x. The signal x is then compared with itself.

The ACF displays peaks at intervals corresponding to the period (and integral multiples thereof) of any periodic or repetitive pattern present in the signal. This property facilitates the detection of rhythms in signals such as the EEG: the presence of the α rhythm would be indicated by a peak in the neighborhood of 0.1 s. The ACF of most signals decays and reaches negligible values after delays of a few milliseconds, except for periodic signals of infinite or indefinite duration for which the ACF will also exhibit periodic peaks. The ACF will also exhibit multiple peaks when the same event repeats itself at regular or irregular intervals. One may need to compute the ACF only up to certain delay limits depending upon the expected characteristics of the signal being analyzed.

The CCF displays peaks at the period of any periodic pattern present in *both* of the signals being analyzed. The CCF may therefore be used to detect common rhythms present between two signals, for example, between two channels of the EEG. When one of the functions being used to compute the CCF is a template representing an event, such as an ECG cycle as in the illustration in Section 3.3.1 or an EEG spike-and-wave complex as in Section 4.4.2, the procedure is known as *template matching*.

Illustration of application: Figure 4.8 shows, in the upper trace, the ACF of a segment of the p4 channel of the EEG in Figure 1.22 over the time interval $4.67 - 5.81$ s. The ACF displays peaks at time delays of 0.11 s and its integral multiples. The inverse of the delay of the first peak corresponds to 9 Hz, which is within the α rhythm range. (The PSD in the lower trace of Figure 4.8 and the others to follow will be described in Section 4.5.) It is therefore obvious that the signal segment analyzed contains the α rhythm. A simple peak-search algorithm may be applied to the ACF to detect the presence of peaks at specific delays of interest or over the entire range of the ACF.

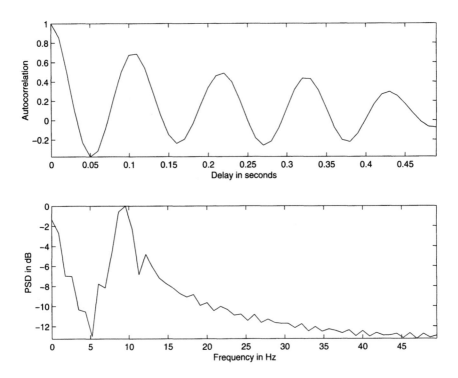

Figure 4.8 Upper trace: ACF of the $4.67 - 5.81$ s portion of the p4 channel of the EEG signal shown in Figure 1.22. Lower trace: The PSD of the signal segment in dB, given by the Fourier transform of the ACF.

To contrast with the preceding example, the upper trace of Figure 4.9 shows the ACF of the $4.2 - 4.96$ s segment of the f3 channel of the EEG in Figure 1.22. The ACF shows no peak in the $0.08 - 1.25$ s region, indicating absence of the α rhythm in the segment analyzed.

Figures 4.10, 4.11, and 4.12 illustrate the CCF results comparing the following portions of the EEG signal shown in Figure 1.22 in order: the p3 and p4 channels over the duration $4.72 - 5.71$ s when both channels exhibit the α rhythm; the o2 and c4 channels over the duration $5.71 - 6.78$ s when the former has the α rhythm but not the latter channel; and the f3 and f4 channels over the duration $4.13 - 4.96$ s when neither channel has α activity. The relative strengths of the peaks in the α range, as described earlier, agree with the joint presence, singular presence, or absence of the α rhythm in the various segments (channels) analyzed.

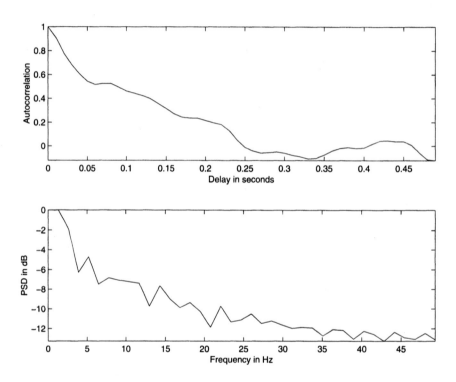

Figure 4.9   Upper trace: ACF of the $4.2 - 4.96$ s portion of the f3 channel of the EEG signal shown in Figure 1.22. Lower trace: The PSD of the signal segment in dB.

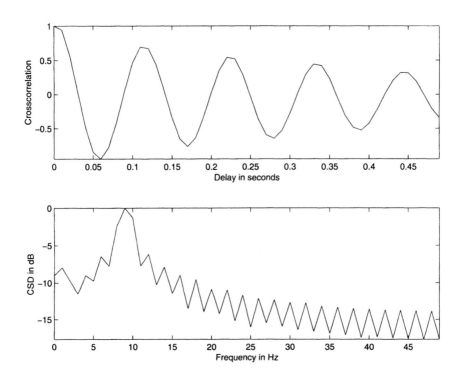

Figure 4.10 Upper trace: CCF between the $4.72 - 5.71$ s portions of the p3 and p4 channels of the EEG signal shown in Figure 1.22. Lower trace: The CSD of the signal segments in dB, computed as the Fourier transform of the CCF.

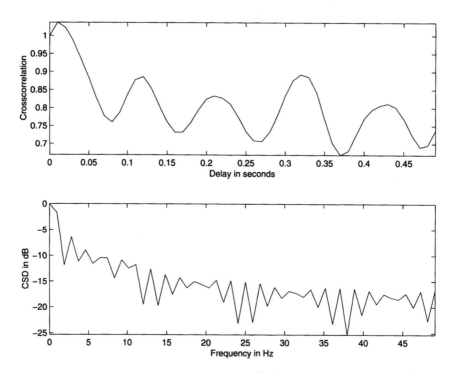

Figure 4.11 Upper trace: CCF between the $5.71 - 6.78$ s portions of the o2 and c4 channels of the EEG signal shown in Figure 1.22. Lower trace: The CSD of the signal segments in dB.

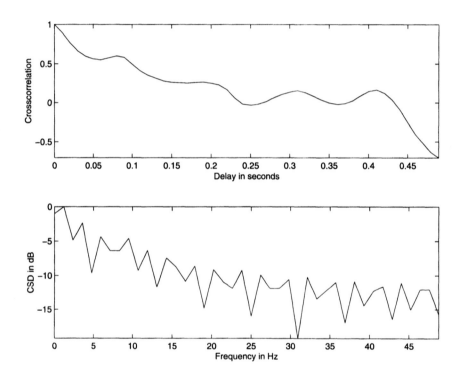

Figure 4.12   Upper trace: CCF between the $4.13 - 4.96$ s portions of the f3 and f4 channels of the EEG signal shown in Figure 1.22. Lower trace: The CSD of the signal segments in dB.

4.4.2 Template matching for EEG spike-and-wave detection

We have already seen the use of template matching for the extraction of ECG cycles for use in synchronized averaging in Section 3.3.1. We shall now consider another application of template matching.

Problem: *Propose a method to detect spike-and-wave complexes in an EEG signal. You may assume that a sample segment of a spike-and-wave complex is available.*

Solution: A spike-and-wave complex is a well-defined event in an EEG signal. The complex is composed of a sharp spike followed by a wave with a frequency of about 3 Hz; the wave may contain a half period or a full period of an almost-sinusoidal pattern. One may therefore extract an epoch of a spike-and-wave complex from an EEG channel and use it for template matching with the same formula as in Equation 3.18 (see also Barlow [97]). The template may be correlated with the same channel from which it was extracted to detect similar events that appear at a later time, or with another channel to search for similar events. A simple threshold on the result should yield the time instants where the events appear.

Illustration of application: The c3 channel of the EEG signal in Figure 1.23 is shown in the upper trace of Figure 4.13. The spike-and-wave complex between 0.60 s and 0.82 s in the signal was selected for use as the template, and template matching was performed with the same channel signal using the formula in Equation 3.18. The result in the lower trace of Figure 4.13 demonstrates strong and clear peaks at each occurrence of the spike-and-wave complex in the EEG signal. The peaks in the result occur at the same instants of time as the corresponding spike-and-wave complexes.

Figure 4.14 shows the f3 channel of the EEG signal in Figure 1.23, along with the result of template matching, using the same template that was used in the previous example from channel c3. The result shows that the f3 channel also has spike-and-wave complexes that match the template.

4.5 CROSS-SPECTRAL TECHNIQUES

The multiple peaks that arise in the ACF or CCF functions may cause confusion in the detection of rhythms; the analyst may be required to discount peaks that appear at integral multiples of the delay corresponding to a fundamental frequency. The Fourier-domain equivalents of the ACF or the CCF permit easier and more intuitive analysis in the frequency domain than in the time domain. The notion of rhythms would be easier to associate with frequencies in *cps* or Hz than with the corresponding inversely related periods (see also the introductory section of Chapter 6).

4.5.1 Coherence analysis of EEG channels

Problem: *Describe a frequency-domain approach to study the presence of rhythms in multiple channels of an EEG signal.*

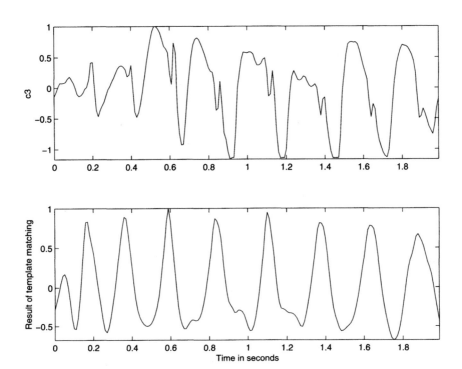

Figure 4.13 Upper trace: the c3 channel of the EEG signal shown in Figure 1.23. Lower trace: result of template matching. The spike-and-wave complex between 0.60 *s* and 0.82 *s* in the signal was used as the template.

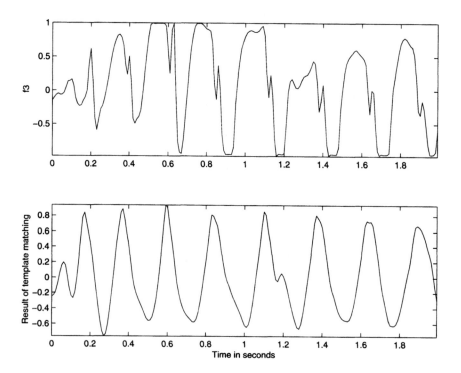

Figure 4.14   Upper trace: the f3 channel of the EEG signal shown in Figure 1.23. Lower trace: result of template matching. The spike-and-wave complex between 0.60 *s* and 0.82 *s* in the c3 channel (see Figure 4.13) was used as the template.

Solution: The Fourier-domain equivalents of the ACF and CCF are the PSD (also known as the autospectrum) and the cross-spectrum (or cross-spectral density — CSD), respectively. The PSD $S_{xx}(f)$ of a signal is related to its ACF via the Fourier transform:

$$S_{xx}(f) = FT[\phi_{xx}(\tau)] = X(f)X^*(f) = |X(f)|^2. \qquad (4.26)$$

The Fourier transform of the CCF between two signals gives the CSD:

$$S_{xy}(f) = FT[\theta_{xy}(\tau)] = X(f)Y^*(f). \qquad (4.27)$$

(For the sake of simplicity, the double-symbol subscripts xx and yy may be replaced by their singular versions, or dropped entirely when not relevant in subsequent discussions.)

The PSD displays peaks at frequencies corresponding to periodic activities in the signal. This property facilitates the detection of rhythms in signals such as the EEG: the presence of the α rhythm would be indicated by a peak or multiple peaks in the neighborhood of $8 - 13\ Hz$. The PSD may also be studied to locate the presence of activity spread over specific bands of frequencies, such as formants in the speech signal or murmurs in the PCG.

The CSD exhibits peaks at frequencies that are present in both of the signals being compared. The CSD may be used to detect rhythms present in common between two channels of the EEG.

The normalized *coherence spectrum* of two signals is given by [5, 32]

$$\Gamma_{xy}(f) = \left[\frac{|S_{xy}(f)|^2}{S_{xx}(f)S_{yy}(f)} \right]^{1/2}. \qquad (4.28)$$

The phase of the coherence spectrum is given by $\psi_{xy}(f) = \angle S_{xy}(f)$, which represents the average phase difference (related to the time delay) between common frequency components in the two signals.

Illustration of application: The coherence between EEG signals recorded from different positions on the scalp depends upon the structural connectivity or functional coupling between the corresponding parts of the brain. Investigations into the neurophysiology of seizure discharges and behavior attributable to disorganization of cerebral function may be facilitated by coherence analysis [32]. The symmetry, or lack thereof, between two EEG channels on the left and right sides of the same position (for example, c3 and c4) may be analyzed via the CSD or the coherence function.

The lower traces in Figures 4.8 and 4.9 illustrate the PSDs of EEG segments with and without the α rhythm, respectively. The former shows a strong and clear peak at about $9\ Hz$, indicating the presence of the α rhythm. Observe that the PSD displays a single peak although the corresponding ACF has multiple peaks at two, three, and four times the delay corresponding to the fundamental period of the α wave in the signal. The PSD in Figure 4.9 exhibits no peak in the α range, indicating the absence of the α rhythm in the signal.

The lower traces in Figures 4.10, 4.11, and 4.12 illustrate the CSDs corresponding to the CCFs in the respective upper traces. Once again, it is easier to deduce the common presence of strong α activity between channels p3 and p4 from the CSD rather than the CCF in Figure 4.10. The single peak at 9 Hz in the CSD is more easily interpretable than the multiple peaks in the corresponding CCF. The CSD in Figure 4.11 lacks a clear peak in the α range, even though the corresponding CCF shows a peak at about 0.1 s, albeit less significant than that in Figure 4.10. The results agree with the fact that one channel has α activity while the other does not. Finally, the CSD in Figure 4.12 is clearly lacking a peak in the α range; the two signal segments have no α activity. Further methods for the analysis of α activity will be presented in Sections 6.4.3 and 7.5.2.

4.6   THE MATCHED FILTER

When a sample observation or template of a typical version of a signal event is available, it becomes possible to design a filter that is *matched* to the characteristics of the event. If a signal that contains repetitions of the event with almost the same characteristics is passed through the *matched filter*, the output should provide peaks at the time instants of occurrence of the event. Matched filters are commonly used for the detection of signals of known characteristics that are buried in noise [105, 106]. They are designed to perform a correlation between the input signal and the signal template, and hence are also known as *correlation filters*.

4.6.1   Detection of EEG spike-and-wave complexes

Problem: *Design a matched filter to detect spike-and-wave complexes in an EEG signal. A reference spike-and-wave complex is available.*

   Solution: Let $x(t)$ be the given reference signal, representing an ideal observation of the event of interest. Let $X(f)$ be the Fourier transform of $x(t)$. Consider passing $x(t)$ through a linear time-invariant filter whose impulse response is $h(t)$; the transfer function of the filter is $H(f) = FT[h(t)]$. The output is given by $y(t) = x(t) * h(t)$ or $Y(f) = X(f)H(f)$.

   It may be shown that the output energy is maximized when

$$H(f) = KX^*(f)\exp(-j2\pi ft_0), \qquad (4.29)$$

where K is a scale factor and t_0 is a time instant or delay [105]. This corresponds to the impulse response being

$$h(t) = Kx(t_0 - t). \qquad (4.30)$$

Thus the transfer function of the matched filter is proportional to the complex conjugate of the Fourier transform of the signal event to be detected. In the time domain, the impulse response is simply a *reversed* or *reflected* version of the reference signal that is scaled and delayed. A suitable delay will have to be added to make the filter causal, as determined by the duration of the reference signal.

As the impulse response is a reversed version of $x(t)$, the convolution operation performed by the matched filter is equivalent to correlation: the output is then equal to the cross-correlation between the input and the reference signal. When a portion of an input signal that is different from $x(t)$ matches the reference signal, the output approximates the ACF ϕ_{xx} of the reference signal at the corresponding time delay. The corresponding frequency domain result is

$$Y(f) = X(f)H(f) = X(f)X^*(f) = S_{xx}(f), \qquad (4.31)$$

which is the PSD of the reference signal (ignoring the time delay and scale factors). The output is therefore maximum at the time instant of occurrence of an approximation to the reference signal. (See also Barlow [97].)

Illustration of application: To facilitate comparison with template matching, the spike-and-wave complex between 0.60 s and 0.82 s in the c3 channel of the EEG in Figure 1.23 was used as the reference signal to derive the matched filter. Figure 4.15 shows the extracted reference signal in the upper trace. The lower trace in the same figure shows the impulse response of the matched filter, which is simply a time-reversed version of the reference signal. The matched filter was implemented as an FIR filter using the MATLAB *filter* command.

Figures 4.16 and 4.17 show the outputs of the matched filter applied to the c3 and f3 channels of the EEG in Figure 1.23, respectively. The upper trace in each plot shows the signal, and the lower trace shows the matched-filter output. It is evident that the matched filter provides a large output for each spike-and-wave complex. Comparing the matched-filter outputs in Figures 4.16 and 4.17 with those of template matching in Figure 4.13 and 4.14, respectively, we observe that they are similar, with the exception that the matched-filter results peak with a delay of 0.22 s after the corresponding spike-and-wave complex. The delay corresponds to the duration of the impulse response of the filter. (*Note:* MATLAB provides the command *filtfilt* for zero-phase forward and reverse digital filtering; this method is not considered in the book.)

4.7 DETECTION OF THE P WAVE

Detection of the P wave is difficult, as it is small, has an ill-defined and variable shape, and could be placed in a background of noise of varying size and origin.

Problem: *Propose an algorithm to detect the P wave in the ECG signal.*

Solution 1: In the method proposed by Hengeveld and van Bemmel [107], VCG signals are processed as follows:

1. The QRS is detected, deleted, and replaced with the base-line. The base-line is determined by analyzing a few samples preceding the QRS complex.

2. The resulting signal is bandpass filtered with -3 dB points at 3 Hz and 11 Hz.

3. A search interval is defined as $QT_{\mathrm{max}} = \frac{2}{9}RR + 250$ ms, where RR is the interval between two successive QRS complexes.

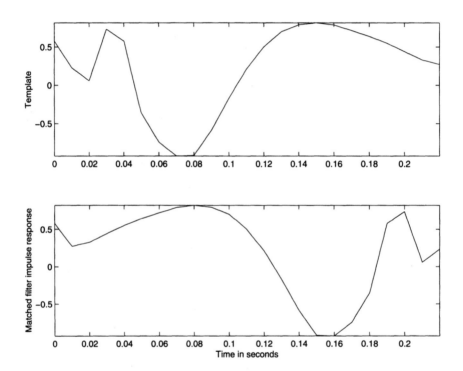

Figure 4.15 Upper trace: The spike-and-wave complex between 0.60 *s* and 0.82 *s* in the c3 channel of the EEG signal shown in Figure 1.23. Lower trace: Impulse response of the matched filter derived from the signal segment in the upper trace. Observe that the latter is a time-reversed version of the former.

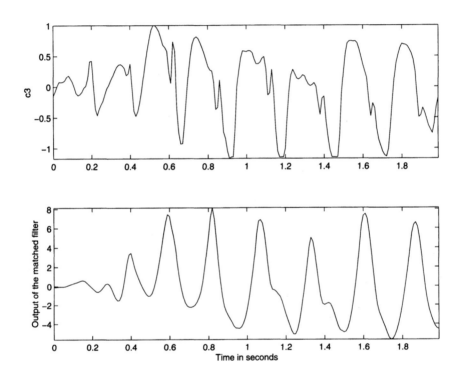

Figure 4.16 Upper trace: The c3 channel of the EEG signal shown in Figure 1.23, used as input to the matched filter in Figure 4.15. Lower trace: Output of the matched filter. See also Figure 4.13.

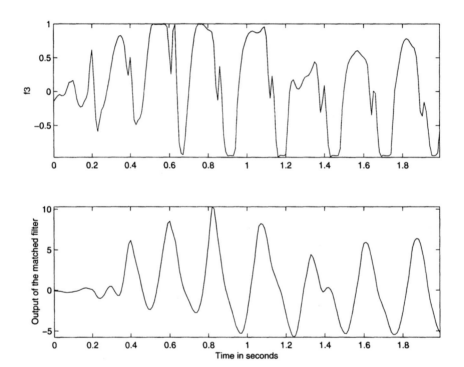

Figure 4.17 Upper trace: The f3 channel of the EEG signal shown in Figure 1.23, used as input to the matched filter in Figure 4.15. Lower trace: Output of the matched filter. See also Figure 4.14.

4. The maximum and minimum values are found in all three VCG leads from the end of the preceding T wave to the onset of the QRS.

5. The signal is rectified and thresholded at 50% and 75% of the maximum to obtain a ternary (three-level) signal.

6. The cross-correlation of the result is computed with a ternary template derived in a manner similar to the procedure in the previous step from a representative set of P waves.

7. The peak in the cross-correlation corresponds to the P location in the original ECG.

The algorithm overcomes the dominance of the QRS complex by first detecting the QRS and then deleting it. Observe that the cross-correlation is computed not with an original P wave, which we have noted could be rather obscure and variable, but with a ternary wave derived from the P wave. The ternary wave represents a simplified template of the P wave.

Figure 4.18 illustrates the results of the various stages of the P-finding algorithm of Hengeveld and van Bemmel [107]. Observe that the original ECG signal shown in part (a) of the figure has a P wave that is hardly discernible. The processed versions of the signal after deleting the QRS, filtering, and rectification are shown in parts (b), (c), and (d). The ternary version in part (e) shows that the P wave has been converted into two pulses corresponding to its upstroke and return parts. The result of cross-correlation with the template in part (f) is shown in part (g). A simple peak-picking algorithm with search limits may be used to detect the peak in the result, and hence determine the P wave position.

Note that the result in part (d) has other waves preceding those related to the P wave. An appropriate search interval should be used so as to disregard the unwanted components.

Solution 2: Gritzali et al. [108] proposed a common approach to detect the QRS, T, and P waves in multichannel ECG signals based upon a transformation they labeled as the "length" transformation. Given a collection of ECG signals from N simultaneous channels $x_1(t), x_2(t), \ldots, x_N(t)$, the length transformation was defined as

$$L(N, w, t) = \int_t^{t+w} \sqrt{\sum_{j=1}^{N} \left(\frac{dx_j}{dt}\right)^2} \, dt, \qquad (4.32)$$

where w is the width of the time window over which the integration is performed. In essence, the procedure computes the total squared derivative of the signals across the various channels available, and integrates the summed quantity over a moving time window. The advantage of applying the derivative-based operator across multiple channels of an ECG signal is that the P and T waves may be well-defined in at least one channel.

In the procedure for waveform detection proposed by Gritzali et al., the QRS is first detected by applying a threshold to $L(N, w, t)$, with w set equal to the average

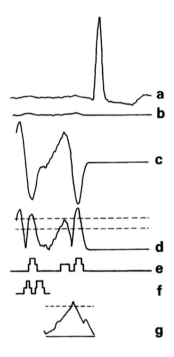

Figure 4.18 Illustration of the results at various stages of the Hengeveld and van Bemmel method for P wave detection. From top to bottom: (a) the original ECG signal; (b) after replacement of the QRS with the base-line; (c) after bandpass filtering; (d) after rectification, with the dashed lines indicating the thresholds; (e) the thresholded ternary signal; (f) the ternary P wave template; and (g) result of cross-correlation between the signals in (e) and (f). Reproduced with permission from S.J. Hengeveld and J.H. van Bemmel, Computer detection of P waves, *Computers and Biomedical Research*, 9:125–132, 1976. ©Academic Press.

QRS width. The onset and offset (end) points of the QRS are represented by a pulse waveform, as indicated in Figure 4.19. The QRS complexes in the signals are then replaced by the iso-electric base-line of the signals, the procedure is repeated with w set equal to the average T duration, and the T waves are detected. The same steps are repeated to detect the P waves. Figure 4.19 illustrates the detection of the QRS, T, and P waves in a three-channel ECG signal. Gritzali et al. also proposed a procedure based upon correlation analysis and least-squares modeling to determine the thresholds required, which will not be described here.

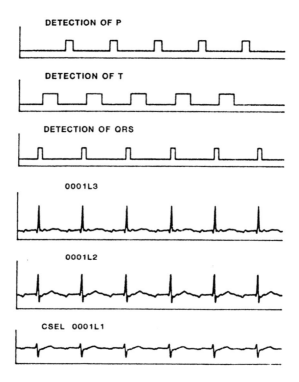

Figure 4.19 Detection of the P, QRS, and T waves in a three-channel ECG signal using the length transformation. The lower three traces show the three ECG channels. The upper three traces indicate the onset and end of the P, QRS, and T waves detected by the procedure in the form of pulse trains. The first P and the last T waves have not been processed. Reproduced with permission from F. Gritzali, G. Frangakis, G. Papakonstantinou, Detection of the P and T waves in an ECG, *Computers and Biomedical Research*, 22:83–91, 1989. ©Academic Press. See Willems et al. [109, 110] for details on the ECG database used by Gritzali et al.

4.8   HOMOMORPHIC FILTERING AND THE COMPLEX CEPSTRUM

In Chapter 3, we have seen linear filters designed to separate signals that were added together. The question asked has been, given $y(t) = x(t) + \eta(t)$, how could one extract $x(t)$ only. Given that the Fourier transform is linear, we know that the Fourier transforms of the signals are also combined in an additive manner: $Y(\omega) = X(\omega) + \eta(\omega)$. Therefore, a linear filter will facilitate the separation of $X(\omega)$ and $\eta(\omega)$, with the assumption that they have significant portions of their energies in different frequency bands.

Suppose now that we are presented with a signal that contains the product of two signals, say, $y(t) = x(t)\,p(t)$. From the multiplication or convolution property of the Fourier transform we have $Y(\omega) = X(\omega) * P(\omega)$, where $*$ represents convolution in the frequency domain. How would we be able to separate $x(t)$ from $p(t)$?

Furthermore, suppose we have $y(t) = x(t) * h(t)$, where $*$ stands for convolution as in the case of the passage of the glottal pulse train or random excitation $x(t)$ through the vocal-tract system with the impulse response $h(t)$. The Fourier transforms of the signals are related as $Y(\omega) = X(\omega)\,H(\omega)$. How would we attempt to separate $x(t)$ and $h(t)$?

4.8.1   Generalized linear filtering

Given that linear filters are well established and understood, it is attractive to consider extending their application to signals that have been combined by operations other than addition, especially by multiplication and convolution as indicated in the preceding paragraphs. An interesting possibility to achieve this is via conversion of the operation combining the signals into addition by one or more transformations. Under the assumption that the transformed signals occupy different portions of the transform space, linear filters may be applied to separate them. The inverses of the transformations used initially would then take us back to the original space of the signals. This approach was proposed in a series of papers by Bogert et al. [111] and Oppenheim et al. [112, 113]. As the procedure extends the application of linear filters to multiplied and convolved signals, it has been referred to as *generalized linear filtering*. Furthermore, as the operations can be represented by algebraically linear transformations between the input and output vector spaces, they have been called *homomorphic systems*.

As a simple illustration of a homomorphic system for multiplied signals, consider again the signal

$$y(t) = x(t)\,p(t). \tag{4.33}$$

Given the goal of converting the multiplication operation to addition, it is evident that a simple logarithmic transformation is appropriate:

$$\log[y(t)] = \log[x(t)\,p(t)] = \log[x(t)] + \log[p(t)]; \quad x(t) \neq 0,\ p(t) \neq 0\ \forall t. \tag{4.34}$$

The logarithms of the two signals are now combined in an additive manner. Taking the Fourier transform, we get

$$Y_l(\omega) = X_l(\omega) + P_l(\omega), \tag{4.35}$$

where the subscript l indicates that the Fourier transform has been applied to a log-transformed version of the signal.

Assuming that the logarithmic transformation has not affected the separability of the Fourier components of the two signals $x(t)$ and $p(t)$, a linear filter (lowpass, highpass, etc.) may now be applied to $Y_l(\omega)$ to separate them. An inverse Fourier transform will yield the filtered signal in the time domain. An exponential operation will complete the reversal procedure (if required).

Figure 4.20 illustrates the operations involved in a multiplicative homomorphic system (or filter). The symbol at the input or output of each block indicates the operation that combines the signal components at the corresponding step. A system of this type is useful in image processing, where an image may be treated as the product of an illumination function and a transmittance or reflectance function. The homomorphic filter facilitates separation of the illumination function and correction for nonuniform lighting. The method has been used to achieve simultaneous dynamic range compression and contrast enhancement [86, 114, 112].

4.8.2 Homomorphic deconvolution

Problem: *Propose a homomorphic filter to separate two signals that have been combined through the convolution operation.*

Solution: Consider the case expressed by the relation

$$y(t) = x(t) * h(t). \tag{4.36}$$

As in the case of the multiplicative homomorphic system, our goal is to convert the convolution operation to addition. From the convolution property of the Fourier transform, we know that

$$Y(\omega) = X(\omega) H(\omega). \tag{4.37}$$

Thus application of the Fourier transform converts convolution to multiplication. Now, it is readily seen that the multiplicative homomorphic system may be applied to convert the multiplication to addition. Taking the complex logarithm of $Y(\omega)$, we have

$$\log[Y(\omega)] = \log[X(\omega)] + \log[H(\omega)]; \quad X(\omega) \neq 0, \ H(\omega) \neq 0 \ \forall \omega. \tag{4.38}$$

(*Note:* $\log[X(\omega)] = \hat{X}(\omega) = |X(\omega)| + j\angle X(\omega)$.)

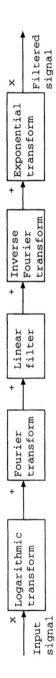

Figure 4.20 Operations involved in a multiplicative homomorphic system or filter. The symbol at the input or output of each block indicates the operation that combines the signal components at the corresponding step.

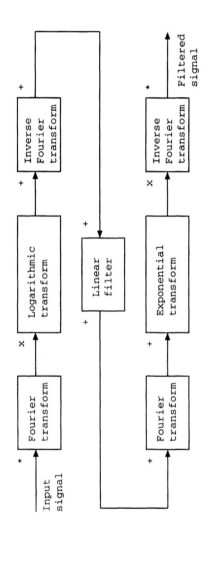

Figure 4.21 Operations involved in a homomorphic filter for convolved signals. The symbol at the input or output of each block indicates the operation that combines the signal components at the corresponding step.

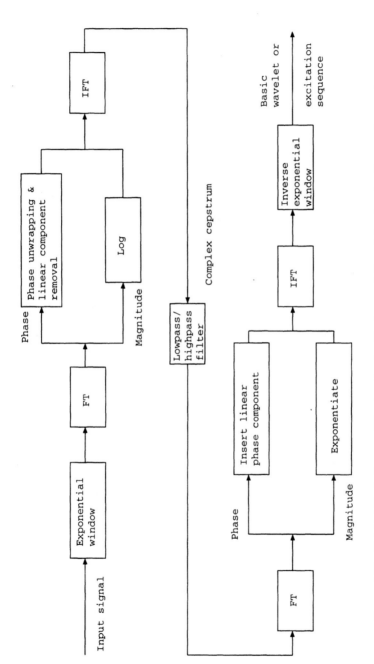

Figure 4.22   Detailed block diagram of the steps involved in deconvolution of signals using the complex cepstrum.

A linear filter may now be used to separate the transformed components of x and h, with the assumption as before that they are separable in the transform space. A series of the inverses of the transformations applied initially will take us back to the original domain.

While the discussion here has been in terms of application of the Fourier transform, the general formulation of the homomorphic filter by Oppenheim and Schafer [86] is in terms of the z-transform. However, the Fourier transform is equivalent to the z-transform evaluated on the unit circle in the z-plane, and the Fourier transform is more commonly used in signal processing than the z-transform.

Figure 4.21 gives a block diagram of the steps involved in a homomorphic filter for convolved signals. Observe that the path formed by the first three blocks (the top row) transforms the convolution operation at the input to addition. The set of the last three blocks (the bottom row) performs the reverse transformation, converting addition to convolution. The filter in between thus deals with (transformed) signals that are combined by simple addition.

4.8.3   Extraction of the vocal-tract response

Problem: *Design a homomorphic filter to extract the basic wavelet corresponding to the vocal-tract response from a voiced-speech signal.*

Solution: We noted in Section 1.2.11 that voiced speech is generated by excitation of the vocal tract, as it is held in a particular form, with a glottal waveform that may be approximated as a series of pulses. The voiced-speech signal may therefore be expressed in discrete-time terms as $y(n) = x(n) * h(n)$, where $y(n)$ is the speech signal, $x(n)$ is the glottal waveform (excitation sequence), and $h(n)$ is the impulse response of the vocal tract (basic wavelet). The $*$ symbol represents convolution, with the assumption that the vocal-tract filter may be approximated by a linear, shift-invariant filter. We may therefore use the homomorphic filter for convolved signals as introduced in the preceding section to separate $h(n)$ and $x(n)$.

The glottal excitation sequence may be further expressed as $x(n) = p(n) * g(n)$, where $p(n)$ is a train of ideal impulses (Dirac delta functions) and $g(n)$ is a smoothing function, to indicate that the physical vocal-cord system cannot produce ideal impulses but rather pulses of finite duration and slope [86]. This aspect will be neglected in our discussions.

Practical application of the homomorphic filter is not simple. Figure 4.22 gives a detailed block diagram of the procedure [86, 115]. Some of the finer details and practical techniques are explained in the following paragraphs.

The complex cepstrum: The formal definition of the complex cepstrum states that it is the inverse z-transform of the complex logarithm of the z-transform of the input signal [115, 86]. (The name "cepstrum" was derived by transposing the syllables of the word "spectrum"; other transposed terms [111, 86, 115] are less commonly used.) If $y(n)$ is the input signal and $Y(z)$ is its z-transform, the complex cepstrum $\hat{y}(n)$ is defined as

$$\hat{y}(n) = \frac{1}{2\pi j} \oint \log[Y(z)] \, z^{n-1} \, dz. \tag{4.39}$$

The contour integral performs the inverse z-transform, and should be evaluated within an annular region in the complex z-plane where $\hat{Y}(z) = \log[Y(z)]$ is single-valued and analytic [86, 2]. The unit of n in $\hat{y}(n)$, that is, in the cepstral domain, is often referred to as *quefrency*, a term obtained by switching the syllables in the term frequency.

Given $y(n) = x(n) * h(n)$, it follows that

$$\hat{Y}(z) = \hat{X}(z) + \hat{H}(z) \ \text{ or } \ \hat{Y}(\omega) = \hat{X}(\omega) + \hat{H}(\omega), \tag{4.40}$$

and further that the complex cepstra of the signals are related simply as

$$\hat{y}(n) = \hat{x}(n) + \hat{h}(n). \tag{4.41}$$

Here, the $\hat{\ }$ symbol over a function of z or ω indicates the complex logarithm of the corresponding function of z or ω, whereas the same symbol over a function of time (n) indicates the complex cepstrum of the corresponding signal. It should be noted that if the original signal $y(n)$ is real, its complex cepstrum $\hat{y}(n)$ is *real*; the prefix *complex* is used to indicate the fact that the preceding z and logarithmic transformations are computed as complex functions. Furthermore, it should be noted that the complex cepstrum is a function of time.

An important consideration in the evaluation of the complex logarithm of $Y(z)$ or $Y(\omega)$ relates to the phase of the signal. The phase spectrum computed as its principal value in the range $0 - 2\pi$, given by $\tan^{-1}\left[\frac{\text{imaginary}\{Y(\omega)\}}{\text{real}\{Y(\omega)\}}\right]$, will almost always have discontinuities that will conflict with the requirements of the inverse z-transformation or inverse Fourier transform to follow later. Thus $Y(\omega)$ needs to be separated into its magnitude and phase components, the logarithmic operation applied to the magnitude, the phase corrected to be continuous by adding correction factors of $\pm 2\pi$ at discontinuities larger than π, and the two components combined again before the subsequent inverse transformation. Correcting the phase spectrum as above is referred to as *phase unwrapping* [115, 86]. It has been shown that a linear phase term, if present in the spectrum of the input signal, may cause rapidly decaying oscillations in the complex cepstrum [115]. It is advisable to remove the linear phase term, if present, during the phase-unwrapping step. The linear phase term may be added to the filtered result (as a time shift) if necessary.

Exponential signals are defined as signals that have a rational z-transform, that is, their z-transforms may be expressed as ratios of polynomials in z. Such signals are effectively represented as weighted sums of exponentials. A few important properties of the complex cepstrum of an exponential signal are summarized below [86].

- $\hat{y}(n)$ will be of infinite duration even if $y(n)$ is of finite duration, and exists for $-\infty < n < \infty$ in general.

- The complex cepstrum $\hat{y}(n)$ decays at least as fast as $\frac{1}{n}$.

- If $y(n)$ is a minimum-phase signal (that is, it has all of its poles or zeros inside the unit circle in the z-plane), then $\hat{y}(n) = 0$ for $n < 0$.

- If $y(n)$ is a maximum-phase signal (that is, it has no poles or zeros inside the unit circle in the z-plane), then $\hat{y}(n) = 0$ for $n > 0$.

Limiting ourselves to causal signals of finite energy, we need not consider the presence of poles on or outside the unit circle in the z-plane. However, the z-transform of a finite-energy signal may have zeros outside the unit circle. Such a composite signal may be separated into its minimum-phase component and maximum-phase component by extracting the causal part ($n > 0$) and anti-causal part ($n < 0$), respectively, of its complex cepstrum, followed by the inverse procedures. The composite signal is equal to the convolution of its minimum-phase component and maximum-phase component. (See also Section 5.4.2.)

Effect of echoes or repetitions of a wavelet: Let us consider a simplified signal $y(n) = x(n) * h(n)$, where

$$x(n) = \delta(n) + a\,\delta(n - n_0), \tag{4.42}$$

with a and n_0 being two constants. (The sampling interval T is ignored, or assumed to be normalized to unity in this example.) The signal may also be expressed as

$$y(n) = h(n) + a\,h(n - n_0). \tag{4.43}$$

The signal thus has two occurrences of the basic wavelet $h(n)$ at $n = 0$ and $n = n_0$. The coefficient a indicates the magnitude of the second appearance of the basic wavelet (called an *echo* in seismic applications), and n_0 indicates its delay (pitch in the case of a voiced-speech signal). The top-most plot in Figure 4.23 shows a synthesized signal with a wavelet and an echo at half the amplitude (that is, $a = 0.5$) of the first wavelet arriving at $n_0 = 0.01125\ s$.

Taking the z-transform of the signal, we have

$$Y(z) = (1 + az^{-n_0})H(z). \tag{4.44}$$

If the z-transform is evaluated on the unit circle, we get the Fourier-transform-based expression

$$Y(\omega) = [1 + a\,\exp(-j\omega n_0)]H(\omega). \tag{4.45}$$

Taking the logarithm, we have

$$\hat{Y}(\omega) = \hat{H}(\omega) + \log[1 + a\,\exp(-j\omega n_0)]. \tag{4.46}$$

If $a < 1$, the log term may be expanded in a power series, to get

$$\hat{Y}(\omega) = \hat{H}(\omega) + a\,\exp(-j\omega n_0) - \frac{a^2}{2}\,\exp(-2j\omega n_0) + \frac{a^3}{3}\,\exp(-3j\omega n_0) - \cdots. \tag{4.47}$$

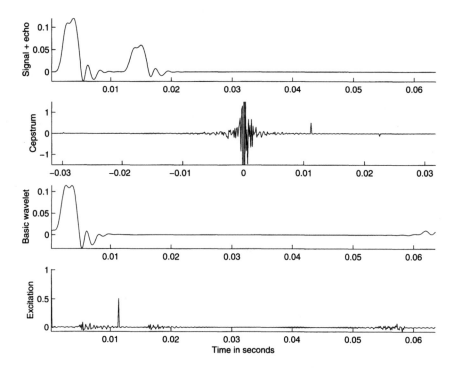

Figure 4.23   From top to bottom: a composite signal with a wavelet and an echo; the complex cepstrum of the signal (the amplitude axis has been stretched to make the peaks at the echo time and its multiples more readily visible; values outside the range ± 1.5 have been clipped); the basic wavelet extracted by shortpass filtering the cepstrum; and the excitation sequence extracted by longpass filtering the cepstrum.

Taking the inverse Fourier transform, we get

$$\hat{y}(n) = \hat{h}(n) + a\,\delta(n - n_0) - \frac{a^2}{2}\,\delta(n - 2n_0) + \frac{a^3}{3}\,\delta(n - 3n_0) - \cdots. \qquad (4.48)$$

The derivation above shows that the complex cepstrum of a signal with a basic wavelet and an echo is equal to the complex cepstrum of the basic wavelet plus a series of impulses at the echo delay and integral multiples thereof [115, 86]. The amplitudes of the impulses are proportional to the echo amplitude (the factor a) and decay for the higher-order repetitions (if $a < 1$). It may be readily seen that if the signal has multiple echoes or repetitions of a basic wavelet, the cepstrum will possess multiple impulse trains, with an impulse at the arrival time of each wavelet and integral multiples thereof. In the case of a voiced-speech signal, the location of the first peak will give the pitch. The second plot in Figure 4.23 shows the complex cepstrum of the signal in the first plot of the same figure. It is seen that the cepstrum has a peak at $0.01125\ s$, the echo arrival time; a smaller (negative) peak is also seen at twice the echo arrival time.

Under the assumption that the complex cepstrum of the basic wavelet decays to negligible values before the first impulse $a\,\delta(n - n_0)$ related to the echo, $\hat{h}(n)$ may be extracted from the complex cepstrum $\hat{y}(n)$ of the composite signal by a simple window that has unit value for $|n| < n_c$, n_c being the cutoff point. (This filter is sometimes referred to as a "shortpass" filter as the cepstrum is a function of time; it might also be called a lowpass filter.) The inverse procedures will yield $h(n)$. The remaining portion of the cepstrum (obtained by "longpass" or highpass filtering) will give $\hat{x}(n)$, which upon application of the inverse procedures will yield $x(n)$. The third and fourth plots in Figure 4.23 show the basic wavelet $h(n)$ and the excitation sequence $x(n)$ extracted by filtering the cepstrum with the cutoff point at $n_c = 0.005\ s$.

In the case where $a \geq 1$, it can be shown that the complex cepstrum will have a train of impulses on its negative time axis, that is, at $(n + kn_0)$, $k = 1, 2, \ldots$ [115, 86]. An appropriate exponential weighting sequence may be used to achieve the condition $a < 1$, in which case the impulse train will appear on the positive axis of the cepstrum. If the weighted signal satisfies the minimum-phase condition, the cepstrum will be causal.

The power cepstrum: Several variants of the cepstrum have been proposed in the literature; Childers [115] provides a review of the related techniques. One variant that is commonly used is the *real cepstrum* or the *power cepstrum*, which is defined as the square of the inverse z-transform of the logarithm of the squared magnitude of the z-transform of the given signal. In practice, the z-transform in the definition stated is replaced by the FFT. The power cepstrum has the computational advantage of not requiring phase unwrapping, but does not facilitate separation of the components of the signal.

By evaluating the inverse z-transform on the unit circle in the z-plane, the power cepstrum $\hat{y}_p(n)$ of a signal $y(n)$ may be defined as

$$\hat{y}_p(n) = \left\{ \frac{1}{2\pi j} \oint \log |Y(z)|^2 z^{n-1} dz \right\}^2. \qquad (4.49)$$

If, as before, we consider $y(n) = x(n) * h(n)$, we have $|Y(z)|^2 = |X(z)|^2 |H(z)|^2$, and it follows that $\log |Y(z)|^2 = \log |X(z)|^2 + \log |H(z)|^2$. Applying the inverse z-transform to this relationship, we get

$$\hat{y}_p(n) = \hat{x}_p(n) + \hat{h}_p(n), \tag{4.50}$$

where $\hat{h}_p(n)$ is the power cepstrum of the basic wavelet and $\hat{x}_p(n)$ is the power cepstrum of the excitation signal. Note that in the above equation the cross-product term was neglected; the cross-term will be zero if the two component power cepstra occupy non-overlapping quefrency ranges. The final squaring operation in Equation 4.49 is omitted in some definitions of the power cepstrum; in such a case, the cross-term does not arise, and Equation 4.50 is valid.

The power cepstrum does not retain the phase information of the original signal. However, it is useful in the identification of the presence of echoes in the signal, and in the estimation of their arrival times. The power cepstrum is related to the complex cepstrum as [115]

$$\hat{y}_p(n) = [\hat{y}(n) + \hat{y}(-n)]^2. \tag{4.51}$$

Let us again consider the situation of a signal with two occurrences of a basic wavelet $h(n)$ and $n = 0$ and $n = n_0$ as in Equations 4.42 and 4.43. Then [115],

$$|Y(z)|^2 = |H(z)|^2 |1 + az^{-n_0}|^2. \tag{4.52}$$

By taking the logarithm of both sides of the equation and substituting $z = \exp(j\omega)$, we get

$$\begin{aligned}
\log |Y(\omega)|^2 &= \log |H(\omega)|^2 + \log[1 + a^2 + 2a \cos(\omega n_0)] \\
&= \log |H(\omega)|^2 + \log(1 + a^2) \\
&+ \log \left(1 + \frac{2a}{1 + a^2} \cos(\omega n_0) \right).
\end{aligned} \tag{4.53}$$

It is now seen that the logarithm of the PSD of the signal will have sinusoidal components (ripples) due to the presence of an echo. The amplitudes and frequencies of the ripples are related to the amplitude a of the echo and its time delay n_0.

Illustration of application: A voiced-speech signal $y(n)$ is the result of convolution of a slowly varying vocal-tract response $h(n)$ with a relatively fast-varying glottal pulse train $x(n)$: $y(n) = x(n) * h(n)$. Under these conditions, it may be demonstrated that the contributions of $h(n)$ to the complex cepstrum $\hat{y}(n)$ will be limited to low values of n within the range of the pitch period of the speech signal [86]. The complex cepstrum $\hat{y}(n)$ will possess impulses at the pitch period and integral multiples thereof. Therefore, a filter that selects a portion of the complex cepstrum for low values of n, followed by the inverse transformations, will yield an estimate of $h(n)$.

When the repetitions of the basic wavelet have magnitudes almost equal to (or even greater than) that of the first wavelet in the given signal record, the contributions of the pulse-train component to the complex cepstrum may not decay rapidly and may cause aliasing artifacts when the cepstrum is computed over a finite duration. A similar situation is caused when the delay between the occurrences of the multiple versions of the basic wavelet is a significant portion of the duration of the given signal. The problem may be ameliorated by applying an exponential weight α^n to the data sequence, with $\alpha < 1$. Childers et al. [115] recommended values of α in the range $0.99 - 0.96$, depending upon the signal characteristics as listed above. Furthermore, they recommend appending or padding zeros to the input signal to facilitate computation of the cepstrum to a longer duration than the signal in order to avoid aliasing errors and ambiguities in time-delay estimates. (See Figure 4.22 for an illustration of the various steps in homomorphic filtering of convolved signals.)

Figure 4.24 illustrates a segment of a voiced-speech signal (extracted from the signal shown in Figure 1.30) and the basic wavelet extracted by shortpass filtering of its complex cepstrum with $n_c = 0.003125$ s. The signal was padded with zeros to twice its duration; exponential weighting with $\alpha = 0.99$ was used. It is seen that the basic vocal-tract response wavelet has been successfully extracted. Extraction of the vocal-tract response facilitates spectral analysis without the effect of the quasi-periodic repetitions in the speech signal.

The fourth trace in Figure 4.24 shows the glottal (excitation) waveform extracted by longpass filtering of the cepstrum with the same parameters as in the preceding paragraph. The result shows impulses at the time of arrival of each wavelet in the composite speech signal. The peaks are decreasing in amplitude due to the use of exponential weighting (with $\alpha = 0.99$) prior to computation of the cepstrum. Inverse exponential weighting can restore the pulses to their original levels; however, the artifact at the end of the excitation signal gets amplified to much higher levels than the desired pulses due to progressively higher values of α^{-n} for large n. Hence the inverse weighting operation was not applied in the present illustration. Regardless, the result indicates that pitch information may also be recovered by homomorphic filtering of voiced-speech signals.

4.9 APPLICATION: ECG RHYTHM ANALYSIS

Problem: *Describe a method to measure the heart rate and average RR interval from an ECG signal.*

Solution: Algorithms for QRS detection such as the Pan-Tompkins method described in Section 4.3.2 are useful for ECG rhythm analysis or heart-rate monitoring. The output of the final smoothing filter could be subjected to a peak-searching algorithm to obtain a time marker for each QRS or ECG beat. The search procedure proposed by Pan and Tompkins was explained in Section 4.3.2. The intervals between two such consecutive markers gives the RR interval, which could be averaged over a number of beats to obtain a good estimate of the inter-beat interval. The heart rate may be computed in *bpm* as 60 divided by the average RR interval in seconds. The heart rate may also be obtained by counting the number of beats detected over a certain period, say 10 s, and multiplying the result with the required factor (6 in the present case) to get the number of beats over one minute.

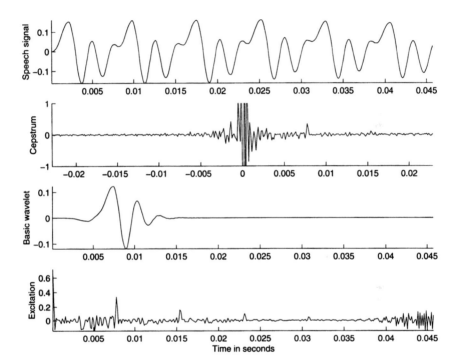

Figure 4.24 From top to bottom: a segment of a voiced-speech signal over six pitch periods (extracted from the signal shown in Figure 1.30 and lowpass filtered); the complex cepstrum of the signal (the amplitude axis has been stretched to make the peaks at the echo time and its multiples more readily visible; values outside the range ±1.0 have been clipped); the (shifted) basic wavelet extracted by shortpass filtering the cepstrum; and the excitation sequence extracted by longpass filtering the cepstrum.

The upper plot in Figure 4.25 shows a filtered version of the noisy ECG signal shown in Figure 3.5. The noisy signal was filtered with an eighth-order Butterworth lowpass filter with a cutoff frequency of 90 Hz, and the signal was down-sampled by a factor of five to an effective sampling rate of 200 Hz. The lower plot shows the output of the Pan-Tompkins method. The Pan-Tompkins result was normalized by dividing by its maximum over the data record available (as the present example was computed off-line). A fixed threshold of 0.1 and a blanking interval of 250 ms was used in a simple search procedure, which was successful in detecting all the beats in the signal. (The blanking interval indicates the period over which threshold checking is suspended once the threshold has been crossed.) The average RR interval was computed as 716 ms, leading to an effective heart rate of 84 bpm.

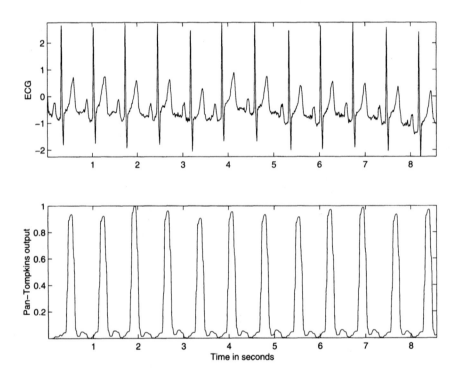

Figure 4.25   Results of the Pan-Tompkins algorithm. Top: lowpass-filtered version of the ECG signal shown in Figure 3.5. Bottom: normalized result of the final integrator.

Results at the various stages of the Pan-Tompkins algorithm for a noisy ECG signal sampled at 200 Hz are shown in Figure 4.26. The bandpass filter has efficiently removed the low-frequency artifact in the signal. The final output has two peaks that are much larger than the others: one at the beginning of the signal due to filtering artifacts, and one at about 7.5 s due to an artifact in the signal. Furthermore, the output has two peaks for the beat with an artifact at 7.5 s. The simple peak-searching procedure as explained in the previous paragraph was applied, which resulted in the detection of 46 beats: one more than the 45 present in the signal due to the artifact at about 7.5 *s*. The average *RR* interval was computed to be 446.6 *ms*, leading to an effective heart rate of 137 *bpm*.

The illustration demonstrates the need for prefiltering the ECG signal to remove artifacts and the need to apply an adaptive threshold to the output of the Pan-Tompkins algorithm for QRS detection. It is readily seen that direct thresholding of the original ECG signal will not be successful in detecting all of the QRS complexes in the signal.

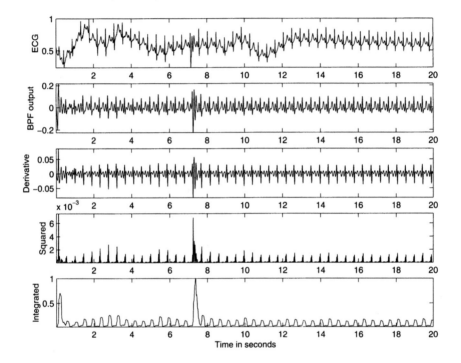

Figure 4.26   Results of the Pan-Tompkins algorithm with a noisy ECG signal. From top to bottom: ECG signal sampled at 200 *Hz*; output of the bandpass filter (BPF); output of the derivative-based operator; the result of squaring; and normalized result of the final integrator.

4.10   APPLICATION: IDENTIFICATION OF HEART SOUNDS

Problem: *Outline a signal processing algorithm to identify S1 and S2 in a PCG signal, and further segment the PCG signal into its systolic and diastolic parts. The ECG and carotid pulse signals are available for reference.*

Solution: We saw in Section 2.3 how the ECG and carotid pulse signals could be used to demarcate the onset of S1 and S2 in the PCG; the procedure, however, was not based upon signal processing but upon visual analysis of the signals. We have, in the present chapter, developed signal processing techniques to detect the QRS complex in the ECG and the dicrotic notch in the carotid pulse signal. We may therefore use these methods to transfer the timing information from the ECG and carotid pulse signals to the PCG signal. In order to perform this task, we need to recognize a few timing relationships between the signals [41, 66].

The beginning of S1 may be taken to be the same instant as the beginning of the QRS. The QRS itself may be detected using one of the three methods described in the present chapter, such as the Pan-Tompkins method.

Detection of the beginning of S2 is more involved. Let the heart rate be HR *bpm*. The pre-ejection period PEP is defined as the interval from the beginning of the QRS to the onset of the corresponding carotid upstroke. The rate-corrected PEP is defined as $PEPC = PEP + 0.4HR$, with the periods in *ms*. $PEPC$ is in the range of 131 ± 13 *ms* for normal adults [41].

The ejection time ET is the interval from the onset of the carotid upstroke to the dicrotic notch. The rate-corrected ejection time in *ms* is $ETC = ET + 1.6HR$, and is in the ranges of 395 ± 13 *ms* for normal adult males and 415 ± 11 *ms* for normal adult females.

Using $PEPC_{max} = 144$ *ms* and $HR_{min} = 60$ *bpm*, we get $PEP_{max} = 120$ *ms*. With $HR_{min} = 60$ *bpm* and $ETC_{max} = 425$ *ms*, we get $ET_{max} = 325$ *ms*. With these parameters, the maximum interval between the QRS and the dicrotic notch is 380 *ms*. The procedure proposed by Lehner and Rangayyan [66] for detection of the dicrotic notch recommends searching the output of the derivative-based method described in Section 4.3.3 in a 500 *ms* interval after the QRS. After the dicrotic notch is detected, we need to subtract the time delay between the beginning of S2 and D to get the time instant where S2 begins. Lehner and Rangayyan [66] measured the average S2 – D delay over the PCG and carotid pulse signals of 60 patients to be 42.6 *ms*, with a standard deviation of 5.0 *ms*.

The following procedure may be used to segment a PCG signal into its systolic and diastolic parts.

1. Use the Pan-Tompkins method described in Section 4.3.2 to locate the QRS complexes in the ECG.

2. Identify one period of the PCG as the interval between two successive QRS locations. Note that any delay introduced by the filters used in the Pan-Tompkins method needs to be subtracted from the detected peak locations to obtain the starting points of the QRS complexes.

3. Use the Lehner and Rangayyan method described in Section 4.3.3 to detect the dicrotic notch in the carotid pulse signal.

4. Let the standardized S2 – D delay be the mean plus two standard deviation values as reported by Lehner and Rangayyan [66], that is, 52.6 *ms*. Subtract the standardized S2 – D delay from the detected dicrotic notch location to obtain the onset of S2.

5. The S1 – S2 interval gives the systolic part of the PCG cycle.

6. The interval between the S2 point and the next detected S1 gives the diastolic part of the PCG cycle.

Figures 4.27 and 4.28 illustrate the results of application of the procedure described above to the PCG, ECG, and carotid pulse signals of a normal subject and a patient with a split S2, systolic ejection murmur, and opening snap of the mitral valve. (Clinical diagnosis indicated the possibility of ventricular septal defect, pulmonary stenosis, or pulmonary hypertension for the 14-month-old female patient with murmur.) The peak positions detected in the output of the Pan-Tompkins method (the third trace in each figure) and the output of the Lehner and Rangayyan method (the fifth trace) have been marked with the ∗ symbol. A simple threshold of 0.75 times the maximum value was used as the threshold to detect the peaks in the output of the Pan-Tompkins method, with a blanking interval of 250 ms.

The QRS and D positions have been marked on the ECG and carotid pulse traces with the triangle and diamond symbols, respectively. Finally, the S1 and S2 positions are marked on the PCG trace with triangles and diamonds, respectively. The filter delays and timing relationships between the three channels of signals described previously have been accounted for in the process of marking the events. Note how the results of event detection in the ECG and carotid pulse signals have been transferred to locate the corresponding events in the PCG. Lehner and Rangayyan [66] used a similar procedure to break PCG signals into systolic and diastolic segments; the segments were then analyzed separately in the time and frequency domains. (See also Sections 6.4.5 and 7.9.)

4.11 APPLICATION: DETECTION OF THE AORTIC COMPONENT OF THE SECOND HEART SOUND

Heart sounds are preferentially transmitted to different locations on the chest. The aortic and pulmonary components A2 and P2 of S2 are best heard at the aortic area (to the right of the sternum, in the second right-intercostal space) and the pulmonary area (left parasternal line in the third left-intercostal space), respectively (see Figure 1.17). A2 is caused by the closure of the aortic valve at the end of systole, and is usually louder than P2 at all locations on the chest. Earlier theories on the genesis of heart sounds attributed the sounds to the opening and closing actions of the valve leaflets *per se*. The more commonly accepted theory at the present time is that described by Rushmer [23]; see Section 1.2.8.

The relative timing of A2 and P2 depends upon the pressure differences across the corresponding valves in the left and right ventricular circulatory systems. In a normal individual, the timing of P2 with reference to A2 varies with respiration; the timing of A2 itself is independent of respiration. The pulmonary pressure (intrathoracic pressure) is decreased during inspiration, leading to a delayed closure of the pulmonary valve and hence an increased (audible and visible) gap between A2 and P2 [23, 41, 42]. The gap is closed and A2 and P2 overlap during expiration in normal individuals. A2 and P2 have individual durations of about 50 ms. The

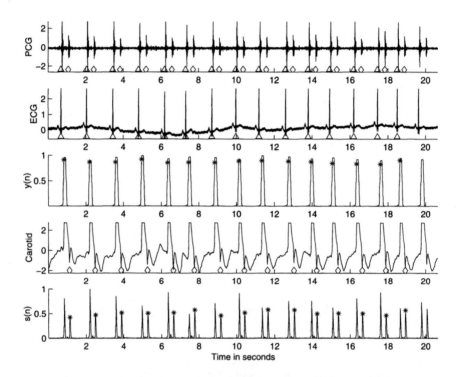

Figure 4.27 Results of segmentation of a PCG signal into systolic and diastolic parts using the ECG and carotid pulse signals for reference. From top to bottom: the PCG signal of a normal subject (male subject, 23 years); the ECG signal; $y(n)$, the output of the Pan-Tompkins method for detection of the QRS after normalization to the range $(0, 1)$; the carotid pulse signal; $s(n)$, the output of the Lehner and Rangayyan method for detection of the dicrotic notch, normalized to the range $(0, 1)$. The peaks detected in the outputs of the two methods have been identified with $*$ marks. The QRS and D positions have been marked with the triangle and diamond symbols, respectively. The S1 and S2 positions are marked on the PCG trace with triangles and diamonds, respectively. The last cardiac cycle was not processed.

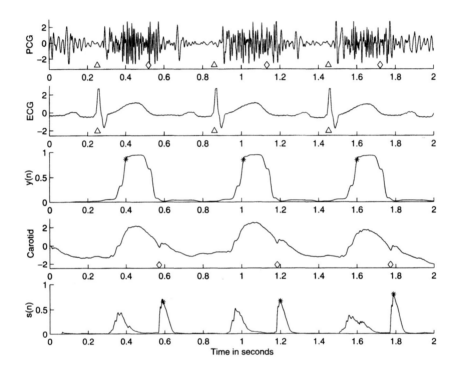

Figure 4.28   Results of segmentation of a PCG signal into systolic and diastolic parts using the ECG and carotid pulse signals for reference. From top to bottom: the PCG signal of a patient with a split S2, systolic ejection murmur, and opening snap of the mitral valve (female patient, 14 months); the ECG signal; $y(n)$, the output of the Pan-Tompkins method for detection of the QRS; the carotid pulse signal; $s(n)$, the output of the Lehner and Rangayyan method for detection of the dicrotic notch. The peaks detected in the outputs of the two methods have been identified with * marks. The QRS and D positions have been marked with the triangle and diamond symbols, respectively. The S1 and S2 positions are marked on the PCG trace with triangles and diamonds, respectively.

normal inspiratory gap between A2 and P2 is of the order of $30 - 40\ ms$, although splits as long as $100\ ms$ have been recorded [41].

A split in S2 longer than $40\ ms$ during sustained expiration is considered to be abnormal [41]. Complete right bundle-branch block could cause delayed activation of the right ventricle, therefore delayed pulmonary valve closure, a delayed P2, and hence a widely split S2. Some of the other conditions that cause a wide split in S2 are atrial septal defect, ventricular septal defect, and pulmonary stenosis. Left bundle-branch block leads to delayed left-ventricular contraction and aortic valve closure (with reference to the right ventricle and the pulmonary valve), causing A2 to appear after P2, and *reversed splitting* of the two components. Some of the other conditions that could cause reversed splitting of S2 are aortic insufficiency and abnormally early pulmonary valve closure. It is thus seen that identification of A2 and P2 and their temporal relationships could assist in the diagnosis of several cardiovascular defects and diseases.

MacCanon et al. [116] conducted experiments on a dog for direct detection and timing of aortic valve closure. They developed a catheter with an electrical contacting device that could be placed at the aortic valve to detect the exact moment of closure of the aortic valve. They also measured the aortic pressure and the PCG at the third left-intercostal space. It was demonstrated that the aortic valve closes at least $5 - 13\ ms$ before the incisura appears in the aortic pressure wave (see Figure 1.27 and 1.28 for illustrations of the aortic pressure waves recorded from a dog). The conclusion reached was that S2 is caused not by the collision of the valve leaflets themselves, but due to the rebound of the aortic blood column and walls after valve closure. MacCanon et al. also hypothesized that the relative high-frequency characteristics of the incisura and S2 result from elastic recoil of the aortic wall and valve in reaction to the distention by the rebounding aortic blood column.

Stein et al. [117, 118] conducted experiments in which intracardiac and intra-arterial sounds were recorded and analyzed. Their experiments indicated that S2 begins *after* the aortic valve closes. They argued that the intensity of S2 depends upon, among other factors, the distensibility of the aortic and pulmonary valves; hemodynamic factors that cause the valves to distend and vibrate; viscosity of the blood and its ability to inhibit diastolic valve motion; and the configuration of the aorta, the pulmonary artery, and the ventricles. It was demonstrated that the pulmonary valve, due to its larger surface area than that of the aortic valve, is more distensible and hence produces a larger sound than the aortic valve even for the same pressure gradient across the valve. In the case of pulmonary hypertension, it was argued that the pulmonary valve would distend further at a higher speed: the rate of development of the diastolic pressure gradient across the closed pulmonary valve would be higher than in normal cases.

Problem: *Given that the second heart sound S2 is made up of an aortic component A2 and a pulmonary component P2 with variable temporal relationships, propose a method to detect only A2.*

Solution: We have seen in the preceding section how the dicrotic notch in the carotid pulse signal may be used to detect the beginning of S2. The technique is based upon the direct relationship between aortic valve closure and the aortic incisura, and consequently the dicrotic notch, as explained above. Now, if we were to detect and segment S2 over several cardiac cycles and several respiratory cycles, we could perform synchronized averaging of S2. A2 should appear at the same instant in every S2 segment, and should be strengthened by the synchronized averaging process. P2, on the other hand, would appear at different times, and should be cancelled out (suppressed) by the averaging process.

Figure 4.29 shows segments of duration $300 \ ms$ containing S2 segmented from nine successive cardiac cycles of the PCG of a patient with atrial septal defect. The PCG signal was segmented using the ECG and carotid pulse signals for reference in a method similar to that illustrated in Figures 4.27 and 4.28. The PCG signal was recorded at the second left-intercostal space, which is closer to the pulmonary area than to the aortic area. The nine S2 segments clearly show the fixed timing of A2 and the variable timing of P2. The last plot is the average of S2 segments extracted from 21 successive cardiac cycles. The averaged signal displays A2 very well, while P2 has been suppressed.

The detection of A2 would perhaps have been better, had the PCG been recorded at the aortic area, where A2 would be stronger than P2. Once A2 is detected, it could be subtracted from each S2 record to obtain individual estimates of P2. Sarkady et al. [119], Baranek et al. [120], and Durand et al. [121] proposed averaging techniques as above with or without envelope detection (but without the use of the carotid pulse); the methods were called aligned ensemble averaging to detect wavelets or coherent detection and averaging.

4.12 REMARKS

We have now established links between the characteristics of certain epochs in a number of biomedical signals and the corresponding physiological or pathological events in the biomedical systems of concern. We have seen how derivative-based operators may be applied to detect QRS complexes in the ECG signal as well as the dicrotic notch in the carotid pulse signal. The utility of correlation and spectral density functions in the detection of rhythms and events in EEG signals was also demonstrated. We have studied how signals with repetitions of a certain event or wavelet, such as a voiced-speech signal, may be analyzed using the complex cepstrum and homomorphic filtering. Finally, we also saw how events detected in one signal may be used to locate the corresponding events in another signal: the task of detecting S1 and S2 in the PCG was made simpler by using the ECG and carotid pulse signals, where the QRS and D waves can be detected more readily than the heart sounds themselves.

Event detection is an important step that is required before we may attempt to analyze the corresponding waves or wavelets in more detail. After a specific wave of interest has been detected, isolated, and extracted, methods targeted to the expected characteristics of the event may be applied for directed analysis of the corresponding physiological or pathological event. Analysis of the event is then not hindered or obscured by other events or artifacts in the acquired signal.

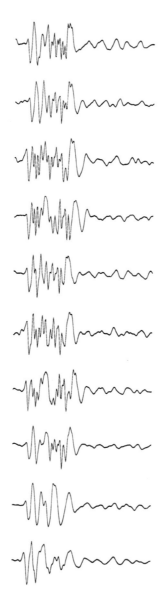

Figure 4.29 Synchronized averaging of S2 to detect A2 and suppress P2. The figure displays nine consecutive segments of S2 (duration = 300 ms) of a patient (female, 7 years) with atrial septal defect leading to a variable split in S2. The trace at the bottom is the average of S2 over 21 consecutive beats.

4.13 STUDY QUESTIONS AND PROBLEMS

1. Prove that the autocorrelation function (ACF) $\phi_{xx}(\tau)$ of any function $x(t)$ is maximum at $\tau = 0$.

   (*Hint:* Start with $E[\{x(t + \tau) \pm x(t)\}^2] \geq 0$.)

2. For a stationary process x, prove that the ACF is even symmetric, that is, $\phi_{xx}(\tau) = \phi_{xx}(-\tau)$. You may use the expectation or time-average definition of the ACF.

3. Starting with the continuous time-average definition of the ACF, prove that the Fourier transform of the ACF is the PSD of the signal.

4. What are the Fourier-domain equivalents of the autocorrelation function and the cross-correlation function? Describe their common features and differences. List their applications in biomedical signal analysis.

5. A signal $x(t)$ is transmitted through a channel. The received signal $y(t)$ is a scaled, shifted, and noisy version of $x(t)$ given as $y(t) = \alpha x(t - t_0) + \eta(t)$ where α is a scale factor, t_0 is the time delay, and $\eta(t)$ is noise. Assume that the noise process has zero mean and is statistically independent of the signal process, and that all processes are stationary.

   Derive expressions for the mean and the ACF of $y(t)$ in terms of the statistics of x and η.

6. Derive an expression for the ACF of the signal $x(t) = \sin(\omega_0 t)$. Use the time-average definition of the ACF.

   From the ACF, derive an expression for the PSD of the signal. Show all steps.

7. A rhythmic episode of a theta wave in an EEG signal is approximated by a researcher to be a sine wave of frequency 5 Hz. The signal is sampled at 100 Hz.

   Draw a schematic representation of the ACF of the episode for delays up to 0.5 s. Label the time axis in samples and in seconds.

   Draw a schematic representation of the PSD of the episode. Label the frequency axis in Hz.

8. The values of a signal sampled at 100 Hz are given by the series
   $\{0, 0, 0, 0, 10, 10, 10, 0, 0, 0, 0, 0, 0, 5, 5, 5, 0, 0, 0, 0, 0, -3, -3, -3, 0, 0, 0\}$.
   An investigator performs template matching with the pattern $\{0, 5, 5, 5, 0\}$. The first sample in each array stands for zero time.

   Plot the output of the template-matching operation and interpret the result. Label the time axis in seconds.

9. A biphasic signal $x(n)$ is represented by the series of samples
   $x(n) = \{0, 1, 2, 1, 0, -1, -2, -1, 0\}$ for $n = 0, 1, 2, \ldots, 8$.

   a) Draw a plot of $x(n)$.

   b) Compose a signal $y(n)$ defined as $y(n) = 3x(n) + 2x(n - 12) - x(n - 24)$. Draw a plot of $y(n)$.

   c) Design a matched filter to detect the presence of $x(n)$ in $y(n)$. Explain how the impulse response $h(n)$ and the frequency response $H(\omega)$ of the filter are related to $x(n)$. Plot $h(n)$.

   d) Compute the output of the filter and plot it. Interpret the output of the filter.

10. A researcher uses the derivative operator (filter) specified as $w(n) = x(n) - x(n - 1)$, where $x(n)$ is the input and $w(n)$ is the output. The result is then passed through the

moving-average filter $y(n) = \frac{1}{3}[w(n) + w(n-1) + w(n-2)]$, where $y(n)$ is the final output desired.

(a) Derive the transfer functions (in the z-domain) of the two filters individually as well as that of the combination.

(b) Does it matter which of the two filters is placed first? Why (not)?

(c) Derive the impulse response of each filter and that of the combination. Plot the three signals.

(d) The signal described by the samples
$\{0, 0, \ldots, 0, 6, 6, 6, 6, 6, 6, 6, 6, 0, 0, \ldots\}$
is applied to the system. Derive the values of the final output signal. Explain the effect of the operations on the signal.

4.14 LABORATORY EXERCISES AND PROJECTS

Note: Data files related to the exercises are available at the site
ftp://ftp.ieee.org/uploads/press/rangayyan/

1. Implement the Pan-Tompkins method for QRS detection in MATLAB. You may employ a simple threshold-based method to detect QRS complexes as the procedure will be run off-line.

   Apply the procedure to the signals in the files ECG3.dat, ECG4.dat, ECG5.dat, and ECG6.dat, sampled at a rate of 200 Hz (see the file ECGS.m). Compute the averaged heart rate and QRS width for each record. Verify your results by measuring the parameters visually from plots of the signals.

2. The files eeg1-xx.dat (where xx indicates the channel name) give eight simultaneously recorded channels of EEG signals with the alpha rhythm. (You may read the signals using the MATLAB program in the file eeg1.m.) The sampling rate is 100 Hz per channel. Cut out a portion of a signal with a clear presence of the alpha rhythm for use as a template or reference signal. Perform cross-correlation of the template with running (short-time) windows of various channels and study the use of the results for the detection of the presence of the alpha rhythm.

3. The files eeg2-xx.dat (where xx indicates the channel name) give ten simultaneously recorded channels of EEG signals with spike-and-wave complexes. (You may read the signals using the MATLAB program in the file eeg2.m.) The sampling rate is 100 Hz per channel. Cut out one spike-and-wave complex from any EEG channel and use it as a template. Perform template matching by cross-correlation or by designing a matched filter. Apply the procedure to the same channel from which the template was selected as well as to other channels. Study the results and explain how they may be used to detect spike-and-wave complexes.

4. The files pec1.dat, pec33.dat, and pec52.dat give three-channel recordings of the PCG, ECG, and carotid pulse signals (sampled at 1,000 Hz; you may read the signals using the program in the file plotpec.m). The signals in pec1.dat (adult male) and pec52.dat (male subject, 23 years) are normal; the PCG signal in pec33.dat has systolic murmur, and is of a patient suspected to have pulmonary stenosis, ventricular septal defect, and pulmonary hypertension (female, 14 months).

Apply the Pan-Tompkins method for QRS detection to the ECG channel and the Lehner and Rangayyan method to detect the dicrotic notch in the carotid pulse channel. Extrapolate the timing information from the ECG and carotid pulse channels to detect the onset of S1 and S2 in the PCG channel. What are the corrections required to compensate the delays between the corresponding events in the three channels?

5

Analysis of Waveshape and Waveform Complexity

Certain biomedical signals such as the ECG and carotid pulse have simple waveshapes (although the QRS wave is often referred to as a "complex"!). The readily identifiable signatures of the ECG and carotid pulse are modified by abnormal events and pathological processes. Hence analysis of waveshapes could be useful in diagnosis.

Signals such as the EMG and the PCG do not have waveshapes that may be identified easily. EMG signals are indeed complex interference patterns of innumerable SMUAPs. PCG signals represent vibration waves that do not possess specific waveshapes. Regardless, even the complexity of the waveforms in signals such as the EMG and the PCG does vary in relation to physiological and pathological phenomena. Analyzing the waveform complexity of such signals may assist in gaining an understanding of the processes they reflect.

5.1 PROBLEM STATEMENT

Explain how waveshapes and waveform complexity in biomedical signals relate to the characteristics of the underlying physiological and pathological phenomena. Propose techniques to parameterize and analyze the signal features you identify.

As in the preceding chapters, the problem statement given above is generic and represents the theme of the present chapter. The following section presents illustrations of the problem with case-studies that provide more specific definitions of the problem with a few signals of interest. The remaining sections of the chapter describe techniques to address the stated problems. It should be noted again that although signal analysis techniques are proposed in the context of specific signals and applications, they should find applications in other fields where signals with comparable characteristics and behavior are encountered.

5.2   ILLUSTRATION OF THE PROBLEM WITH CASE-STUDIES

5.2.1   The QRS complex in the case of bundle-branch block

We saw in Section 1.2.4 that the His bundle and its branches conduct the cardiac excitation pulse from the AV node to the ventricles. A block in one of the bundle branches causes asynchrony between the contraction of the left and the right ventricles. This, in turn, causes a staggered summation of the action potentials of the myocytes of the left and the right ventricles over a longer-than-normal duration. The result is a longer and possibly jagged QRS complex, as illustrated by the ECG of a patient with right bundle-branch block in Figure 1.15.

5.2.2   The effect of myocardial ischemia and infarction on QRS waveshape

Occlusion of a coronary artery or a branch thereof due to deposition of fat, calcium, and so on, results in reduced blood supply to a portion of the cardiac musculature. The part of the myocardium served by the affected artery then suffers from ischemia, that is, lack of blood supply. Prolonged ischemia leads to myocardial infarction, when the affected tissue dies. The deceased myocytes cannot contract any more, and no longer produce action potentials.

The action potential of an under-nourished ventricular myocyte reflects altered repolarization characteristics: the action potential is of smaller amplitude and shorter duration [10, 122]. The result of the summation of the action potentials of all of the active ventricular myocytes will thus be different from the normal QRS complex. The primary change reflected in the ECG is a modified ST segment that is either elevated or depressed, depending upon the lead used and the position of the affected region; the T wave may also be inverted. Chronic myocardial infarction causes a return to a normal ST segment, and a pronounced Q wave [23].

5.2.3   Ectopic beats

Ectopic beats are generated by cardiac tissue that possess abnormal pacing capabilities. Ectopic beats originating from focal points on the atria could cause altered P waveshapes due to different paths of propagation of the excitation pulse and hence different activation sequences of atrial muscle units. However, the QRS complex of atrial ectopic beats will appear normal as the conduction of the excitation past the AV node would be normal.

Ectopic beats originating on the ventricles (that are necessarily premature beats, that is, PVCs) typically possess bizarre QRS waveshapes due to widely differing paths of conduction and excitation of the ventricular muscle units. Figure 1.14 illustrates an ECG signal with PVCs. PVCs typically lack a preceding P wave; however, an ectopic beat triggered during the normal AV node delay will demonstrate a normal preceding P wave. PVCs triggered by ectopic foci close to the AV node may possess near-normal QRS shape as the path of conduction may be almost the same as in the case of a normal impulse from the AV node. On the other hand, beats triggered by ectopic foci near the apex could take a widely different path of propagation, resulting in a far-from-normal QRS waveshape. In addition to waveshape, the preceding and succeeding RR intervals play important roles in determining the nature of ectopic beats.

5.2.4 EMG interference pattern complexity

We saw in Section 1.2.3 that motor units are recruited by two mechanisms — spatial and temporal recruitment — in order to produce increasing levels of contraction and muscular force output. As more and more motor units are brought into action and their individual firing rates increase (within certain limits), the SMUAPs of the active motor units overlap and produce a complex interference pattern. Figures 1.9 and 1.10 illustrate an EMG signal obtained from the crural diaphragm of a dog during one normal breath cycle. The increasing complexity of the waveform with increasing level of the breath is clearly seen in the expanded plot in Figure 1.10.

Although a surface EMG interference pattern is typically too complex for visual analysis, the general increase in the level of activity ("busy-ness") may be readily seen. It is common practice in EMG laboratories to feed EMG signals to an amplified speaker: low levels of activity when the SMUAPs are not overlapping (that is, separated in time) result in discrete "firing" type of sounds; increasing levels of contraction result in increased "chatter" in the sound produced. EMG signals may be analyzed to derive parameters of waveform complexity that increase with increasing muscular contraction, thereby providing a correlate to mechanical activity that is derived from its electrical manifestation.

5.2.5 PCG intensity patterns

Although the vibration waves in a PCG signal may not be amenable to direct visual analysis, the general intensity pattern of the signal over a cardiac cycle may be readily appreciated either by auscultation or visual inspection. Certain cardiovascular diseases and defects alter the relative intensity patterns of S1 and S2, cause additional sounds or murmurs, and/or split S2 into two distinct components, as already described in Section 1.2.8. While many diseases may cause systolic murmurs, for example, the intensity pattern or envelope of the murmur could assist in arriving at a specific diagnosis. It should also be noted that definitive diagnosis based on the PCG would usually require comparative analysis of PCG signals from a few positions on the chest. Figures 1.24, 1.26, 2.4, 4.27, and 4.28 illustrate PCG signals of a normal subject and patients with systolic murmur, split S2, and opening snap of the mitral valve. The differences in the overall intensity patterns of the signals are obvious. However, signal processing techniques are desirable to convert the signals into positive-valued envelopes that could be treated as distributions of signal energy over time. Such a transformation permits the treatment of signal intensity patterns as PDFs, which lends to the computation of various statistical measures and moments.

5.3 ANALYSIS OF EVENT-RELATED POTENTIALS

The most important parameter extracted from a visual ERP is the timing or latency of the first major positivity; since the average of this latency is about 120 ms for normal adults, it is referred to as P120 (see Figure 3.12). The latencies of the troughs before and after P120, called N80 and N145, respectively, are also of interest. The amplitudes of the ERP at the corresponding instants are of lesser importance. Delays in the latencies that are well beyond the normal range could indicate problems in the visual system. Asymmetries in the latencies of the left and right parts of the visual system could also be indicative of disorders.

The lowest trace in Figure 3.12 is an averaged flash visual ERP recorded from a normal adult male subject. The signal has been labeled to indicate the N80, P120, and N145 points, the corresponding actual latencies for the subject being 85, 100.7, and 117 ms, respectively.

Auditory ERPs are weaker and more complex than visual ERPs, requiring averaging over several hundred or a few thousand stimuli. Auditory ERPs are analyzed for the latencies and amplitudes of several peaks and troughs. Clinical ERP analysis is usually performed manually, there being no pressing need for signal processing techniques beyond synchronized averaging.

5.4 MORPHOLOGICAL ANALYSIS OF ECG WAVES

The waveshape of an ECG cycle could be changed by many different abnormalities, including myocardial ischemia or infarction, bundle-branch block, and ectopic beats. It is not possible to propose a single analysis technique that can assist in categorizing all possible abnormal causes of change in waveshape. The following subsections address a few illustrative cases.

5.4.1 Correlation coefficient

Problem: *Propose a general index to indicate altered QRS waveshape. You are given a normal QRS template.*

Solution: Jenkins et al. [67] applied the correlation coefficient γ_{xy} as defined in Equation 4.21 to classify ECG cycles as normal beats or beats with abnormal morphology. A normal beat was used as a template to compute γ_{xy} for each detected beat. They found that most normal beats possessed γ_{xy} values above 0.9, and that

PVCs and beats with abnormal shape had considerably lower values. A threshold of 0.9 was used to assign a code to each beat as 0:abnormal or 1:normal in terms of waveshape. Figure 2.2 shows an ECG signal with five abnormal beats that have the first symbol in the 4-symbol code as 0, indicating an abnormal shape due to generation by an ectopic focus or due to aberrant conduction of a pulse generated by the SA node. The normal beats have the first symbol of the code as 1, indicating a high correlation with the normal template.

5.4.2 The minimum-phase correspondent and signal length

The normal ECG signal contains epochs of activity where the signal's energy is concentrated. Discounting the usually low-amplitude P and T waves, most of the energy of a normal ECG signal is concentrated within an interval of about 80 ms that is spanned by the QRS complex. The normally iso-electric PQ, ST, and TP segments contain no energy as the signal amplitude is zero over the corresponding intervals. We have observed that certain abnormal conditions cause the QRS to widen or the ST segment to bear a nonzero value. In such a case, it could be said that the energy of the signal is being spread over a longer duration. Let us now consider how we may capture this information, and investigate if it may be used for waveshape analysis.

Problem: *Investigate the effect of the distribution of energy over the time axis on a signal's characteristics. Propose measures to parameterize the effects and study their use in the classification of ECG beats.*

Solution: A signal $x(t)$ may be seen as a distribution of the amplitude of a certain variable over the time axis. The square of the signal, that is, $x^2(t)$, may be interpreted as the instantaneous energy of the signal-generating process. The function $x^2(t)$, $0 \leq t \leq T$, may then be viewed as an energy distribution or density function, with the observation that the total energy of the signal is given by

$$E_x = \int_0^T x^2(t)\, dt. \tag{5.1}$$

Such a representation facilitates the definition of moments of the energy distribution, leading to a centroidal time

$$t_{\bar{x}} = \frac{\int_0^T t\, x^2(t)\, dt}{\int_0^T x^2(t)\, dt}, \tag{5.2}$$

and dispersion of energy about the centroidal time

$$\sigma_{t_{\bar{x}}}^2 = \frac{\int_0^T (t - t_{\bar{x}})^2\, x^2(t)\, dt}{\int_0^T x^2(t)\, dt}. \tag{5.3}$$

Observe the similarity between the equations above and Equations 3.1 and 3.3: the normalized function

$$p_x(t) = \frac{x^2(t)}{\int_0^T x^2(t)\, dt} \tag{5.4}$$

is now treated as a PDF. Other moments may also be defined to characterize and study the distribution of $x^2(t)$ over the time axis. The preceding equations have been stated in continuous time for the sake of generality; they are valid for discrete-time signals, with a simple change of t to n and $\int dt$ to $\sum_n$.

Minimum-phase signals: The distribution of the energy of a signal over its duration is related to its amplitude spectrum and, more importantly, to its phase spectrum. The notion of minimum phase is useful in analyzing related signal characteristics. The minimum-phase property of signals may be explained in both the time and frequency domains [86, 123, 124, 125, 126, 127, 102].

In the time domain, a signal $x(n)$ is a minimum-phase signal if both the signal and its inverse $x_i(n)$ are one-sided (that is, completely causal or anti-causal) signals with finite energy, that is, $\sum_{n=0}^{\infty} x^2(n) < \infty$ and $\sum_{n=0}^{\infty} x_i^2(n) < \infty$. (*Note:* The inverse of a signal is defined such that $x(n) * x_i(n) = \delta(n)$; equivalently, we have $X_i(z) = \frac{1}{X(z)}$.)

Some of the important properties of a minimum-phase signal are:

- For a given amplitude spectrum there exists one and only one minimum-phase signal.

- Of all finite-energy, one-sided signals with identical amplitude spectra, the energy of the minimum-phase signal is optimally concentrated toward the origin, and the signal has the smallest phase lag and phase-lag derivative at each frequency.

- The z-transform of a minimum-phase signal has all of its poles and zeros inside the unit circle in the z-plane.

- The complex cepstrum of a minimum-phase signal is causal (see also Section 4.8.3).

The extreme example of a minimum-phase signal is the delta function $\delta(t)$, which has all of its energy concentrated at $t = 0$. The magnitude spectrum of the delta function is real and equal to unity for all frequencies; the phase lag at every frequency is zero.

Minimum-phase and maximum-phase components: A signal $x(n)$ that does not satisfy the minimum-phase condition, referred to as a composite signal or a mixed-phase signal, may be split into its minimum-phase component and maximum-phase component by filtering its complex cepstrum $\hat{x}(n)$ [86, 115, 128]. To obtain the minimum-phase component, the causal part of the complex cepstrum (see Section 4.8.3) is chosen as follows:

$$\hat{x}_{\min}(n) = \begin{cases} 0 & n < 0 \\ 0.5\hat{x}(n) & n = 0 \\ \hat{x}(n) & n > 0 \end{cases}. \qquad (5.5)$$

Application of the inverse procedures yields the minimum-phase component $x_{\min}(n)$. Similarly, the maximum-phase component is obtained by application of the inverse procedures to the anti-causal part of the cepstrum, selected as

$$\hat{x}_{\text{max}}(n) = \begin{cases} \hat{x}(n) & n < 0 \\ 0.5\hat{x}(n) & n = 0 \\ 0 & n > 0 \end{cases} . \tag{5.6}$$

The minimum-phase and maximum-phase components of a signal satisfy the following relationships:

$$\hat{x}(n) = \hat{x}_{\text{min}}(n) + \hat{x}_{\text{max}}(n), \tag{5.7}$$

and

$$x(n) = x_{\text{min}}(n) * x_{\text{max}}(n). \tag{5.8}$$

The minimum-phase correspondent (MPC): A mixed-phase signal may be converted to a minimum-phase signal that has the same spectral magnitude as the original signal by filtering the complex cepstrum of the original signal as

$$\hat{x}_{\text{MPC}}(n) = \begin{cases} 0 & n < 0 \\ \hat{x}(n) & n = 0 \\ \hat{x}(n) + \hat{x}(-n) & n > 0 \end{cases} \tag{5.9}$$

and applying the inverse procedures [86, 115, 128]. The result is known as the *minimum-phase correspondent* or MPC of the original signal [102]. The MPC will possess optimal concentration of energy around the origin under the constraint imposed by the specified magnitude spectrum (of the original mixed-phase signal).

Observe that $\hat{x}_{\text{MPC}}(n)$ is equal to twice the even part of $\hat{x}(n)$ for $n > 0$. This leads to a simpler procedure to compute the MPC, as follows: Let us assume $\hat{X}(z) = \log X(z)$ to be analytic over the unit circle in the z-plane. We can write $\hat{X}(\omega) = \hat{X}_R(\omega) + j\hat{X}_I(\omega)$, where the subscripts R and I indicate the real and imaginary parts, respectively. $\hat{X}_R(\omega)$ and $\hat{X}_I(\omega)$ are the log-magnitude and phase spectra of $x(n)$, respectively. Now, the inverse Fourier transform of $\hat{X}_R(\omega)$ is equal to the even part of $\hat{x}(n)$, defined as $\hat{x}_e(n) = [\hat{x}(n) + \hat{x}(-n)]/2$. Thus we have

$$\hat{x}_{\text{MPC}}(n) = \begin{cases} 0 & n < 0 \\ \hat{x}_e(n) & n = 0 \\ 2\hat{x}_e(n) & n > 0 \end{cases} . \tag{5.10}$$

This result means that we do not need to compute the complex cepstrum, which requires the unwrapped phase spectrum of the signal, but need only to compute a *real cepstrum* using the log-magnitude spectrum. Furthermore, given that the PSD is the Fourier transform of the ACF, we have $\log[\text{FT}[\phi_{xx}(n)]] = 2\hat{X}_R(\omega)$. It follows that, in the cepstral domain, $\hat{\phi}_{xx}(n) = 2\hat{x}_e(n)$, and therefore [128]

$$\hat{x}_{\text{MPC}}(n) = \begin{cases} 0 & n < 0 \\ 0.5\hat{\phi}_{xx}(n) & n = 0 \\ \hat{\phi}_{xx}(n) & n > 0 \end{cases} , \tag{5.11}$$

where $\hat{\phi}_{xx}(n)$ is the cepstrum of the ACF $\phi_{xx}(n)$ of $x(n)$.

Signal length: The notion of signal length (SL), as introduced by Berkhout [124], is different from signal duration. The duration of a signal is the extent of time over which the signal exists, that is, has nonzero values (neglecting periods within the total signal duration where the signal could be zero). SL relates to how the energy of a signal is distributed over its duration. SL depends upon both the magnitude and phase spectra of the signal. For one-sided signals, minimum SL implies minimum phase; the converse is also true [124].

The general definition of SL of a signal $x(n)$ is given as [124]

$$SL = \frac{\sum_{n=0}^{N-1} w(n)\, x^2(n)}{\sum_{n=0}^{N-1} x^2(n)}, \tag{5.12}$$

where $w(n)$ is a nondecreasing, positive weighting function with $w(0) = 0$. The definition of $w(n)$ depends upon the application and the desired characteristics of SL. It is readily seen that samples of the signal away from the origin $n = 0$ receive progressively heavier weighting by $w(n)$. The definition of SL as above may be viewed as a normalized moment of $x^2(n)$. If $w(n) = n$, we get the centroidal time instant of $x^2(n)$ as in Equation 5.2.

For a given amplitude spectrum and hence total energy, the minimum-phase signal has its energy optimally concentrated near the origin. Therefore, the minimum-phase signal will have the lowest SL of all signals with the specified amplitude spectrum. Signals with increasing phase lag have their energy spread over a longer time duration, and will have larger SL due to the increased weighting by $w(n)$.

Illustration of application: The QRS-T wave is the result of the spatio-temporal summation of the action potentials of ventricular myocytes. The duration of normal QRS-T waves is in the range of $350 - 400\ ms$, with the QRS itself limited to about $80\ ms$ due to rapid and coordinated depolarization of the ventricular motor units via the Purkinje fibers. However, PVCs, in general, have QRS-T complexes that are wider than normal, that is, they have their energy distributed over longer time spans within their total duration. This is due to different and possibly slower and disorganized excitation sequences triggering the ventricular motor units: ectopic triggers may not get conducted through the Purkinje system, and may be conducted through the ventricular muscle cells themselves. Furthermore, PVCs do not, in general, display separate QRS and T waves, that is, they lack an iso-electric ST segment.

Regardless of the above distinctions, normal ECG beats and PVCs have similar amplitude spectra, indicating that the difference between the signals may lie in their phase. SL depends upon both the amplitude spectrum and the phase spectrum of the given signal, and parameterizes the distribution of energy over the duration of the signal. Based upon the arguments above, Murthy and Rangaraj [102] proposed the application of SL to classify ECG beats as normal or ectopic (or PVC, along with the use of the RR interval to indicate prematurity). Furthermore, to overcome ambiguities in the determination of the onset of each beat, they computed the SL of the MPC of the ECG signals (segmented so as to include the P, QRS, and T waves of each cycle). Use of the MPC resulted in a "rearrangement" of the waves such that the dominant QRS wave appeared at the origin in the MPC.

Figure 5.1 illustrates a normal ECG signal and three PVCs of a patient with multiple ectopic foci generating PVCs of widely differing shapes [102]. The figure also illustrates the corresponding MPCs and lists the SL values of all the signals. The SL values of the MPCs of the abnormal waves are higher than the SL of the MPC of the normal signal (see the right-hand column of signals in Figure 5.1). The SL values of the original PVCs do not exhibit such a separation from the SL of the normal signal (see the left-hand column of signals in Figure 5.1). Ambiguities due to the presence of base-line segments of variable lengths at the beginning of the signals have been overcome by the use of the MPCs. The MPCs have the most-dominant wave in each case at the origin, reflecting a rearrangement of energy or waves so as to meet the minimum-phase criteria.

Figure 5.2 shows plots of the RR intervals and SL values computed using the original ECG signals and their MPCs for several beats of the same patient whose representative ECG waveforms are illustrated in Figure 5.1 [102]. The SL values of the normal signals and the ectopic beats exhibit a significant overlap in the range $28 - 35$ (plot (a) in Figure 5.2). However, the SL values of the MPCs of the PVCs are higher than those of the normal beats, which facilitates their classification (plot (b) in Figure 5.2).

Murthy and Rangaraj [102] applied their QRS detection method (described in Section 4.3.1) to ECG signals of two patients with ectopic beats, and used the SL of MPC to classify the beats with a linear discriminant function (described in Section 9.4.1). They analyzed 208 beats of the first patient (whose signals are illustrated in Figures 5.1 and 5.2): 132 out of 155 normals and 48 out of 53 PVCs were correctly classified; one beat was missed by the QRS detection algorithm. Misclassification of normal beats as PVCs was attributed to wider-than-normal QRS complexes and depressed ST segments in some of the normal beats of the patient (see Figure 5.2). The signal of the second patient included 89 normals and 18 PVCs, all of which were detected and classified correctly. It was observed that computation of the MPC was not required in the case of the second patient: the SL values of the original signals provided adequate separation between normal and ectopic beats. The segments of normal ECG cycles used by Murthy and Rangaraj included the P wave; better results could perhaps be obtained by using only the QRS and T waves since most PVCs do not include a distinct P wave and essentially correspond to the QRS and T waves in a normal ECG signal.

It should be noted that the QRS width may be increased by other abnormal conditions such as bundle-branch block; the definition of SL as above would lead to higher SL for wider-than-normal QRS complexes. Furthermore, ST segment elevation or depression would be interpreted as the presence of energy in the corresponding time interval in the computation of SL. Abnormally large T waves could also lead to SL values that are larger than those for normal signals. More sophisticated logic and other parameters in addition to SL could be used to rule out these possibilities and affirm the classification of a beat as an ectopic beat.

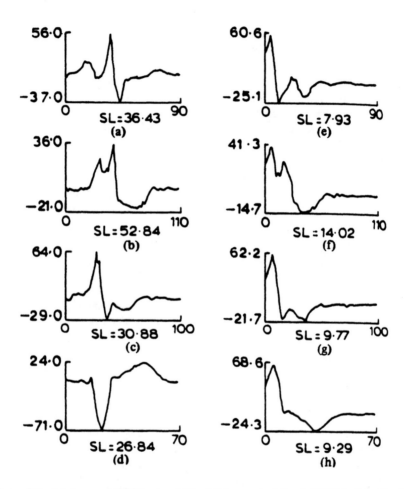

Figure 5.1   (a) A normal ECG beat and (b) – (d) three ectopic beats (PVCs) of a patient with multiple ectopic foci. (e) – (h) MPCs of the signals in (a) – (d). The SL values of the signals are also indicated [102]. Note that the abscissa is labeled in samples, with a sampling interval of 10 ms. The ordinate is not calibrated. The signals have different durations and amplitudes although plotted to the same size. Reproduced with permission from I.S.N. Murthy and M.R. Rangaraj, New concepts for PVC detection, *IEEE Transactions on Biomedical Engineering*, 26(7):409–416, 1979. ©IEEE.

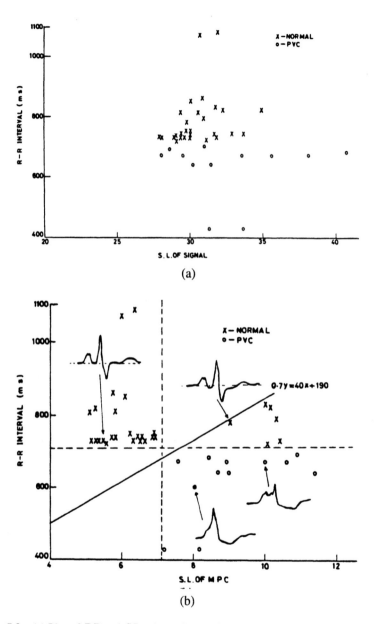

Figure 5.2 (a) Plot of RR and SL values of several beats of a patient with multiple ectopic foci (as in Figure 5.1). (b) Same as (a) but with the SL of the MPCs of the signals. A few representative ECG cycles are illustrated. The linear discriminant (decision) function used to classify the beats is also shown [102]. Reproduced with permission from I.S.N. Murthy and M.R. Rangaraj, New concepts for PVC detection, *IEEE Transactions on Biomedical Engineering*, 26(7):409–416, 1979. ©IEEE.

5.4.3 ECG waveform analysis

Measures such as the correlation coefficient and SL described in the preceding sub-sections provide general parameters that could assist in comparing waveforms. The representation, however, is in terms of gross features, and many different waveforms could possess the same or similar feature values. Detailed analysis of ECG waveforms will require the use of several features or measurements for accurate categorization of various QRS complex shapes and correlation with cardiovascular diseases. Since the ECG waveform depends upon the lead system used, sets of features may have to be derived for multiple-lead ECGs, including as many as 12 leads that are commonly used in clinical practice.

The steps required for ECG waveform analysis may be expressed as [31]:

1. Detection of ECG waves, primarily the QRS complex, and possibly the P and T waves.

2. Delimitation of wave boundaries, including the P, QRS, and T waves.

3. Measurement of inter-wave intervals, such as RR, PQ, QT, ST, QQ, and PP intervals.

4. Characterization of the morphology (shape) of the waves.

The last step above may be achieved using parameters such as the correlation coefficient and SL as described earlier, or via detailed measurements of the peaks of the P, Q, R, S, and T waves (some could be negative); the durations of the P, Q, R, S, QRS, and T waves; and the inter-wave intervals defined above [31]. The nature of the PQ and ST segments, in terms of their being iso-electric or not (in case of the latter, as being positive or negative, or elevated or depressed), should also be documented. However, a large number of such features would make the development of further pattern classification rules difficult.

Cox et al. [31, 129] proposed four measures to characterize QRS complexes, defined as follows:

1. *Duration* — the duration or width of the QRS complex.

2. *Height* — the maximum amplitude minus the minimum amplitude of the QRS complex.

3. *Offset* — the positive or negative vertical distance from the midpoint of the base-line to the center of the QRS complex. The base-line is defined as the straight line connecting the temporal boundary points of the QRS complex. The center is defined as the midpoint between the highest and lowest bounds in amplitude of the QRS complex.

4. *Area* — the area under the QRS waveform rectified with respect to a straight line through the midpoint of the base-line.

Since the measures are independent of time, they are less sensitive to the preceding procedures for the detection of fiducial markers.

The measures were used to develop a system for arrhythmia monitoring, known as "Argus" for Arrhythmia Guard System, for use in coronary-care units. Figure 5.3 shows the grouping of more than 200 QRS complexes of a patient with multi-focal PVCs into 16 dynamic families by Argus using the four features defined above [31]. The families labeled $00, 01, 02, 04, 06$, and 10 were classified as normal beats by Argus (163 beats which were all classified as normals by a cardiologist; 91% of the normals were correctly labeled by Argus). PVCs of different shapes from more than two ectopic foci form the remaining families, with some of them having shapes close to those of the patient's normal sinus beats. Of the 52 beats in the remaining families, 96% were labeled as PVCs by the cardiologist; Argus labeled 85% of them as PVCs, 13% as not PVCs, and 2% as border-line beats [129]. Cox et al. [31] summarize one of the clinical tests of Argus with over $50,000$ beats, some noteworthy points being as follows: 85% of $45,364$ normal beats detected and classified correctly, with 0.04% beats missed; 78% of $4,010$ PVCs detected and classified correctly, with 5.3% beats missed; and 38 normals (less than 0.1% of the beats) falsely labeled as PVCs.

5.5 ENVELOPE EXTRACTION AND ANALYSIS

Signals with complex patterns such as the EMG and PCG may not permit direct analysis of their waveshape. In such cases, the intricate high-frequency variations may not be of interest; rather, the general trends in the level of the overall activity might convey useful information. Considering, for example, the EMG in Figure 1.9, observe that the general signal level increases with the level of activity (breathing). As another example, the PCG in the case of aortic stenosis, as illustrated in Figure 1.26, demonstrates a diamond-shaped systolic murmur: the *envelope* of the overall signal carries important information. Let us therefore consider the problem of extraction of the envelope of a seemingly complex signal.

Problem: *Formulate algorithms to extract the envelope of an EMG or PCG signal to facilitate analysis of trends in the level of activity or energy in the signal.*

Solution: The first step required in order to derive the envelope of a signal with positive and negative deflections is to obtain the absolute value of the signal at each time instant, that is, perform full-wave rectification. This procedure will create abrupt discontinuities at time instants when the original signal values change sign, that is, at zero-crossings. The discontinuities create high-frequency components of significant magnitude. This calls for the application of a lowpass filter with a relatively low bandwidth in the range of $0 - 10$ or $0 - 50$ Hz to obtain smooth envelopes of EMG and PCG signals. A moving-average filter may be used to perform lowpass filtering, leading to the basic definition of a time-averaged envelope as

$$y(t) = \frac{1}{T_a} \int_{t-T_a}^{t} |x(t)| \, dt, \tag{5.13}$$

where T_a is the duration of the moving-average window.

In a procedure similar in principle to that described above, Lehner and Rangayyan [66] applied a weighted MA filter to the squared PCG signal to obtain a

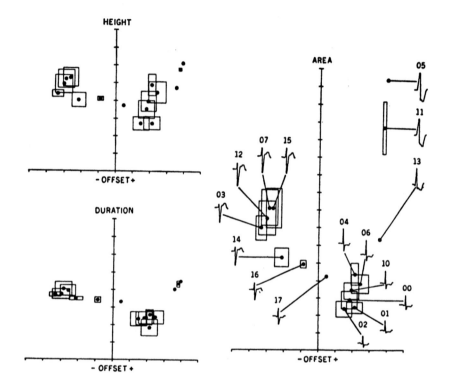

Figure 5.3   Use of four features to catalog QRS complexes into one of 16 dynamic families of similar complexes enclosed by four-dimensional boxes. The waveforms of typical members of each family are shown in the area-versus-offset feature plane. The family numbers displayed are in the octal (base eight) system. The families labeled 00, 01, 02, 04, 06, and 10 were classified as normal beats, with the others being PVCs or border-line beats. Reproduced with permission from J.R. Cox, Jr., F.M. Nolle, and R.M. Arthur, Digital analysis of the electroencephalogram, the blood pressure wave, and the electrocardiogram, *Proceedings of the IEEE*, 60(10):1137–1164, 1972. ©IEEE.

smoothed energy distribution curve $E(n)$ as

$$E(n) = \sum_{k=1}^{M} x^2(n - k + 1)w(k), \qquad (5.14)$$

where $x(n)$ is the PCG signal, $w(k) = M - k + 1$, and $M = 32$ with the signal sampled at $1,024\ Hz$. Observe that the difference between energy and power is simply a division by the time interval being considered, which may be treated as a scale factor or ignored.

The envelope represents the total averaged activity (electrical, acoustic, and so on) within the averaging window. An improved filter such as a Bessel filter [26] may be required if a smooth envelope is desired. The filter should strike a balance between the need to smooth discontinuities in the rectified signal and the requirement to maintain good sensitivity to represent relevant changes in signal level or amplitude. This procedure is known as envelope detection or amplitude demodulation. A few related procedures and techniques are described in the following subsections.

5.5.1 Amplitude demodulation

Amplitude modulation (AM) of signals for radio transmission involves multiplication of the signal $x(t)$ to be transmitted by an RF carrier $\cos(w_c t)$, where w_c is the carrier frequency. The AM signal is given as $y(t) = x(t)\cos(w_c t)$ [1, 2]. If the exact carrier wave used at the transmitting end were available at the receiving end as well (including the phase), *synchronous demodulation* becomes possible by multiplying the received signal $y(t)$ with the carrier. We then have the demodulated signal as

$$x_d(t) = y(t)\cos(w_c t) = x(t)\cos^2(w_c t) = \frac{1}{2}x(t) + \frac{1}{2}x(t)\cos(2w_c t). \qquad (5.15)$$

The AM component at $2w_c$ may be removed by a lowpass filter, which will leave us with the desired signal $x(t)$.

If $x(t)$ is always positive, or a DC bias is added to meet this requirement, it becomes readily apparent that the envelope of the AM signal is equal to $x(t)$. An extremely simple *asynchronous demodulation* procedure that does not require the carrier then becomes feasible: we just need to follow the envelope of $y(t)$. Given also that the carrier frequency w_c is far greater than the maximum frequency present in $x(t)$, the positive envelope of $y(t)$ may be extracted by performing half-wave rectification. A lowpass filter with an appropriate time constant to "fill the gaps" between the peaks of the carrier wave will give a good estimate of $x(t)$. The difference between the use of a full-wave rectifier or a half-wave rectifier (that is, the larger gaps between the peaks of the carrier wave available after either type of rectification) can be easily made up by increasing the time constant of the filter. The main differences between various envelope detectors lie in the way the rectification operation is performed, and in the lowpass filter used [1, 2].

In a related procedure known as complex demodulation , a given arbitrary signal is demodulated to derive the time-varying amplitude and phase characteristics of the signal for each frequency (band) of interest [130, 131, 132]. In this approach, an arbitrary signal $x(t)$ is expressed as

$$x(t) = a(t) \cos[f_o t + \psi(t)] + x_r(t), \tag{5.16}$$

where f_o is the frequency of interest, $a(t)$ and $\psi(t)$ are the time-varying amplitude and phase of the component at f_o, and $x_r(t)$ is the remainder of the signal $x(t)$ after the component at f_o has been removed. It is assumed that $a(t)$ and $\psi(t)$ vary slowly in relation to the frequencies of interest. The signal $x(t)$ may be equivalently expressed in terms of complex exponentials as

$$x(t) = \frac{1}{2} a(t) \left\{ \exp\{j[f_o t + \psi(t)]\} + \exp\{-j[f_o t + \psi(t)]\} \right\} + x_r(t). \tag{5.17}$$

In the procedure of complex demodulation, the signal is shifted in frequency by $-f_o$ via multiplication with $2 \exp(-j f_o t)$, to obtain the result $y(t)$ as

$$
\begin{aligned}
y(t) &= 2x(t) \exp(-j f_o t) \\
&= a(t) \exp[j\psi(t)] + a(t) \exp\{-j[2f_o t + \psi(t)]\} + 2x_r(t) \exp(-j f_o t).
\end{aligned}
\tag{5.18}
$$

The second term in the expression above is centered at $2f_o$, whereas the third term is centered at f_o; only the first term is placed at DC. Therefore, a lowpass filter may be used to extract the first term, to obtain the final result $y_o(t)$ as

$$y_o(t) \approx a(t) \exp[j\psi(t)]. \tag{5.19}$$

The desired entities may then be extracted as $a(t) \approx |y_o(t)|$ and $\psi(t) \approx \angle y_o(t)$.

The frequency resolution of the method depends upon the bandwidth of the lowpass filter used. The procedure may be repeated at every frequency (band) of interest. The result may be interpreted as the envelope of the signal for the specified frequency (band). The method was applied for the analysis of HRV by Shin et al. [130] and the analysis of heart rate and arterial blood pressure variability by Hayano et al. [131].

In applying envelope detection to biomedical signals such as the PCG and the EMG, it should be noted that there is no underlying RF carrier wave in the signal: the envelope rides on relatively high-frequency acoustic or electrical activity that has a composite spectrum. The difference in frequency content between the envelope and the "carrier activity" will not be comparable to that in AM. Regardless, we could expect at least a ten-fold difference in frequency content: the envelope of an EMG or PCG signal may have an extremely limited bandwidth of $0 - 20 \ Hz$, whereas the underlying signal has components up to at least $200 \ Hz$, if not to $1,000 \ Hz$. Application of envelope detection to the analysis of EMG related to respiration will be illustrated in Section 5.9.

5.5.2 Synchronized averaging of PCG envelopes

The ECG and PCG form a good signal pair for synchronized averaging: the latter could be averaged over several cardiac cycles using the former as the trigger.

However, the PCG is not amenable to direct synchronized averaging as the vibration waves may interfere in a destructive manner and cancel themselves out. Karpman et al. [133] proposed to first rectify the PCG signal, smooth the result using a lowpass filter, and then perform synchronized averaging of the envelopes so obtained using the ECG as the trigger. The PCG envelopes were averaged over up to 128 cardiac cycles to get repeatable averaged envelopes. It should noted that while synchronized averaging can reduce the effects of noise, breathing, coughing, and so on, it can also smudge the time boundaries of cardiac events if the heart rate is not constant during the averaging procedure.

Figure 5.4 illustrates the envelopes obtained for a normal case and seven cases of systolic murmur due to aortic stenosis (AS), atrial septal defect (ASD), hypertrophic subaortic stenosis (HSS), rheumatic mitral regurgitation (MR), ventricular septal defect (VSD), and mitral regurgitation with posterior leaflet prolapse (PLP). The typical diamond-shaped envelope in the case of aortic stenosis results in an envelope shaped like an isosceles triangle due to rectification. Mitral regurgitation results in a rectangular holo-systolic murmur envelope.

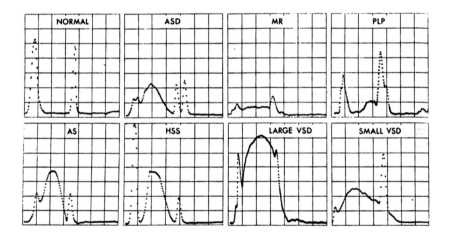

Figure 5.4 Averaged envelopes of the PCG signals of a normal subject and patients with systolic murmur due to aortic stenosis (AS), atrial septal defect (ASD), hypertrophic subaortic stenosis (HSS), rheumatic mitral regurgitation (MR), ventricular septal defect (VSD), and mitral regurgitation with posterior leaflet prolapse (PLP). Reproduced with permission from L. Karpman, J. Cage, C. Hill, A.D. Forbes, V. Karpman, and K. Cohn, Sound envelope averaging and the differential diagnosis of systolic murmurs, *American Heart Journal*, 90(5):600–606, 1975. ©American Heart Association.

Karpman et al. analyzed 400 cases of systolic murmurs due to six types of diseases and defects, and obtained an accuracy of 89% via envelope analysis. They proposed a decision tree to classify systolic murmurs based upon envelope shape and its relation to the envelopes of S1 and S2, which is illustrated in Figure 5.5.

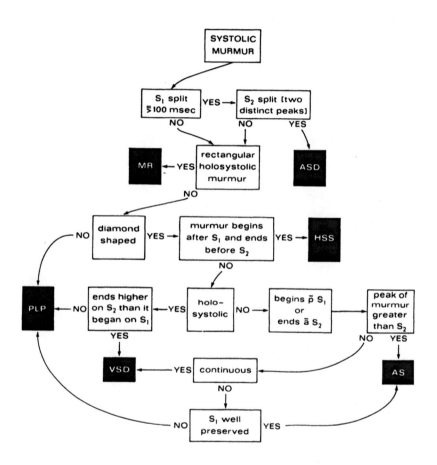

Figure 5.5 Decision tree to classify systolic murmurs based upon envelope analysis. For details on the abbreviations used, refer to the text or the caption of Figure 5.4. Reproduced with permission from L. Karpman, J. Cage, C. Hill, A.D. Forbes, V. Karpman, and K. Cohn, Sound envelope averaging and the differential diagnosis of systolic murmurs, *American Heart Journal*, 90(5):600–606, 1975. ©American Heart Association.

5.5.3 The envelogram

Sarkady et al. [119] proposed a Fourier-domain algorithm to obtain envelopes of PCG signals. They defined the *envelogram estimate* as the magnitude of the analytic signal $y(t)$ formed using the PCG $x(t)$ and its Hilbert transform $x_H(t)$ as

$$y(t) = x(t) + jx_H(t). \tag{5.20}$$

(*Note:* An analytic function is a complex function of time having a Fourier transform that vanishes for negative frequencies [5, 86].) The Hilbert transform of a signal is defined as the convolution of the signal with $\frac{1}{\pi t}$, that is,

$$x_H(t) = \int_{-\infty}^{\infty} \frac{x(\tau)}{\pi(t - \tau)} \, d\tau. \tag{5.21}$$

The Fourier transform of $\frac{1}{\pi t}$ is $-j \, \text{sgn}(\omega)$, where

$$\text{sgn}(\omega) = \begin{cases} -1 & \omega < 0 \\ 0 & \omega = 0 \\ 1 & \omega > 0 \end{cases}. \tag{5.22}$$

Then, we have $Y(\omega) = X(\omega)[1 + \text{sgn}(\omega)]$. $Y(\omega)$ is a one-sided or single-sideband function of ω containing positive-frequency terms only.

Based upon the definitions and properties described above, Sarkady et al. [119] proposed the following algorithm to obtain the envelogram estimate:

1. Compute the DFT of the PCG signal.

2. Set the negative-frequency terms to zero; that is, $X(k) = 0$ for $\frac{N}{2}+2 \le k \le N$, with the DFT indexed $1 \le k \le N$ as in MATLAB.

3. Multiply the positive-frequency terms, that is, $X(k)$ for $2 \le k \le \frac{N}{2} + 1$, by 2; the DC term $X(1)$ remains unchanged.

4. Compute the inverse DFT of the result.

5. The magnitude of the result gives the envelogram estimate.

The procedure described above, labeled also as complex demodulation by Sarkady et al., yields a high-resolution envelope of the input signal. Envelograms and PSDs computed from PCG signals over single cardiac cycles tend to be noisy and are affected by respiration and muscle noise. Sarkady et al. recommended synchronized averaging of both envelograms and PSDs of PCGs over several cycles. A similar method was used by Baranek et al. [120] to obtain the envelopes of PCG signals for the detection of the aortic component A2 of S2.

Illustration of application: The top-most plots in Figures 5.6 and 5.7 show one cycle each of the PCG signals of a normal subject and of a patient with systolic murmur, split S2, and opening snap of the mitral valve. The PCG signals were segmented by using the Pan-Tompkins method to detect the QRS complexes in the ECG signal, as illustrated in Figures 4.27 and 4.28 for the same signals. The envelograms of the PCG cycles illustrated and the averaged envelograms (over 16 beats for the normal and 26 beats for the case with murmur) obtained using the method of Sarkady et al. [119] are shown in the second and third plots of Figures 5.6 and 5.7, respectively. Observe that while a split S2 is visible in the individual signal and envelogram illustrated in Figure 5.6, the split is not seen in the averaged envelogram and envelope, possibly due to breathing-related variations over the duration of the signal record and averaging.

Furthermore, based upon the method of Karpman et al. [133], the averaged envelopes were computed by taking the absolute value of the signal over each cardiac cycle, smoothing with a Butterworth lowpass filter with $N = 8$ and $f_c = 50\ Hz$, and synchronized averaging. The last plots in Figures 5.6 and 5.7 show the averaged envelopes. (The Butterworth filter has introduced a small delay in the envelope; the delay may be avoided by using the *filtfilt* command in MATLAB.) The averaged envelograms and averaged envelopes for the normal case display the envelopes of S1 and S2; the individual components of S1 and S2 have been smoothed over and merged in the averaged results. The averaged envelograms and averaged envelopes for the case with murmur clearly demonstrate the envelopes of S1, the systolic murmur, the split S2, and the opening snap of the mitral valve.

5.6 ANALYSIS OF ACTIVITY

Problem: *Propose measures of waveform complexity or activity that may be used to analyze the extent of variability in signals such as the PCG and EMG.*

Solution: The samples of a given EMG or PCG signal may, for the sake of generality, be treated as a random variable x. Then, the variance $\sigma_x^2 = E[(x - \mu_x)^2]$ represents an averaged measure of the variability or *activity* of the signal about its mean. If the signal has zero mean, or is preprocessed to meet the same condition, we have $\sigma_x^2 = E[x^2]$; that is, the variance is equal to the average power of the signal. Taking the square root, we get the standard deviation of the signal as equal to its root mean-squared (RMS) value. Thus the RMS value could be used as an indicator of the level of activity about the mean of the signal. A much simpler indicator of activity is the number of zero-crossings within a specified interval; the zero-crossing rate (ZCR) increases as the high-frequency energy of the signal increases. A few measures related to the concepts introduced above are described in the following subsections, with illustrations of application.

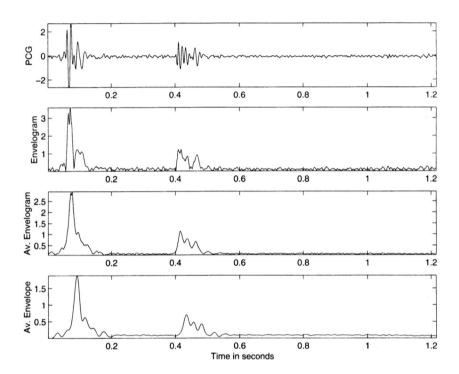

Figure 5.6 Top to bottom: PCG signal of a normal subject (male, 23 years); envelogram estimate of the signal shown; averaged envelogram over 16 cardiac cycles; averaged envelope over 16 cardiac cycles. The PCG signal starts with S1. See Figure 4.27 for an illustration of segmentation of the same signal.

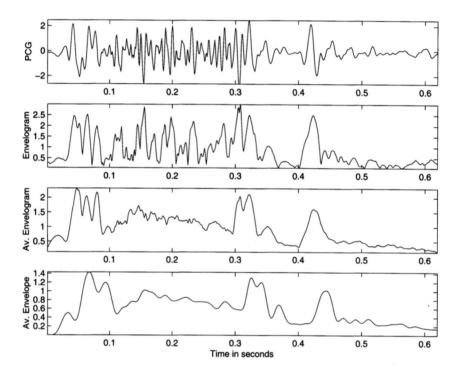

Figure 5.7 Top to bottom: PCG signal of a patient (female, 14 months) with systolic murmur (approximately 0.1 − 0.3 *s*), split S2 (0.3 − 0.4 *s*), and opening snap of the mitral valve (0.4 − 0.43 *s*); envelogram estimate of the signal shown; averaged envelogram over 26 cardiac cycles; averaged envelope over 26 cardiac cycles. The PCG signal starts with S1. See Figure 4.28 for an illustration of segmentation of the same signal.

5.6.1 The root mean-squared value

The RMS value of a signal $x(n)$ over its total duration of N samples is given by

$$RMS = \left[\frac{1}{N} \sum_{n=0}^{N-1} x^2(n) \right]^{\frac{1}{2}}. \tag{5.23}$$

This global measure of signal level (related to power), however, is not useful for the analysis of trends in nonstationary signals. A running estimate of the RMS value of the signal computed over a causal window of M samples, defined as

$$RMS(n) = \left[\frac{1}{M} \sum_{k=0}^{M-1} x^2(n-k) \right]^{\frac{1}{2}}, \tag{5.24}$$

could serve as a useful indicator of the average power of the signal as a function of time. The duration of the window M needs to be chosen in accordance with the bandwidth of the signal, with $M << N$. Such an approach for computing running parameters of signals falls under the general scheme of *short-time analysis* of nonstationary signals [46].

Gerbarg et al. [134, 135] derived power versus time curves of PCG signals by computing the average power in contiguous segments of duration 10 *ms*, and used the curves to identify systolic and diastolic segments of the signals. They noted that within a 10 *s* PCG record, at least one diastolic segment would be longer than the corresponding systolic segment, and that all systolic segments in the record would have approximately the same duration. Innocent (physiological) systolic murmurs in children were observed to be limited to the first and middle thirds of the systolic interval between S1 and S2, whereas pathological systolic murmurs due to mitral regurgitation were noted to be holo-systolic (spanning the entire systolic period). Based upon these observations, Gerbarg et al. computed ratios of the mean power of the *last third of systole* to the mean power of systole and also to a certain "standard" noise level. A ratio was also computed of the mean energy of systole to the mean energy of the PCG over the complete cardiac cycle. Agreement in the range of $78 - 91\%$ was obtained between computer classification based upon the three ratios defined above and clinical diagnosis of mitral regurgitation in different groups of subjects.

Use of the RMS function for the analysis of EMG and VMG signals and thereby muscular activity will be illustrated in Section 5.10.

5.6.2 Zero-crossing rate

An intuitive indication of the "busy-ness" of a signal is provided by the number of times it crosses the zero-activity line or some other reference level. ZCR is defined as the number of times the signal crosses the reference within a specified interval. However, ZCR could be easily affected by DC bias, base-line wander, and low-frequency artifacts. For these reasons, it would be advisable to measure the ZCR of the derivative of the signal, which would be similar to the definition of turning points in the test for randomness described in Section 3.1.1. Saltzberg and Burch [136] discuss the relationship between ZCR and moments of PSDs, and their application to EEG analysis.

In spite of its simplicity, ZCR has been used in practical applications such as speech signal analysis to perform speech versus silence decision and to discriminate between voiced and unvoiced sounds [46] (see also Figure 3.1), and PCG analysis for the detection of murmurs. Jacobs et al. [137] used ZCR to perform normal versus abnormal classification of PCG signals using the ECG as a trigger, and obtained correct-classification rates of 95% for normals (58/61) and 94% for abnormals (77/82). They indicated a decision limit of 20 zero-crossings in a cardiac cycle. Yokoi et al. [138] proposed a mass-screening system based upon measurements of the maximum amplitude and ZCR in 8 ms segments of PCG signals (sampled at 2 kHz). They obtained correct-classification rates of 98% with 4,809 normal subjects and 76% with 1,217 patients with murmurs.

5.6.3   Turns count

Willison [139] proposed to analyze the level of activity in EMG signals by determining the number of spikes occurring in the interference pattern (see also [22, 140, 141]). Instead of counting zero-crossings, Willison's method investigates the significance of every change in phase (direction or slope) of the EMG signal called a *turn*. Turns greater than 100 μV are counted, with the threshold selected so as to avoid counting insignificant fluctuations due to noise. The method is similar to counting turning points as in the test for randomness described in Section 3.1.1, but is expected to be robust in the presence of noise due to the threshold imposed. The method is not directly sensitive to SMUAPs, but significant phase changes caused by superimposed SMUAPs are counted. Willison [139] found that EMG signals of subjects with myopathy possessed higher turns counts than those of normal subjects at comparable levels of volitional effort.

Illustration of application: The top-most plot in Figure 5.8 illustrates the EMG signal over two breath cycles from the crural diaphragm of a dog recorded via implanted fine-wire electrodes [26]. The subsequent plots illustrate, in top-to-bottom order, the short-time RMS values, the turns count by Willison's procedure, and the smoothed envelope of the signal. The RMS and turns count values were computed using a causal moving window of duration 70 ms (210 samples). The window duration needs to be chosen to strike a balance between the extent of smoothing desired in the turns count series and the accuracy in reflecting the nonstationary nature of the signal (increasing level of activity with inspiration in the present example). The envelope was obtained by taking the absolute value of the signal (equivalent to full-wave rectification) followed by a Butterworth lowpass filter of order $N = 8$ and cutoff frequency $f_c = 8$ Hz. It is seen that all three of the derived features demonstrate the expected increasing trend with the level of contraction (breath), and can serve as correlates or indicators of muscle contraction and the concomitant EMG complexity. The results may be further smoothed (lowpass filtered) if desired.

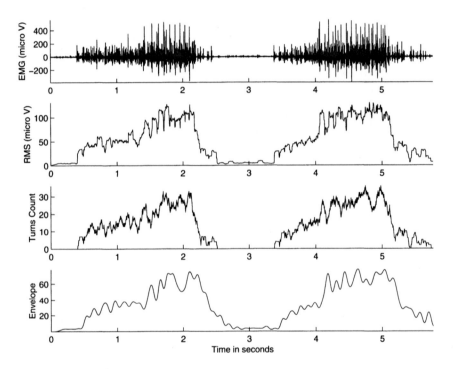

Figure 5.8 Top to bottom: EMG signal over two breath cycles from the crural diaphragm of a dog recorded via implanted fine-wire electrodes; short-time RMS values; turns count using Willison's procedure; and smoothed envelope of the signal. The RMS and turns count values were computed using a causal moving window of **70** *ms* duration. EMG signal courtesy of R.S. Platt and P.A. Easton, Department of Clinical Neurosciences, University of Calgary.

Figure 5.9 illustrates one 70 *ms* segment of the EMG signal in Figure 5.8 with the boundary points of the significant turns as detected by Willison's procedure marked by the '*' symbol. The procedure was implemented by first computing the derivative of the EMG signal and detecting points of change in its sign. A turn was marked wherever the EMG signal differed by at least 100 μV between successive points of sign change in the derivative. Observe from Figure 5.9 that the EMG signal need not cross the zero line to cause a turns count, and that zero-crossings with voltage swings of less than 100 μV are not counted as turns.

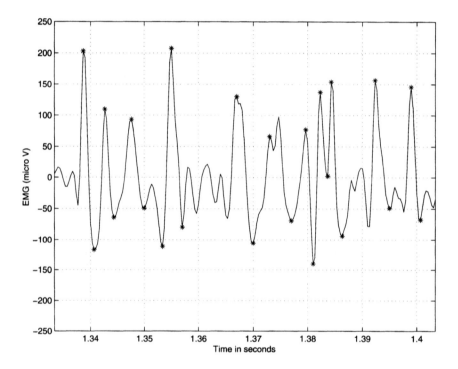

Figure 5.9 Illustration of the detection of turns in a 70 *ms* window of the EMG signal in Figure 5.8. The segments of the signal between pairs of '*' marks have been identified as significant turns.

5.6.4 Form factor

Based upon the notion of variance as a measure of signal activity, Hjorth [142, 143, 144] (see also [32]) proposed a method for the analysis of EEG waves. In this method, short-time segments of duration 1 *s* or longer are analyzed and three parameters are computed. The first parameter is called *activity* and is simply the variance σ_x^2 of the signal segment $x(n)$. The second parameter, called *mobility* M_x, is computed as the square root of the ratio of the activity of the first derivative of the signal to the activity of the (original) signal:

$$M_x = \left[\frac{\sigma_{x'}^2}{\sigma_x^2}\right]^{\frac{1}{2}} = \frac{\sigma_{x'}}{\sigma_x}, \tag{5.25}$$

where x' stands for the first derivative of x. The third parameter, called *complexity* or the *form factor* FF, is defined as the ratio of the mobility of the first derivative of the signal to the mobility of the signal itself:

$$FF = \frac{M_{x'}}{M_x} = \frac{\sigma_{x''}/\sigma_{x'}}{\sigma_{x'}/\sigma_x}, \tag{5.26}$$

where x'' stands for the second derivative of the signal. The complexity of a sinusoidal wave is unity; other waveforms have complexity values increasing with the extent of variations present in them.

Hjorth [143, 144] described the mathematical relationships between the activity, mobility, complexity, and PSD of a signal, and applied them to model EEG signal generation. Binnie et al. [145, 146] describe the application of FF and spectrum analysis to EEG analysis for the detection of epilepsy. However, because the computation of FF is based upon the first and second derivatives of the signal and their variances, the measure is sensitive to noise. A complex and relatively wide-band signal such as the EMG is not amenable to analysis via FF. Application of FF to discriminate between normal and ectopic ECG beats will be illustrated in Section 5.7.

We have explored a few measures to characterize waveform complexity in this section. Many authors have proposed several other diverse measures and interpretations of waveform or system complexity in the literature, examples of which include features based upon nonlinear dynamics and the correlation dimension [147], and the embedding dimension of time-varying dynamic systems [148].

5.7   APPLICATION: PARAMETERIZATION OF NORMAL AND ECTOPIC ECG BEATS

Problem: *Develop a parameter to discriminate between normal ECG waveforms and ectopic beats (PVCs).*

Solution: We have observed several times that ectopic beats, due to the abnormal propagation paths of the associated excitation pulses, typically possess waveforms that are significantly different from those of the normal QRS waveforms of the same subject. More often than not, ectopic beats have bizarre and complex waveshapes. The form factor FF described in Section 5.6.4 parameterizes the notion of waveform complexity, providing a value that increases with complexity. Therefore, FF appears to be a suitable measure to discriminate between normal and ectopic beats. Note that the RR interval by itself cannot indicate ectopic beats, as the RR interval could vary due to sinus arrhythmia and conduction problems, as well as due to heart-rate variations.

Figure 5.10 displays a segment of the ECG of a patient with ectopic beats; the segment illustrates the initiation of an episode of *ventricular bigeminy* where every normal beat is followed by an ectopic beat [23]. The ECG of the patient was processed using the Pan-Tompkins algorithm for QRS detection (see Section 4.3.2). QRS marker points were detected using a simple threshold applied to the output of the Pan-Tompkins algorithm. Each beat was segmented at points 160 ms before and 240 ms after the detected marker point; the diamond and circle symbols on the ECG in Figure 5.10 indicate the starting and ending points of the corresponding beats. The FF value was computed for each segmented beat. The RR interval (in ms) and FF value are printed for each beat in Figure 5.10. It can be readily seen that the FF values for the PVCs are higher than those for the normal beats.

Note from Figure 5.10 that the RR intervals for the PVCs are lower than those for the normal beats, and that the normal beats that follow the PVCs have higher-than-normal RR intervals due to the compensatory pause. Pattern classification of the ECG beats in this example as normal or PVCs using RR and FF will be described in Section 9.12.

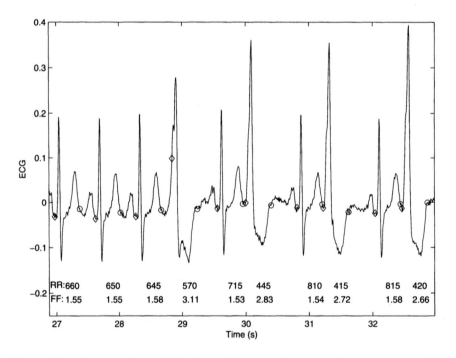

Figure 5.10 Segment of the ECG of a patient (male, 65 years) with ectopic beats. The diamond and circle symbols indicate the starting and ending points, respectively, of each beat obtained using the Pan-Tompkins algorithm for QRS detection. The RR interval (in ms) and form factor FF values are printed for each beat.

5.8 APPLICATION: ANALYSIS OF EXERCISE ECG

Problem: *Develop an algorithm to analyze changes in the ST segment of the ECG during exercise.*

Solution: Hsia et al. [149] developed a method to analyze changes in the ST segment of the ECG signal as the subject performed exercises. The analysis was performed as part of a radionuclide ventriculography (gated blood-pool imaging) procedure. In this procedure, nuclear medicine images are obtained of the left ventricle before and after exercising by the patient on a treadmill or bicycle ergometer. Images are obtained at different phases of the cardiac cycle by gating the radionuclide (gamma ray) emission data with reference to the ECG; image data for each phase are averaged over several cardiac cycles. Analysis of exercise ECG is complicated due to base-line artifacts caused by the effects of respiration, skin resistance changes due to perspiration, and soft tissue movement affecting electrode contact. Detection of changes in the ST segment in the presence of such artifacts poses a major challenge.

One of the main parameters used by Hsia et al. is related to the correlation coefficient as defined in Equation 3.18. The measure, however, is affected by base-line variations. To address this, a modified correlation coefficient was defined as

$$\gamma_{xy} = \frac{\sum_{n=0}^{N-1}[x(n)][y(n) - \Delta]}{\sqrt{\sum_{n=0}^{N-1}[x(n)]^2 \sum_{n=0}^{N-1}[y(n) - \Delta]^2}}. \tag{5.27}$$

Here, $x(n)$ is the template, $y(n)$ is the ECG signal being analyzed, Δ is a base-line correction factor defined as the difference between the base-line of $y(n)$ and the base-line of $x(n)$, and N is the duration (number of samples) of the template and the signal being analyzed. The template was generated by averaging up to 20 QRS complexes that met a specified RR interval constraint.

Hsia et al. proposed a method to establish the base-line of each ECG beat by searching for the PQ segment by backtracking from the R point detected (trigger for gating the image data). The region of three consecutive samples with the minimum change (maximum flatness) preceding the QRS was taken to represent the base-line of the beat. (*Note:* The PQ segment is almost always iso-electric, whereas the ST segment is variable in the case of cardiac diseases.) The search procedure also established the width of the QRS complex to be used in template matching (N in Equation 5.27). Beats with $\gamma_{xy} < 0.85$ were considered to be abnormal. The base-line correction factor in Equation 5.27 provided the robustness required.

Groups of 16 successive normal beats were aligned and averaged to obtain a representative waveform. The ST segment level was computed as the difference between a reference ST point and the iso-electric level of the current averaged beat. The averaging procedure included a condition to reject beats with abnormal morphology, such as PVCs. The ST reference point was defined as $R + 64\ ms + \max(4, \frac{200 - HR}{16}) \times 4\ ms$ or $S + 44\ ms + \max(4, \frac{200 - HR}{16}) \times 4\ ms$, where R or S indicates the position of the R or S wave of the present beat in ms, and HR is the heart rate in bpm. ST level differences of more than $2\ mV$ were reported by the program. Furthermore, the slope of the ST segment was computed by using two samples before and two samples after the ST point detected as described above (a duration of 16 ms with the sampling rate being 250 Hz).

In addition to the analysis of the ST segment, the method of Hsia et al. performed rhythm analysis, identification of PVCs and other abnormal beats, and assisted in the rejection of radionuclide emission data related to abnormal beats from the imaging procedure. The combined use of nuclear medicine imaging and ECG analysis was expected to improve the accuracy of the diagnosis of myocardial ischemia.

5.9 APPLICATION: ANALYSIS OF RESPIRATION

Problem: *Propose a method to relate EMG activity to airflow during inspiration.*

Solution: Platt et al. [26] recorded EMG signals from the parasternal intercostal and crural diaphragm muscles of dogs. One EMG signal was obtained from a pair of electrodes mounted at a fixed distance of 2 mm placed between fibers in the third left parasternal intercostal muscle about 2 cm from the edge of the sternum. The crural diaphragm EMG was obtained via fine-wire electrodes sewn in-line with the muscle fibers and placed 10 mm apart. During the signal acquisition experiment, the dog breathed through a snout mask, and a pneumo-tachograph was used to measure airflow. Figures 1.9, 1.10, and 5.8 show samples of the crural EMG signal.

Although the EMG signal is commonly used in many physiological studies including analysis of respiration, the intricate variations in the signal are often not of interest. A measure of the total or integrated electrical activity, ideally reflecting the global activity in the pool of active motor units of the muscle, would serve the purposes of most analyses [26]. As the EMG signal is nonstationary, short-time measures are called for. The smoothed envelope of the EMG signal is commonly used under these circumstances.

Platt et al. observed that the filters commonly used for smoothing rectified EMG signals had poor high-frequency attenuation, resulting in noisy envelopes. They proposed a modified Bessel filter for application to the EMG signal after full-wave rectification; the filter severely attenuated frequencies beyond 20 Hz with gain < -70 dB, and yielded EMG envelopes that were much smoother than those given by other filters.

The EMG envelopes derived by Platt et al. agreed very well with the inspiratory airflow pattern. Figure 5.11 shows plots of the parasternal intercostal EMG signal over two breath cycles, the corresponding filtered envelope, and the airflow pattern. Figure 5.12 shows the correlation between the filtered EMG envelope amplitude and the airflow in liters per second. It is evident that the envelope extracted by this method is an excellent correlate of inspiratory airflow.

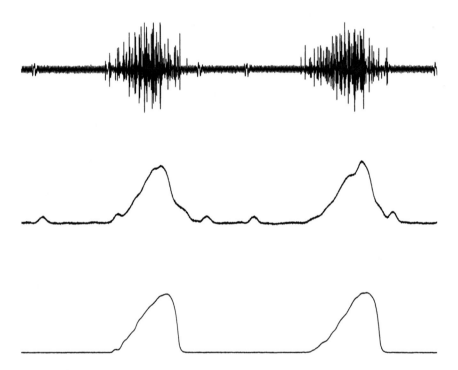

Figure 5.11   Top to bottom: EMG signal over two breath cycles from the parasternal inter-costal muscle of a dog recorded via implanted electrodes; EMG envelope obtained with the modified Bessel filter with a time constant of 100 ms; and inspiratory airflow. The duration of the signals plotted is 5 s. The several minor peaks appearing in the envelope are related to the ECG which appears as an artifact in the EMG signal. Data courtesy of R.S. Platt and P.A. Easton, Department of Clinical Neurosciences, University of Calgary [26].

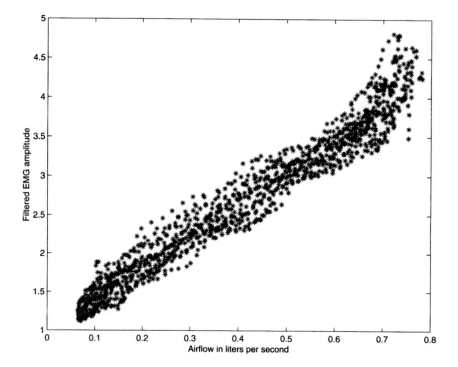

Figure 5.12   Correlation between EMG amplitude from Bessel-filtered envelope versus inspiratory airflow. The EMG envelope was filtered using a modified Bessel filter with a time constant of 100 ms. Data courtesy of R.S. Platt and P.A. Easton, Department of Clinical Neurosciences, University of Calgary [26].

5.10 APPLICATION: ELECTRICAL AND MECHANICAL CORRELATES OF MUSCULAR CONTRACTION

Problem: *Derive parameters from the electrical and mechanical manifestations of muscular activity that correlate with the level of contraction or force produced.*

Solution: Zhang et al. [47, 48] studied the usefulness of simultaneously recorded EMG and VMG signals in the analysis of muscular force produced by contraction. In their experimental procedure, the subjects performed isometric contraction (that is, with no movement of the associated leg) of the rectus femoris (thigh) muscle to different levels of torque with a Cybex II dynamometer. Four levels of contraction were performed from 20% to 80% of the maximal voluntary contraction (MVC) level of the individual subject. The experiments were performed at three knee-joint angles of 30°, 60°, and 90°. Each contraction was held for a duration of about 6 s, and the subjects rested in between experiments to prevent the development of muscle fatigue. The VMG signal was recorded using a Dytran 3115a accelerometer, and surface EMG signals were recorded using disposable $Ag - AgCl$ electrodes. The VMG signals were filtered to the bandwidth $3 - 100 \ Hz$ and the EMG signals were filtered to $10 - 300 \ Hz$. The VMG and EMG signals were sampled at $250 \ Hz$ and $1,000 \ Hz$, respectively. Figure 2.3 illustrates sample recordings of the VMG and EMG signals at two levels of contraction.

RMS values were computed for each contraction level over a duration of 5 s. Figure 5.13 shows the variation of the RMS values of the EMG and VMG signals acquired at a knee-joint angle of 60° and averaged over four subjects. The almost-linear trends of the RMS values of both the signals with muscular contraction indicate the usefulness of the parameter in the analysis of muscular activity. It should, however, be noted that the relationship between RMS values and contraction may not follow the same (linear) pattern for different muscles. Figure 5.14 shows the RMS versus %MVC relationships for three muscles: the relationship is linear for the first dorsal interosseus (FDI), whereas it is nonlinear for the biceps and deltoid muscles [150].

5.11 REMARKS

We have now reached the stage in our study where we can derive parameters from segments of biomedical signals. We focused our attention on characteristics that could be observed or derived in the time domain. The parameters considered were designed with the aim of discriminating between different types of waveshapes, or of representing change in waveform complexity through the course of a physiological or pathological process. We have seen how the various parameters explored in the present chapter can help in distinguishing between normal and ectopic ECG beats, and how certain measures can serve as correlates of physiological activity such as respiration.

It should be borne in mind that, in most practical applications, a single parameter or a couple of measures may not adequately serve the purposes of signal analysis or

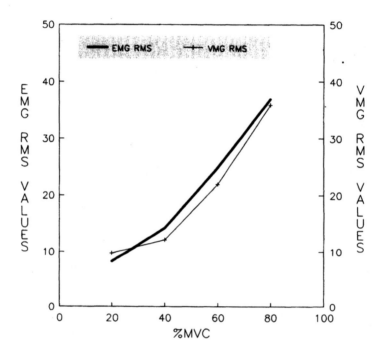

Figure 5.13 RMS values of the VMG and EMG signals for four levels of contraction of the rectus femoris muscle at **60°** knee-joint angle averaged over four subjects. Reproduced with permission from Y.T. Zhang, C.B. Frank, R.M. Rangayyan, and G.D. Bell, Relationships of the vibromyogram to the surface electromyogram of the human rectus femoris muscle during voluntary isometric contraction, *Journal of Rehabilitation Research and Development*, 33(4): 395–403, 1996. ©Department of Veterans Affairs.

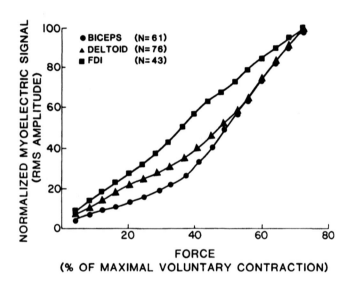

Figure 5.14 EMG RMS value versus level of muscle contraction expressed as a percentage of the maximal voluntary contraction level (%MVC) for each subject. The relationship is displayed for three muscles. FDI: first dorsal interosseus. N: number of muscles in the study. Reproduced with permission from J.H. Lawrence and C.J. de Luca, Myoelectric signal versus force relationship in different human muscles, *Journal of Applied Physiology,* 54(6):1653–1659, 1983. ©American Physiological Society.

diagnostic decision making. A single parameter such as the form factor or signal length may assist in distinguishing some types of PVCs from normal ECG beats; however, several cardiovascular diseases and defects may cause changes in the ECG signal that may lead to similar variations in the FF or SL values. A practical application would need to maintain a broad scope of analysis and use several parameters to detect various possible abnormalities. As always, an investigator should consider the possibility that a parameter observed to be useful in, say, ECG analysis in the time domain, may serve the needs in the analysis of some other signal, such as the PCG or EMG, in a different domain.

5.12 STUDY QUESTIONS AND PROBLEMS

1. Prove that the form factor FF of a sinusoidal wave is equal to unity.

2. The following discrete-time signals are defined over the interval 0 to 10 s with the sampling frequency being 1 Hz:

   - $x_1(n) = u(n) - u(n - 5)$.
   - $x_2(n) = 2u(n - 3) - 2u(n - 8)$.
   - $x_3(n) = u(n - 2) - u(n - 9)$.
   - $x_4(n) = u(n - 2) - u(n - 10)$.

   $u(n)$ is the discrete-time unit step function.

   The signal length SL of a signal $x(n)$ is defined as

   $$SL = \frac{\sum_{n=0}^{N-1} w(n)\, x^2(n)}{\sum_{n=0}^{N-1} x^2(n)},$$

   where $w(n)$ is a nondecreasing weighting function, and N is the number of samples in the signal. Let $w(n) = n$, $n = 0, 1, 2, \ldots, N - 1$.

   Draw sketches of each signal with the weighting function $w(n)$ superimposed. Compute the SL values for the four signals given. Interpret your results and compare the characteristics of the four signals in terms of their SL values.

3. You are given a signal with the samples $\{0, 0, 2, 2, 3, -3, 2, 0, 0\}$ and a template with the samples $\{1, -1\}$. Perform the template matching operation and derive the sample values for the output. Provide an interpretation of the result.

4. Discuss the similarities and differences between the problems of

   (i) detection of spike transients in EEG signals, and

   (ii) the detection of QRS complexes in ECG signals.

5. You have been hired to develop a heart-rate monitor for use in a coronary-care unit. Design a system to accept a patient's ECG signal, filter it to remove artifacts and noise, sample the signal, measure the heart rate, and set off alarms as appropriate. Provide a block diagram of the system, with details (in point form) of the signal processing steps to be performed in each block. Specify the important parameters for each processing step.

6. A needle EMG signal under low levels of muscle contraction was observed to contain a mixture of three trains of MUAPs. One of the trains contains quasi-periodic occurrences of a monophasic MUAP, the second contains occurrences of a biphasic MUAP, and the third contains occurrences of a triphasic MUAP. It was also observed that the MUAPs do not overlap in the EMG signal.

Propose a signal analysis procedure to:

(a) detect the occurrence (location in time) of each MUAP of each type individually, and

(b) determine the firing rate of each motor unit.

Note that each MUAP needs to be detected and labeled as being one of monophasic, biphasic, or triphasic type.

Your solution should include:

(i) plots of the EMG signal (make up one according to the description above) with labels for the components;

(ii) plots of the signal at various stages of your analysis procedure;

(iii) equations for important steps of your signal analysis procedure; and

(iv) point-form statements describing the reason or logic behind each step you propose.

7. A researcher is attempting to develop a digital signal processing system for the ac-quisition and analysis of heart sound signals (PCG signals). Assist the researcher in addressing the following concerns and problems:

(a) What are the typical bandwidths of normal PCG signals and those with murmurs? What is the recommended sampling frequency?

(b) What are the sources of artifacts that one has to consider in recording PCG signals? Name one physiological source and one other source, and recommend techniques to limit or eliminate both.

(c) How can one identify the locations of the first and second heart sounds (S1 and S2)? Which other biomedical signals would you recommend for assistance in this problem? Draw schematic diagrams of the signals and identify the corresponding cardiac events and timing relationships.

(d) Propose a technique to obtain the envelope of the PCG signal. List all steps of the method you propose and provide the required parameters.

(e) Draw schematic PCG signals and their envelopes over one cardiac cycle for a normal case, a case with systolic murmur, and a case with diastolic murmur. Identify each event in each case.

8. You are given a database of single-motor-unit action potentials (SMUAPs) containing several types of normal and abnormal patterns. Each signal record has one SMUAP. The patterns and features of interest are:

(i) Monophasic SMUAPs.

(ii) Biphasic SMUAPs.

(iii) Triphasic SMUAPs.

(iv) Polyphasic SMUAPs with more than three phases.

(a) Propose two parameters (computed features) to help in separating the four classes of SMUAPs. Give the required equations or procedures and explain their relationship to the signal characteristics described above. Describe conditions or preprocessing steps that are required in order for your methods to work well.

(b) Draw a schematic plot of the feature-vector space and demarcate regions where you expect features of the four SMUAP types to lie.

(c) State decision rules to classify the four SMUAP types using the two measures you propose.

9. Why is the ST segment of the ECG relevant in diagnosis? Recommend signal analysis techniques for the analysis of ST segment variations in clinical applications.

5.13 LABORATORY EXERCISES AND PROJECTS

Note: Data files related to the exercises are available at the site
ftp://ftp.ieee.org/uploads/press/rangayyan/

1. The signal in the file emg_dog2.dat was recorded from the crural diaphragm of a dog using fine-wire electrodes sewn in-line with the muscle fibers and placed 10 *mm* apart. The signal represents two cycles of breathing, and has been sampled at 10 *kHz*. (See also the file emg_dog2.m.)

   Write a MATLAB program to perform full-wave rectification (absolute value) or half-wave rectification (threshold at zero, with the mean value of the signal being zero). Apply a lowpass Butterworth filter of order eight and cutoff frequency in the range 10 to 20 *Hz* to the result. Analyze and evaluate the results with the two methods of rectification and at least two different lowpass cutoff frequencies. Compare the results with the envelope provided in the file emg_dog2_env.dat.

2. The root mean squared (RMS) value of a signal within a specific duration is related to the average power level of the signal. Write a MATLAB program to compute the RMS value at each instant for the EMG signal in the file emg_dog2.dat by using a causal short-time analysis window of duration in the range 50 − 150 *ms*. Use at least two different window durations and analyze the results. (See also the file emg_dog2.m.)

3. Develop a MATLAB program to compute the turns count in causal moving windows of duration in the range 50 − 150 *ms*. Apply the method to the EMG signal in the file emg_dog2.dat. (See also the file emg_dog2.m.) Study the results for different thresholds in the range 0 − 200 *μV*.

   Compare the envelope, RMS, and turns count curves in terms of their usefulness as representatives of inspiratory airflow (data provided in the file emg_dog2_flo.dat).

4. The file safety.wav contains the speech signal for the word "safety" uttered by a male speaker, sampled at 8 *kHz*. (See also the file safety.m.) The signal has a significant amount of background noise (as it was recorded in a normal computer laboratory). Develop procedures to derive short-time RMS, turns count, and ZCR in moving windows of duration in the range 10 − 100 *ms*. Study the variations in the parameters in relation to the voiced, unvoiced, and silence (background noise) portions of the signal.

   What do you expect the results to be if the procedures are applied to the first derivative of the signal? Confirm your assertions or expectations by performing the study.

5. Develop a program to derive the envelogram. Apply the procedure to the PCG signals in the files pec1.dat, pec33.dat, and pec52.dat. (See the file plotpec.m.)

   Extend the procedure to average the envelograms over several cardiac cycles using the ECG as the trigger. How will you handle the variations in the duration (number of samples) of the signals from one beat to another?

6. The ECG signal in the file ecgpvc.dat contains a large number of PVCs, including episodes of bigeminy. (See the file ecgpvc.m.) Apply the Pan-Tompkins procedure to detect and segment each beat. Label each beat as normal or PVC by visual inspection. Record the number of beats missed, if any, by your detection procedure.

   Compute the RR interval and the form factor FF for each beat. Use a duration of 80 samples (400 ms) spanning the QRS – T portion of each beat to compute FF. The P wave need not be considered in the present exercise.

   Compute the mean and standard deviation of the FF and RR values for the normal beats and the PVCs. Evaluate the variation of the two parameters between the two categories of beats.

6

Frequency-domain Characterization of Signals and Systems

Many biomedical systems exhibit innate rhythms and periodicity that is more readily expressed and appreciated in terms of frequency than time units. As a basic example, consider cardiac function: we express cardiac rhythm more conveniently in terms of *beats per minute* — a measure of the frequency of occurrence or the rate of repetition — than in terms of the duration of a beat or the interval between beats in seconds (the *RR* interval). A cardiac rhythm expressed as 72 *bpm* is more easily understood than a statement of the corresponding *RR* interval as 0.833 *s*. By the same token, the notion of an EEG rhythm is conveyed more readily by a description in *cycles per second* in lay terms, or in *Hertz (Hz)* in technical terms. Even engineers would find a frequency-domain expression easier to appreciate than a time-domain description, such as an alpha rhythm having a frequency of 11.5 *Hz* versus the equivalent period of 0.087 *s*.

When the signal being studied is made up of discrete (that is, separate and distinct) events in time, such as the ECG or a train of SMUAPs, the basic rhythm or rate of activity present in the signal can indeed be assessed directly in the time domain. On the other hand, signals such as the PCG display complex or complicated patterns in the time domain that do not facilitate ready appreciation of their frequency-domain characteristics; furthermore, the time-domain waveforms may differ from one occurrence of the signal (one heart beat) to another.

The PCG provides an interesting example of a signal with multiple frequency-domain features: in addition to the beat-to-beat periodicity or rhythm, the heart sounds within a cardiac cycle exhibit *resonance*. Due to the multi-compartmental nature of the cardiac system, we should expect heart sounds to possess multiple resonance frequencies: this leads to the need to describe the PCG, not only in terms of a rhythm (the heart rate) or a single resonance frequency, but also a composite *spectrum* of several dominant or resonance frequencies. Furthermore, constrained flow of blood through an orifice such as a septal defect or across a stenosed valve acting as a baffle could lead to turbulence, resulting in *wide-band noise*. In the case of noise-like murmurs, we would be able to identify neither rhythms nor resonance frequencies: the need arises to consider the distribution of the signal's energy or power over a wide band of frequencies, leading to the notion of the power *spectral density* function.

We have seen in Chapter 3 that it is often more convenient and meaningful to describe filters in terms of their frequency response — $H(z)$, $H(\omega)$, or $H(f)$ — than in terms of their impulse response $h(t)$ or the time-domain input – output relationship (difference equation). Furthermore, we saw in Section 4.4 that it is easier to interpret the PSDs of EEG waves than it is to interpret their theoretically equivalent ACFs. The Fourier and other similar transforms provide an invertible or reversible transformation from the time domain to the frequency domain (and vice-versa). Therefore, it may be argued that no new information is created by taking a given signal from the time domain to the frequency domain. However, the distribution of the energy or power of the signal in the frequency domain that is provided by the Fourier transform — the spectrum or PSD of the signal — facilitates better analysis and description of the frequency-domain characteristics of the signal. The PSD of a signal is not only useful in analyzing the signal, but also in designing amplifiers, filters, data-acquisition and transmission systems, and signal processing systems to treat the signal appropriately. We have seen in Section 3.5 that we need not only the signal PSD but also the noise PSD in order to be able to implement the optimal Wiener filter.

The treatment of biomedical signals as stochastic processes provides flexibility and a sense of generality in analysis, but imposes conditions and requirements in the estimation of their statistics including the ACF and PSD. In the present chapter, we shall investigate methods to estimate the PSD and frequency-domain parameters of biomedical signals and systems. We shall also study methods to derive spectral parameters that can characterize the given signal as well as the system that generated the signal. The motivation for the study, as always, shall be to distinguish between normal and abnormal signals or systems, and the potential use of the methods in diagnosis.

6.1 PROBLEM STATEMENT

Investigate the potential use of the Fourier spectrum and parameters derived thereof in the analysis of biomedical signals. Identify physiological and pathological processes that could modify the frequency content of the corresponding signals. Outline the signal processing tasks needed to perform spectral analysis of biomedical signals and systems.

As in the preceding chapters, the problem statement given above is generic, and represents the theme of the present chapter. The various signal analysis techniques described and the examples used for illustration in the following sections will address the points raised in the problem statement, with attention to specific problems and techniques.

6.2 ILLUSTRATION OF THE PROBLEM WITH CASE-STUDIES

6.2.1 The effect of myocardial elasticity on heart sound spectra

The first and second heart sounds — S1 and S2 — are typically composed of low-frequency components; this is to be expected due to the fluid-filled and elastic nature of the cardiohemic system. Sakai et al. [151] processed recorded heart sound signals by using tunable bandpass filters (with a bandwidth of 20 Hz, tuned over the range $20 - 40\ Hz$ to $400 - 420\ Hz$), and estimated the frequency distributions of S1 and S2. They found the heart sound spectra to be maximum in the $20 - 40\ Hz$ band; that S1 had a tendency to demonstrate peaks at lower frequencies than those of S2; and that S2 exhibited a "gentle peaking" between $60\ Hz$ and $220\ Hz$.

Gerbarg et al. [134, 135] developed a computer program to simulate a filter bank, and obtained averaged power spectra of S1 and S2 of $1,000$ adult males, 32 high-school children, and 75 patients in a hospital. The averaged PSDs of S1 and S2 obtained by them indicated peak power in the range $60 - 70\ Hz$, and relative power levels lower than $-10\ dB$ beyond $150\ Hz$. The PSD of S2 displayed slightly more high-frequency energy than that of S1.

Frome and Frederickson [152] applied the FFT to the analysis of first and second heart sounds. They described how segmented S1 and S2 data may be combined into a single complex signal, and how a single FFT may be used to obtain the FFTs of the two signals. Computer data processing techniques were described to obtain smoothed, averaged periodograms (described later in Section 6.4.1) of S1 and S2 separately.

Yoganathan et al. [153] applied the FFT for the analysis of S1 of 29 normal subjects. The FFT spectra of 250 ms windows containing S1 were averaged over 15 beats for each subject. It was found that the frequency spectrum of S1 contains peaks in a low-frequency range ($10 - 50\ Hz$) and a medium-frequency range ($50 - 140\ Hz$) [153]. In a similar study, the spectrum of S2 was observed to contain peaks in low-frequency ($10 - 80\ Hz$), medium-frequency ($80 - 220\ Hz$), and high-frequency ranges ($220 - 400\ Hz$) [154]. It has been suggested that the resonance peaks in the spectra may be related to the elastic properties of the heart muscles and the dynamic events causing the various components of S1 and S2 (see Section 1.2.8).

Adolph et al. [155] used a dynamic spectrum analyzer to study the frequency content of S1 during the iso-volumic contraction period. The center frequency of a filter with 20 Hz bandwidth was initially set to 30 Hz, and then varied in 10 Hz increments up to 70 Hz. The outputs of the filters were averaged over the same (prerecorded) 10 consecutive beats. Finally, the ratios of the average peak voltage of the filtered outputs to that of the total S1 signal during the iso-volumic contraction period were computed.

Adolph et al. hypothesized that the frequency content of S1 during the iso-volumic contraction period should depend on the relative contributions of the mass and elasticity of the left ventricle. The mass of the left ventricle with its blood content remains constant during iso-volumic contraction. Therefore, it was reasoned that the frequency content of S1 should decrease (that is, shift toward lower frequencies) in the case of diseases that reduce ventricular elasticity, such as myocardial infarction.

Figure 6.1 shows averaged S1 spectra for normal subjects and patients with acute or healed myocardial infarction; it is seen that the reduced elasticity due to myocardial infarction has reduced the relative content of power near 40 Hz. However, Adolph et al. also noted that an increase in ventricular mass as in the case of trained athletes, or a reduction in elasticity combined with an increase in the mass as in the case of myocardiopathy, could also cause a similar shift in the frequency content of S1. Regardless, they found that frequency analysis of S1 was of value in differentiating acute pulmonary embolism from acute myocardial infarction. Clarke et al. [156] also found reduction in the spectral energy of S1 to be a common accompaniment of myocardial ischemia.

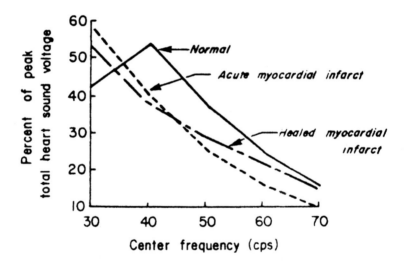

Figure 6.1   First heart sound spectra for normal, acute myocardial infarct, and healed my-ocardial infarct cases. The latter two cases exhibit an increased percentage of low-frequency components. Reproduced with permission from R.J. Adolph, J.F. Stephens, and K. Tanaka, The clinical value of frequency analysis of the first heart sound in myocardial infarction, *Circulation,* 41:1003–1014, 1970. ©American Heart Association.

6.2.2   Frequency analysis of murmurs to diagnose valvular defects

As we noted in Section 1.2.8, cardiovascular valvular defects and diseases cause high-frequency, noise-like sounds known as murmurs. Murmurs are often the only indicators of the early stages of certain cardiovascular diseases; prompt diagnosis could prevent further deterioration of the condition and possible complications.

We noted in Section 5.6.2 that zero-crossing analysis in the time domain was applied to assist in the detection of murmurs by Jacobs et al. [137] and Yokoi et al. [138]. Although ZCR increases with the presence of higher-frequency components, it does not yield a direct measure of the frequency content or the spectrum of the signal.

Application of electronic signal filtering techniques to analyze the frequency content of heart sounds and murmurs was initiated as early as the 1950s. Geckeler et al. [157] and McKusick et al. [158, 159] studied the applicability of the sound spectrograph for the analysis of heart sounds and murmurs. The sound spectrograph was developed in the late 1940s by Bell Telephone Laboratories as a tool to produce what was labeled as *visible speech*. The spectrograph used a bandpass filter (or a bank of bandpass filters) to determine the power of the given signal in each frequency band of interest. The signal was usually recorded and played back repeatedly as the center frequency of the bandpass filter was varied. The output was recorded on heat-sensitive or light-sensitive paper to produce a 2D distribution of frequency content of the signal at every instant of time as a gray-level image (essentially a *time-frequency distribution*, to be discussed in Section 8.4.1). Winer et al. [160] proposed iso-intensity contour plotting of the spectrogram instead of using variations in intensity (gray scale); they reported that, whereas normal heart sounds indicated the presence of regularity in the contours of equal intensity, abnormal sounds and murmurs produced irregular contour line structures with extensive "convolutions" and roughness. It was suggested that the cardio-spectrograms (or spectral phonocardiography) could provide physiologic and pathologic information beyond that provided by auscultation, without suffering from the psychoacoustic impediments that affected human observers.

Yoshimura et al. [161] used a tunable bandpass filter with low and high cutoff frequencies in the range $18 - 1,425\ Hz$ to process recorded PCG signals. They determined that the diastolic *rumble* of mitral stenosis occupied the range $20 - 200\ Hz$, whereas the diastolic *blow* of aortic regurgitation spanned a much higher frequency range of $200 - 1,600\ Hz$ (although the characteristic range was $400 - 800\ Hz$).

Gerbarg et al. [134, 135] developed a computer program to simulate a filter bank and obtain power spectra of heart sounds and murmurs, with the aim of developing a system for mass-screening to detect cardiovascular diseases. They argued that innocent (physiological) systolic murmur in children is limited to the first and middle thirds of the systolic interval between S1 and S2, whereas pathological systolic murmur due to mitral regurgitation is holo-systolic (spans the entire systolic period). Therefore, they computed ratios of the mean power of the *last third of systole* to the mean power of systole and also to a certain "standard" noise level. A ratio was also computed of the mean energy of systole to the mean energy of the PCG over the complete cardiac cycle. Gerbarg et al. obtained $78 - 91\%$ agreement of their computer classification based upon the three ratios defined above with clinical diagnosis of mitral regurgitation in different groups of subjects. Although they would not claim that a fully automated program for the diagnosis of mitral regurgitation had been developed, they indicated that the feasibility of computer-based diagnosis had been established, and that simulation of human auscultation had been partially achieved.

The specific problem of detection of the murmur due to aortic insufficiency in the presence of the murmur due to mitral stenosis was considered by van Vollenhoven et al. [162]. Aortic insufficiency causes an early diastolic murmur (with a blowing or hissing quality) that is best heard in the aortic area (second right-intercostal space, just right of the sternum), whereas the mid-diastolic rumbling murmur of mitral stenosis is best heard at the apex. A tunable bandpass filter with 50 Hz bandwidth and center frequency tunable in steps of 50 Hz was used by van Vollenhoven et al. to study the frequency content in a 100 ms window during the diastolic phase of recorded PCG signals. They found that the murmur of mitral stenosis was limited in frequency content to less than 400 Hz, whereas the murmur in the case of aortic insufficiency combined with mitral stenosis had more high-frequency energy in the range $300 - 1,000 \, Hz$.

Sarkady et al. [119] suggested synchronized averaging of the PSDs of PCG signals over several cardiac cycles computed using the FFT algorithm. Johnson et al. [163, 164] studied FFT-based PSDs of the systolic murmur due to aortic stenosis. They computed the PSDs of systolic windows of duration $86, 170$, and $341 \, ms$, and averaged the results over 10 cardiac cycles. Johnson et al. hypothesized that higher murmur frequencies are generated as the severity of aortic stenosis increases. In their study of patients who underwent catheterization and cardiac fluoroscopy, the trans-valvular systolic pressure gradient was measured during pull-back of the catheter from the left ventricle through the aortic valve, and found to be in the range $10 - 140 \, mm \, of \, Hg$. Spectral power ratios (described in Section 6.5.2) were computed considering the band $25 - 75 \, Hz$ as the constant area (CA) related to normal sounds and the band $75 - 150 \, Hz$ as the predictive area (PA) related to murmurs.

Figure 6.2 illustrates the PSDs of four patients with aortic stenosis of different levels of severity. The PSDs in the figure are segmented into the CA and PA parts as described above; the trans-valvular systolic pressure gradient (in $mm \, of \, Hg$) and the PA/CA spectral power ratio are also shown for each case. Johnson et al. found that the spectral power ratio increased linearly with the trans-valvular systolic pressure gradient, and hence correlated well with the severity of aortic stenosis. The importance of recording the PCG in the aortic area, pre-filtering the PCG to $25 - 1,500 \, Hz$, and the selection of an appropriate systolic murmur window was discussed by Johnson et al. Although there were confounding factors, it was indicated that the noninvasive PCG-based technique could be useful in identifying the need for catheterization as well as follow-up of patients with aortic stenosis.

6.3 THE FOURIER SPECTRUM

The Fourier transform is the most commonly used transform to study the frequency-domain characteristics of signals [1, 2, 14, 86]. This is mainly because the Fourier transform uses sinusoidal functions as its basis functions. Projections are computed of the given signal $x(t)$ onto the complex exponential basis function of frequency ω

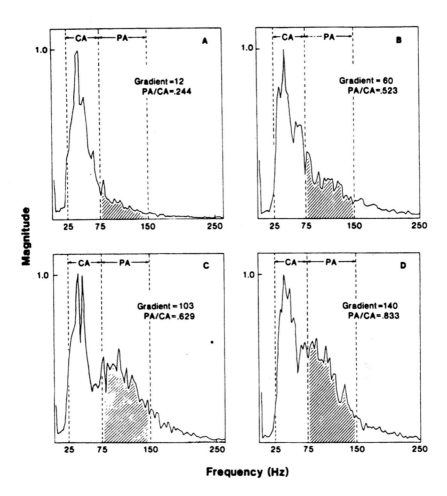

Figure 6.2 Averaged and normalized PSDs of four patients with aortic stenosis of different levels of severity. Each PSD is segmented into two parts: a constant area CA and a predictive area PA. The trans-valvular systolic pressure gradient (measured via catheterization in $mm\ of\ Hg$) and the PA/CA spectral power ratio are shown for each case. Reproduced with permission from the American College of Cardiology: G.R. Johnson, R.J. Adolph, and D.J. Campbell, Estimation of the severity of aortic valve stenosis by frequency analysis of the murmur, *Journal of the American College of Cardiology,* 1(5):1315–1323, 1983 ©Elsevier Science.

radians/s, given by $\exp(j\omega t) = \cos(\omega t) + j\,\sin(\omega t)$, as

$$X(\omega) = \int_{-\infty}^{\infty} x(t)\ \exp(-j\omega t)\ dt, \tag{6.1}$$

or in the frequency variable f in Hz as

$$X(f) = \int_{-\infty}^{\infty} x(t)\ \exp(-j2\pi ft)\ dt. \tag{6.2}$$

(The complex exponential function is conjugated in computing the projection. In some fields, the forward Fourier transform is defined with $\exp(+j\omega t)$ in the integral.) The above equations represent *analysis* of the signal $x(t)$ with reference to the complex exponential basis functions. The lower limit of the integral will be 0 if the signal is causal; the upper limit will be equal to the duration of the signal in the case of a finite-duration signal. The value of $X(\omega)$ or $X(f)$ at each frequency of interest $\omega = 2\pi f$ represents the "amount" of the corresponding cosine and sine functions present in the signal $x(t)$. Note that, in general, $X(\omega)$ is complex for real signals, and includes the magnitude and phase of the corresponding complex exponential.

The inverse transformation, representing *synthesis* of the signal $x(t)$ as a weighted combination of the complex exponential basis functions, is given as

$$x(t) = \frac{1}{2\pi} \int_{-\infty}^{\infty} X(\omega)\ \exp(j\omega t)\ d\omega = \int_{-\infty}^{\infty} X(f)\ \exp(j2\pi ft)\ df. \tag{6.3}$$

The second version of the above equation with the frequency variable f in Hz may be more convenient in some situations than the first one with ω in *radians/s*, due to the absence of the $\frac{1}{2\pi}$ factor. (If the forward Fourier transform is defined with $\exp(+j\omega t)$, the inverse Fourier transform will have $\exp(-j\omega t)$ in the integral; this distinction is not significant.)

In the case of a discrete-time signal $x(n)$, we may still compute the Fourier transform with a continuous frequency variable ω as

$$X(\omega) = \sum_{n=-\infty}^{\infty} x(n)\ \exp(-j\omega n), \tag{6.4}$$

with the normalized-frequency range $0 \le \omega \le 2\pi$ (equivalent to $0 \le f \le 1$). The lower limit of the summation will be 0 if the signal is causal. The upper limit of the summation will be equal to the index $(N-1)$ of the last sample in the case of a finite-duration signal with N samples. The frequency variable ω may also be defined over the range $0 \le \omega \le \omega_s$ (equivalent to $0 \le f \le f_s$), in which case n in the above equation should be multiplied by the sampling interval T in seconds. The Fourier transform is equivalent to the z-transform evaluated on the unit circle with $z = \exp(j\omega)$. Note that the Fourier transform of a discrete-time (sampled) signal is periodic with the period equal to the sampling frequency ω_s or 2π on the normalized-frequency scale.

When processing digital signals on a computer, the frequency variable ω will also have to be sampled, as $\omega = 2\pi \frac{f_s}{N} k$, or in the case of normalized frequency as $\omega = \frac{2\pi}{N} k$, where k is the frequency sample index and N is the number of samples to be used over one period of the periodic spectrum $X(\omega)$. Then, we have the DFT (analysis) relationship

$$X(k) = \sum_{n=0}^{N-1} x(n) \, \exp\left(-j\frac{2\pi}{N}kn\right), \quad k = 0, 1, 2, \ldots, N-1. \tag{6.5}$$

In the above equation, it is assumed that the given signal has N samples; it may be shown that the Fourier transform of a discrete-time signal with N samples is completely determined by N samples of its Fourier transform equally spaced around the unit circle in the z-plane [86]. The inverse DFT (synthesis) relationship is given by the expression

$$x(n) = \frac{1}{N} \sum_{k=0}^{N-1} X(k) \, \exp\left(j\frac{2\pi}{N}kn\right), \quad n = 0, 1, 2, \ldots, N-1. \tag{6.6}$$

Sampling the frequency variable causes the signal to become periodic in the time domain. The equations above define the forward and inverse DFTs over one period.

Note that

$$\exp\left(j\frac{2\pi}{N}kn\right) = \cos\left(\frac{2\pi}{N}kn\right) + j \, \sin\left(\frac{2\pi}{N}kn\right) \tag{6.7}$$

represents the sine and cosine functions of normalized frequency $f = \frac{1}{N}k$, $k = 0, 1, 2, \ldots, N-1$. The normalized frequency lies in the range $0 \leq f \leq 1$, and may be converted to the real frequency in Hz by multiplication with the sampling frequency f_s Hz. Equation 6.5 represents the dot product or projection of the given signal $x(n)$ onto each complex exponential or sinusoid $\exp(j\frac{2\pi}{N}kn)$ (conjugated). Equation 6.6 represents synthesis of the signal $x(n)$ as a linear, weighted combination of the complex exponential basis functions, the weights being the DFT coefficients $X(k)$.

Several important properties of the DFT and their implications are listed below [1, 2, 14, 86].

- A signal $x(n)$ and its DFT $X(k)$ are both periodic sequences.

- If a signal $x(n)$ has N samples, its DFT $X(k)$ must be computed with at least N samples equally spaced over the normalized-frequency range $0 \leq \omega \leq 2\pi$ (or, equivalently, around the unit circle in the z-plane) for complete representation and determination of $X(\omega)$, and hence exact reconstruction of $x(n)$ via the inverse DFT of $X(k)$. Of course, one may use more than N samples to compute $X(k)$ in order to employ an FFT algorithm with $L = 2^M \geq N$ samples, where M is an integer, or to obtain $X(\omega)$ with finer frequency sampling than $\frac{2\pi}{N}$.

- The DFT is linear: the DFT of $ax(n) + by(n)$ is $aX(k) + bY(k)$, where $X(k)$ and $Y(k)$ are the DFTs of $x(n)$ and $y(n)$, respectively.

- The DFT of $x(n - n_o)$ is $\exp(-j\frac{2\pi}{N}kn_o)X(k)$, where $X(k)$ is the DFT of $x(n)$. A time shift leads to a linear component being added to the phase of the original signal. As all sequences in DFT relationships are periodic, the shift operation should be defined as a circular or periodic shift. If at least n_o zeros are present or are padded at the end of the signal before the shift operation, a circular shift will be equivalent to a linear shift.

- The DFT of $x(n) * h(n)$ is $X(k)H(k)$, where $X(k)$ and $H(k)$ are the DFTs of $x(n)$ and $h(n)$, respectively. The inverse DFT of $X(k)H(k)$ is $x(n) * h(n)$. Similarly, $x(n)h(n)$ and $X(k) * H(k)$ form a DFT pair. Convolution in one domain is equivalent to multiplication in the other. It is necessary for all the signals in the above relationships to have the same number of samples N.

  As all sequences in DFT relationships are periodic, the convolution operations in the above relationships are *periodic convolution* and not linear convolution. Note that circular or periodic convolution is defined for periodic signals having the same period, and that the result will also be periodic with the same period as that of the individual input signals.

  The result of linear convolution of two signals $x(n)$ and $h(n)$ with different durations N_x and N_h samples, respectively, will have a duration of $N_x + N_h - 1$ samples. If linear convolution is desired via the inverse DFT of $X(k)H(k)$, the DFTs must be computed with $L \geq N_x + N_h - 1$ samples. The individual signals should be padded with zeros at the end to make their effective durations equal for the sake of DFT computation and multiplication. All signals and their DFTs are then periodic with the augmented period of L samples.

- The DFT of a real signal $x(n)$ will possess conjugate symmetry, that is, $X(-k) = X^*(k)$. As a consequence, the real part and the magnitude of $X(k)$ will be even sequences, whereas the imaginary part and the phase of $X(k)$ will be odd sequences.

- According to Parseval's theorem, the total energy of the signal must remain the same before and after Fourier transformation. We then have the following equalities:

$$\int_{-\infty}^{\infty} |x(t)|^2 \, dt \;=\; \frac{1}{2\pi} \int_{-\infty}^{\infty} |X(\omega)|^2 \, d\omega, \tag{6.8}$$

$$\sum_{n=-\infty}^{\infty} |x(n)|^2 \;=\; \frac{1}{2\pi} \int_{-\pi}^{\pi} |X(\omega)|^2 \, d\omega,$$

$$\sum_{n=0}^{N-1} |x(n)|^2 \;=\; \frac{1}{N} \sum_{k=0}^{N-1} |X(k)|^2.$$

Since the integral of $|X(\omega)|^2$ over all ω or the sum of $|X(k)|^2$ over all k represents the total energy of the signal (or average power, if the quantity is divided by the duration of the signal), $|X(\omega)|^2$ and $|X(k)|^2$ represent the spread or *density* of the power of the signal along the frequency axis.

6.4 ESTIMATION OF THE POWER SPECTRAL DENSITY FUNCTION

We have already encountered the ACF and CCF in Equations 3.9, 3.12, and 4.24: the first two equations cited provided a general definition of the ACF as a statistical expectation or an integral over a duration tending to ∞; the third treated the CCF as the projection of one signal onto another and neglected a scale factor that was of no consequence in the application. We shall now investigate more closely the procedures required to estimate the ACF, and hence the PSD, from finite-length signal records.

Let us consider a signal record of N samples: $x(n)$, $n = 0, 1, 2, \ldots, N - 1$. In order to compute the time-averaged ACF $\phi_{xx}(m)$ for a delay of m samples, we need to form the product $x(n)x(n \pm m)$ and sum over the available range of data samples. The true ACF is given as $\phi_{xx}(m) = E[x(n)x(n + m)]$. Note that one of the copies of the signal entering the computation of the ACF should be conjugated if the signal is complex.

It is readily seen that we may sum from $n = 0$ to $n = N - 1$ when computing $\phi_{xx}(0)$ with $x(n)x(n) = x^2(n)$. However, when computing $\phi_{xx}(1)$ with $x(n)x(n + 1)$, we can only sum from $n = 0$ to $n = N - 2$. As we apply a linear shift of m samples to one copy of the signal to compute $\phi_{xx}(\pm m)$, m samples of one of the copies of the signal drop out of the window of analysis indicated by the overlap between the two copies of the signal. Therefore, only $N - |m|$ pairs of data samples are available to estimate the ACF for the delay of $\pm m$ samples. We then have a sample-mean estimate of the ACF given by

$$\phi_1(m) = \frac{1}{N - |m|} \sum_{n=0}^{N-|m|-1} x(n)x(n + m). \tag{6.9}$$

The subscript xx has been omitted in the above equation; the subscript 1 indicates the use of one type of averaging scale factor in estimating the ACF. Oppenheim and Schafer [86] show that $\phi_1(m)$ is a consistent estimate of $\phi_{xx}(m)$: it has zero bias and has a variance that tends to zero as $N \to \infty$. However, the variance of the estimate becomes exceptionally large as m approaches N: very few non-zero pairs of samples are then available to compute the ACF, and the estimate is useless.

An alternative definition of the ACF ignores the lack of $|m|$ non-zero pairs of samples, and applies the same scale factor for all delays, leading to

$$\phi_2(m) = \frac{1}{N} \sum_{n=0}^{N-|m|-1} x(n)x(n + m). \tag{6.10}$$

Note that the upper limit of summation in the above expression could be stated as $N - 1$ with no effect on the result; the first or the last $|m|$ samples of $x(n)$ will not overlap with $x(n+m)$, and result in zero product terms. Oppenheim and Schafer [86] show that $\phi_2(m)$ has a bias equal to $\frac{m}{N}\phi_{xx}(m)$: the bias tends to the actual value being estimated as m approaches N, although the variance is almost independent of m and tends to zero as $N \to \infty$. Regardless, both the ACF estimates are asymptotically unbiased (the bias of $\phi_2(m)$ tends to zero as $N \to \infty$), and yield good estimates of the ACF as long as the number of samples N is large and $m << N$.

Note that the two ACF estimates $\phi_1(m)$ and $\phi_2(m)$ are inter-related as

$$\phi_2(m) = \frac{N - |m|}{N} \phi_1(m). \tag{6.11}$$

Thus $\phi_2(m)$ is a scaled version of $\phi_1(m)$. However, since the scaling factor is a function of m, it is more commonly referred to as a *window*; more discussion on this interpretation will be presented in Section 6.4.1. It should also be observed that the distinction between $\phi_1(m)$ and $\phi_2(m)$ is comparable to that between the unbiased and biased sample variance measures, where the division is by $N - 1$ or N, respectively, with N being the number of samples available.

6.4.1   The periodogram

Since the PSD and the ACF are a Fourier transform pair, we may compute an estimate of the PSD as

$$S_2(\omega) = \sum_{m=-(N-1)}^{N-1} \phi_2(m) \exp(-j\omega m), \tag{6.12}$$

assuming that, indeed, the ACF is computed or available for $|m|$ up to $N - 1$. The Fourier transform of the signal $x(n)$, $n = 0, 1, 2, \ldots, N - 1$, is given as

$$X(\omega) = \sum_{n=0}^{N-1} x(n) \exp(-j\omega n). \tag{6.13}$$

It can be shown that

$$S_2(\omega) = \frac{1}{N}|X(\omega)|^2. \tag{6.14}$$

The PSD estimate $S_2(\omega)$ is known as the *periodogram* of the signal $x(n)$ [86]. Oppenheim and Schafer [86] show that $S_2(\omega)$ is a biased estimate of the PSD, with

$$E[S_2(\omega)] = \sum_{m=-(N-1)}^{N-1} \frac{N - |m|}{N} \phi_{xx}(m) \exp(-j\omega m). \tag{6.15}$$

If we consider the Fourier transform of $\phi_1(m)$, we get a different estimate of the PSD as

$$S_1(\omega) = \sum_{m=-(N-1)}^{N-1} \phi_1(m) \exp(-j\omega m), \tag{6.16}$$

with the expected value [86]

$$E[S_1(\omega)] = \sum_{m=-(N-1)}^{N-1} \phi_{xx}(m) \exp(-j\omega m). \qquad (6.17)$$

Because of the finite limits of the summation, $S_1(\omega)$ is a biased estimate of the PSD.

The two estimates $S_2(\omega)$ and $S_1(\omega)$ may be seen as the Fourier transforms of windowed ACFs, with the window functions being a triangular function — known as the Bartlett window, $w_B(m)$ — in the first case, and a rectangular window $w_R(m)$ in the second case:

$$w_B(m) = \begin{cases} \frac{N-|m|}{N}, & |m| < N \\ 0, & \text{otherwise} \end{cases}, \qquad (6.18)$$

$$w_R(m) = \begin{cases} 1 & |m| < N \\ 0, & \text{otherwise} \end{cases}. \qquad (6.19)$$

Note that the windows defined above have a (non-zero) duration of $(2N-1)$ samples.

Since the ACF is multiplied with the window function, the PSD is convolved with the Fourier transform of the window function, leading to spectral leakage and loss of resolution (more details on windows will follow in Section 6.4.3). The Fourier transforms of the Bartlett and rectangular windows are, respectively [86],

$$W_B(\omega) = \frac{1}{N} \left[\frac{\sin(\omega N/2)}{\sin(\omega/2)} \right]^2, \qquad (6.20)$$

and

$$W_R(\omega) = \frac{\sin[\omega(2N-1)/2]}{\sin(\omega/2)}. \qquad (6.21)$$

Oppenheim and Schafer [86] show that the periodogram has a variance that does not approach zero as $N \to \infty$; instead, the variance of the periodogram is of the order of σ_x^4 regardless of N. Thus the periodogram is not a consistent estimate of the PSD.

6.4.2 The need for averaging

A common approach to reduce the variance of an estimate is to average over a number of statistically independent estimates. We have seen in Section 3.3.1 how the variance of the noise in noisy signals may be reduced by synchronized averaging over a number of observations of the corrupted signal. In a similar vein, a number of periodograms may be computed over multiple observations of a signal and averaged to obtain a better estimate of the PSD. It is necessary for the process to be stationary, at least during the period over which the periodograms are computed and averaged.

   Problem: *How can we obtain an averaged periodogram when we are given only one signal record of finite duration?*

Solution: Oppenheim and Schafer [86] describe the following procedure, attributed to Bartlett, to average periodograms of segments of the given signal record:

1. Divide the given data sequence $x(n)$, $n = 0, 1, 2, \ldots, N - 1$, into K segments of M samples each. We then have the segments given by

$$x_i(n) = x(n + (i - 1)M), \quad 0 \le n \le M - 1, \quad 1 \le i \le K. \tag{6.22}$$

2. Compute the periodogram of each segment as

$$S_i(\omega) = \frac{1}{M} \left| \sum_{n=0}^{M-1} x_i(n) \exp(-j\omega n) \right|^2, \quad 1 \le i \le K. \tag{6.23}$$

The Fourier transform in the above equation is evaluated as a DFT (using the FFT algorithm) in practice.

3. If the ACF $\phi_{xx}(m)$ is negligible for $|m| > M$, the periodograms of the K segments of duration M samples each may be assumed to be mutually independent. Then, the Bartlett estimate $S_B(\omega)$ of the PSD is obtained as the sample mean of the K independent observations of the periodogram:

$$S_B(\omega) = \frac{1}{K} \sum_{i=1}^{K} S_i(\omega). \tag{6.24}$$

Oppenheim and Schafer [86] show that the expected value of the Bartlett estimate $S_B(\omega)$ is the convolution of the true PSD $S_{xx}(\omega)$ with the Fourier transform of the Bartlett window given in Equation 6.20 (with N replaced by M). The convolution relationship indicates the bias in the estimate, and has the effect of spectral smearing and leakage; the bias may therefore be interpreted as a loss in resolution. Although $S_B(\omega)$ is a biased estimate, its variance tends to zero as the number of segments K increases. Therefore, it is a consistent estimate.

When we have a (stationary) signal of fixed duration of N samples, we will face limitations on the number of segments K that we may obtain. While the variance of the estimate decreases as K is increased, it should be recognized that there is a concomitant decrease in the number of samples M per segment. As M decreases, the main lobe of the Fourier transform of the Bartlett window (see Equation 6.20) widens; frequency resolution is lost because the estimate is the convolution of the true PSD with the window's frequency response. An illustration of application of the Bartlett procedure will be provided at the end of Section 6.4.3.

Cyclo-stationary signals such as the PCG offer a unique and interesting approach to synchronized averaging of periodograms over a number of cycles, without the trade-off between the reduction of variance and the loss of resolution imposed by segmentation as described above. This is presented as an illustration of application in Section 6.4.5.

6.4.3 The use of windows: Spectral resolution and leakage

The Bartlett procedure may be viewed as an ensemble averaging approach to reduce the variance (which may be interpreted as noise) of the periodogram. Another approach to obtain a smooth spectrum is to convolve the periodogram $S(\omega)$ with a filter or smoothing function $W(\omega)$ in the frequency domain (similar to the use of an MA filter in the time domain). The smoothed estimate $S_s(\omega)$ is given by

$$S_s(\omega) = \frac{1}{2\pi} \int_{-\pi}^{\pi} S(\nu)\, W(\omega - \nu)\, d\nu, \qquad (6.25)$$

where ν is a temporary variable for integration.

As the PSD is a nonnegative function, the smoothing function $W(\omega)$ should satisfy $W(\omega) \geq 0$, $-\pi \leq \omega \leq \pi$. The Fourier transform of the Bartlett window $W_B(\omega)$ meets this requirement. Oppenheim and Schafer [86] show that the variance of the smoothed periodogram is reduced approximately by the factor

$$\frac{1}{N} \sum_{m=-(M-1)}^{M-1} w^2(m) = \frac{1}{2\pi N} \int_{-\pi}^{\pi} W^2(\omega)\, d\omega, \qquad (6.26)$$

with reference to the variance of the original periodogram; here N is the total number of samples in the signal and $(2M - 1)$ is the number of samples in the smoothing window function. A rectangular window offers a variance-reduction factor of approximately $\frac{2M}{N}$, whereas the factor for the Bartlett window is $\frac{2M}{3N}$ [86]. It should be noted that smoothing of the spectrum (reduction of variance) is achieved at the price of loss of frequency resolution.

Since the periodogram is the Fourier transform of the ACF estimate $\phi(m)$, the convolution operation in the frequency domain in Equation 6.25 is equivalent to multiplying $\phi(m)$ with $w(m)$, the inverse Fourier transform of $W(\omega)$. This result suggests that the same PSD estimate as $S_s(\omega)$ may be obtained by applying a window to the ACF estimate and then taking the Fourier transform of the result. As the ACF is an even function, the window should also be even.

Based upon the arguments outlined above, Welch [165] (see also Oppenheim and Schafer [86]) proposed a method to average modified periodograms. In Welch's procedure, the given signal is segmented as in the Bartlett procedure, but a window is applied directly to the original signal segments before Fourier transformation. The periodograms of the windowed segments are defined as

$$S_{Wi}(\omega) = \frac{1}{ME_w} \left| \sum_{n=0}^{M-1} x_i(n) w(n) \exp(-j\omega n) \right|^2, \quad i = 1, 2, \dots, K, \qquad (6.27)$$

where E_w is the average power of the window given by

$$E_w = \frac{1}{M} \sum_{n=0}^{M-1} w^2(n). \qquad (6.28)$$

Note that the duration of the window is now M samples. The Welch PSD estimate $S_W(\omega)$ is obtained by averaging the modified periodograms as

$$S_W(\omega) = \frac{1}{K} \sum_{i=1}^{K} S_{Wi}(\omega). \tag{6.29}$$

Welch [165] showed that, if the segments are not overlapping, the variance of the averaged modified periodogram is inversely proportional to K, the number of segments used. Welch also suggested that the segments may be allowed to overlap, in which case the modified periodograms are not mutually independent. The spectral window that is effectively convolved with the PSD in the frequency domain is proportional to the squared magnitude of the Fourier transform of the time-domain data window applied. Therefore, no matter which type of a data window is used, the spectral smoothing function is nonnegative, thereby guaranteeing that the PSD estimate will be nonnegative as well.

Some of the commonly used data windows are defined below [86, 166]; the windows are of length N samples and causal, defined for $0 \leq n \leq N - 1$.

Rectangular:

$$w(n) = 1. \tag{6.30}$$

Bartlett (triangular):

$$w(n) = \begin{cases} \frac{2n}{N-1}, & 0 \leq n \leq \frac{N-1}{2}, \\ 2 - \frac{2n}{N-1}, & \frac{N-1}{2} \leq n \leq N - 1. \end{cases} \tag{6.31}$$

Hamming:

$$w(n) = 0.54 - 0.46 \cos\left(\frac{2\pi n}{N-1}\right). \tag{6.32}$$

Hanning (von Hann):

$$w(n) = \frac{1}{2}\left[1 - \cos\left(\frac{2\pi n}{N-1}\right)\right]. \tag{6.33}$$

Blackman:

$$w(n) = 0.42 - 0.5 \cos\left(\frac{2\pi n}{N-1}\right) + 0.08 \cos\left(\frac{4\pi n}{N-1}\right). \tag{6.34}$$

Figure 6.3 illustrates the rectangular, Bartlett, Hanning, and Hamming windows with $N = 256$ samples. A Hanning window with $N = 128$ samples is also illustrated (centered with reference to the longer-duration windows).

Use of the tapered windows (all of the above, except the rectangular window) provides the advantage that the ends of the given signal are reduced to zero (with the further exception of the Hamming window, for which the end-values are not zero but 0.08). This feature means that there are no discontinuities in the periodic version of

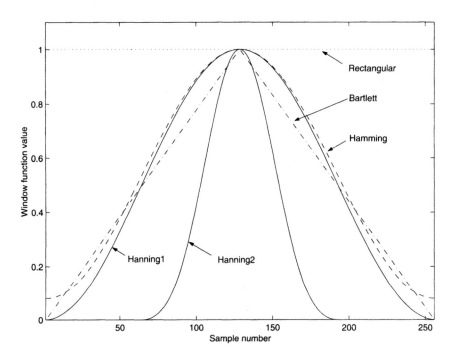

Figure 6.3 Commonly used window functions: rectangular, Bartlett, Hamming, and Hanning windows with $N = 256$ (Hanning1), and Hanning window with $N = 128$ samples (Hanning2). All windows are centered at the 128th sample.

the signal encountered in DFT-based procedures. All of the window functions listed above are symmetric (even) functions, and therefore have a linear phase (or a real spectrum with zero phase if the window is centered at the origin).

Figures 6.4 to 6.8 illustrate the log-magnitude frequency responses of the window functions shown in Figure 6.3. The frequency responses were computed after padding the windows to a total duration of $L = 2,048$ samples for FFT computation. The plots are on an expanded scale over the limited normalized-frequency range of $(0, 0.1)$ in order to illustrate clearly the characteristics of the main-lobe and the side-lobes. The discontinuities in the frequency responses of the rectangular and Bartlett windows in Figures 6.4 and 6.5 are due to the log of the zeros of the responses being $-\infty$.

The rectangular window has the narrowest main lobe of width $\frac{4\pi}{N}$; the main lobe is wider at $\frac{8\pi}{N}$ for the Bartlett, Hanning, and Hamming windows; it is the widest at $\frac{12\pi}{N}$ for the Blackman window [86]. A reduction in window width will lead to an increase in the main-lobe width, as illustrated by the frequency responses of the two Hanning windows in Figures 6.7 and 6.8. Note that the wider the main lobe, the greater is the spectral smoothing, and hence the loss of spectral resolution is more severe.

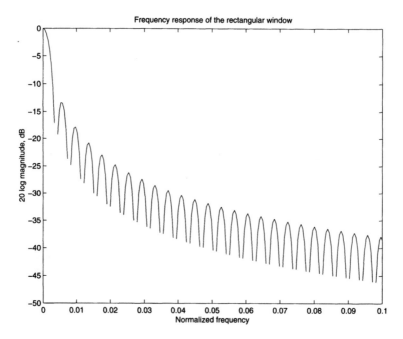

Figure 6.4   Log-magnitude frequency response of the rectangular window illustrated in Figure 6.3. The window width is $N = 256$ samples.

The rectangular window has the highest peak side-lobe levels of all of the windows listed at -13 *dB*, with the Bartlett, Hamming, Hanning, and Blackman windows having their peak side-lobe levels at -25 *dB*, -31 *dB*, -41 *dB*, and -57 *dB*, respectively [86]. Higher levels of the side-lobes will cause increased spectral leakage (weighted summation of spectral components with significant weights over a wide

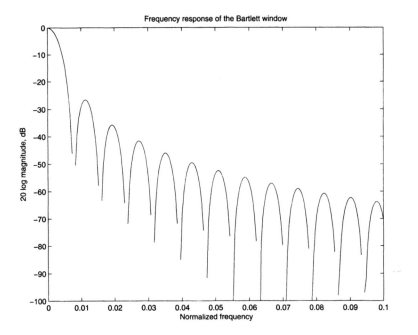

Figure 6.5 Log-magnitude frequency response of the Bartlett window illustrated in Figure 6.3. The window width is $N = 256$ samples.

range of frequencies due to convolution in the frequency domain), resulting in a more distorted spectrum. Note that reduction of leakage through the use of the tapered windows comes at the price of increased main-lobe width, and therefore more severe loss of spectral resolution (more smoothing).

Illustration of application: The Welch method of windowing signal segments and averaging their PSDs was applied to the o2 channel of the EEG signal illustrated in Figure 1.22. The number of samples in the signal is $N = 750$, with the sampling frequency being $f_s = 100 \ Hz$. Note that the specific EEG signal record may be assumed to be stationary over its relatively short duration of 7.5 s. The dominant activity in the signal is the alpha rhythm, which appears throughout the duration of the signal record.

The PSD of the entire signal was first computed using no window (that is, the rectangular window was applied implicitly); the FFT array was computed with $L = 1,024$ samples. The top trace in Figure 6.9 illustrates the PSD of the signal.

For the first averaged periodogram procedure, the EEG signal was segmented with $M = 64$ samples each, with implicit usage of the rectangular window (equivalent to the Bartlett method). A total of $K = 11$ segments were obtained. Each segment was padded with zeros to a length of $L = 1,024$ for the sake of FFT computation. The PSDs of the segments were then averaged, followed by normalization and logarithmic transformation. The second and third plots in Figure 6.9 illustrate the PSD of a sample segment (the 11th segment) and the averaged PSD (the Bartlett estimate), respectively.

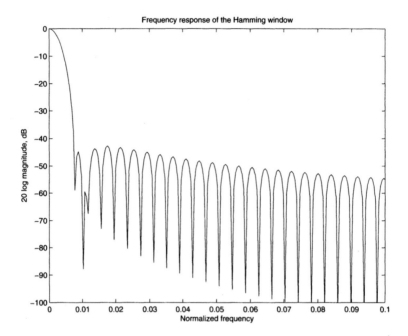

Figure 6.6   Log-magnitude frequency response of the Hamming window illustrated in Figure 6.3. The window width is $N = 256$ samples.

It is seen that the averaged PSD (third trace) provides a smooth spectral estimate with a clearly dominant peak at approximately 10 Hz, representing the alpha rhythm present in the signal. The PSD of the individual segment (middle trace) displays many peaks and valleys that are possibly spurious and not significant, and have been suppressed or smoothed by the averaging process. The single PSD computed from the entire signal (top trace) exhibits numerous variations that may not be relevant and could confound visual or automated analysis. (*Note:* Direct comparison of the PSDs is possible since they have the same number of samples, that is, the same frequency sampling.)

Figure 6.10 illustrates a second set of PSDs similar to that in Figure 6.9, but with the usage of the Hanning window in the Welch procedure. The effect of the Hanning window is not significant in the case of the PSD of the entire signal (top trace), as the window length is reasonably large ($N = 750$). However, the Hanning window has clearly smoothed the multiple (possibly spurious) peaks and valleys in the PSD of the segment illustrated in the middle trace. The wider main-lobe of the Hanning window's frequency response has caused a more severe loss of frequency resolution (smoothing) than the rectangular window in the case of the corresponding PSD in Figure 6.9. Finally, the averaged PSD in the lowest trace of Figure 6.10 clearly illustrates the benefit of the Hanning window in the significantly reduced power levels beyond 30 Hz. The lower side-lobe levels of the Hanning window have resulted in less spectral leakage than in the case of the rectangular window as illustrated by

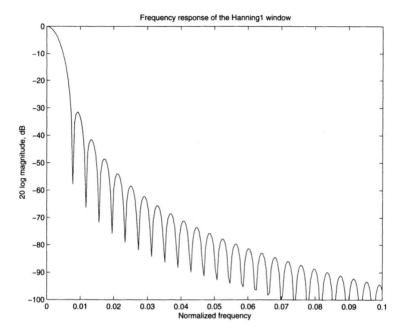

Figure 6.7 Log-magnitude frequency response of the Hanning1 window illustrated in Figure 6.3. The window width is $N = 256$ samples.

the corresponding PSD in Figure 6.9. The price paid, however, is evidenced by the wider peak in the averaged PSD with the Hanning window, which spans the range $5 - 15 \, Hz$ at the $-10 \, dB$ level. The two distinct peaks at about $10 \, Hz$ and $12 \, Hz$ that are evident in the top traces of Figures 6.9 and 6.10 as well as in the smoothed PSD in the bottom trace of Figure 6.9 are no longer seen separately in the bottom trace of Figure 6.10. Regardless, the averaged PSD with the Hanning window appears to be smoother and more amenable to analysis than the corresponding result with the rectangular window.

6.4.4 Estimation of the autocorrelation function

Good estimates of the ACF are required in applications such as the design of the optimal Wiener filter and estimation of the statistics of stochastic processes. Once a PSD estimate has been obtained by a method such as the Bartlett or Welch procedures, we may take the inverse Fourier transform of the result and use the result as an estimate of the ACF. We may also fit a smooth curve or a parametric model (Gaussian, Laplacian, etc.) to the PSD or to the equivalent ACF model.

Let us consider again the expression

$$\phi_2(m) = \frac{1}{N} \sum_{n=0}^{N-|m|-1} x(n)x(n+m). \tag{6.35}$$

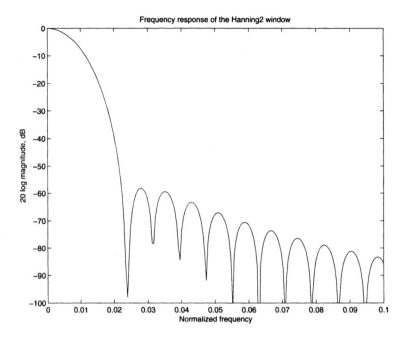

Figure 6.8 Log-magnitude frequency response of the Hanning2 window illustrated in Figure 6.3. The window width is $N = 128$ samples.

As the ACF is an even function, we need to compute it only for positive m. It is evident that the ACF estimate is simply the result of linear convolution of $x(n)$ with $x(-n)$ (with the scale factor $\frac{1}{N}$). If the DFT of $x(n)$ is $X(k)$, the DFT of $x(-n)$ is $X^*(k)$. Since convolution in the time domain is multiplication in the frequency domain, we could compute the DFT $X(k)$ of $x(n)$, obtain $X(k)X^*(k) = |X(k)|^2$, and take its inverse DFT. However, the DFT procedure provides circular convolution and not linear convolution. Therefore, we need to pad $x(n)$ with at least $M - 1$ zeros, where M is the largest lag for which the ACF is desired. The DFT must then be computed with at least $L = N + M - 1$ samples, where N is the number of samples in the original signal. If this requirement is built into the periodogram or averaged periodogram procedure, the inverse DFT of the final PSD estimate may be used as an estimate of the ACF (with the scale factor $\frac{1}{N}$, or division by $\phi_{xx}(0)$ to get the normalized ACF).

6.4.5 Synchronized averaging of PCG spectra

Every individual is familiar with the comforting *lub – dub* sounds of his or her heart beat; every prospective parent would have taken pleasure in listening to the throbbing heart of the yet-to-be-born baby. Use of the heart sounds is extremely common in clinical practice: the stethoscope is the most common sign and tool of a physician.

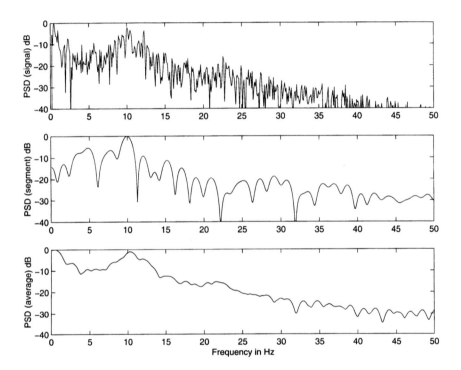

Figure 6.9 Bartlett PSD estimate of the o2 channel of the EEG signal in Figure 1.22. Top trace: PSD of the entire signal. Middle trace: PSD of the 11th segment. Bottom trace: Averaged PSD using $K = 11$ segments of the signal. The rectangular window was (implicitly) used in all cases. Number of samples in the entire signal: $N = 750$. Number of samples in each segment: $M = 64$. All FFT arrays were computed with $L = 1,024$ samples. Sampling frequency $f_s = 100 \ Hz$.

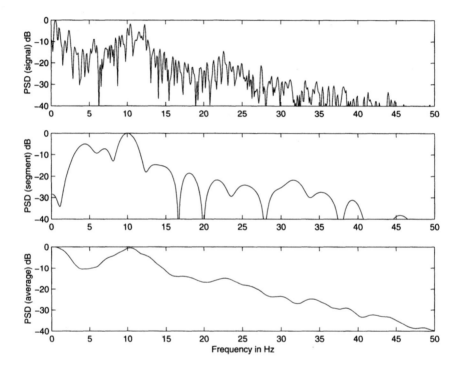

Figure 6.10 Welch PSD estimate of the o2 channel of the EEG signal in Figure 1.22. Top trace: PSD of the entire signal. Middle trace: PSD of the 11th segment. Bottom trace: Averaged PSD using $K = 11$ segments of the signal. The Hanning window was used in all cases. Number of samples in the entire signal and the size of the Hanning window used in computing the PSD of the entire signal: $N = 750$. Number of samples in each segment and the size of the Hanning window used in the averaged periodogram method: $M = 64$. All FFT arrays were computed with $L = 1,024$ samples. Sampling frequency $f_s = 100\ Hz$.

Yet, behind this common signal lie many sophisticated and potentially complicating characteristics.

The PCG is a nonstationary signal due to the fact that the amount of blood in each cardiac chamber and the state of contraction of the muscles change continually during each cardiac cycle. S2 usually has more high-frequency content than S1: the PSD of a normal PCG signal changes within about 300 ms. Valve opening or closing sounds, being of short duration of the order of 10 ms, are of a transient and high-frequency character. The presence of murmurs adds another dimension of nonstationarity, with frequency content well beyond that of the normal heart sounds: the PSD of an abnormal PCG could change every 100 ms or less. Individual epochs of S1, S2, valve snaps, and murmurs are of limited durations of the order of $10 - 300$ ms. These aspects of the PCG preclude segmented averaging as recommended by the Bartlett or Welch procedures.

Over and above all of the factors mentioned in the preceding paragraph, the transmission characteristics of the chest wall change during breathing. (Living systems are dynamic!) The PCG signals recorded at various locations on the chest are also subject to different transmission-path effects. While adult subjects may cooperate in PCG signal acquisition by holding their breath or performing other maneuvers, these possibilities cannot be considered in the case of infants and young children in poor states of health. The PCG signal presents more challenges in acquisition and analysis than most of the other biomedical signals we have encountered [40].

Problem: *Propose a method to obtain averaged PSD estimates of the systolic and diastolic heart sounds.*

Solution: The cyclo-stationarity of the PCG lends itself to a unique approach to averaging PCG segments corresponding to the same phase of the cardiac cycle extracted from multiple beats. If the subject were to hold his/her breath during the period of acquisition of the PCG record, the chest-wall transmission characteristics will be stationary over the multiple cardiac cycles in the record. Therefore, we may segment S1, S2, or any portion of the cardiac cycle of interest from as many beats as are available, and average their PSD estimates in a procedure similar to the Bartlett or Welch procedures. (*Note:* Direct averaging of the PCG signals themselves or of their complex Fourier transforms could lead to undesired cancellation of noise-like murmurs or asynchronous frequency components and their disappearance from the result! Refer to Sections 4.11 and 6.6 for discussions on intentional cancellation of asynchronous components in the PCG via synchronized averaging.)

We saw in Sections 5.5.2 and 5.5.3 how the envelope or the envelogram of the PCG may be averaged over several cardiac cycles. However, there was no need to segment parts of a cardiac cycle in envelope analysis: nonstationarity of the signal within a cardiac cycle was not a concern. In the present application of PSD analysis, there is a need to segment the PCG further.

A procedure was described in Section 4.10 for segmentation of the systolic and diastolic parts of PCG signals based upon the detection of the QRS complex in the ECG and the detection of the dicrotic notch in the carotid pulse signal. Further segmentation of the systolic or diastolic parts into S1 and systolic murmur or S2 and diastolic murmur, respectively, would require more sophisticated methods, which will be the topics of Chapter 8. For now, let us consider the task of obtaining averaged PSDs of the systolic and diastolic parts of a PCG signal.

Figure 6.11 shows the PCG signal over one cardiac cycle of a normal subject segmented using the procedure described in Section 4.10 and illustrated in Figure 4.27. The periodograms of the systolic and diastolic parts of the PCG cycle illustrated are also shown in the figure. In order to obtain better PSD estimates, the periodogram of each systolic or diastolic segment was computed separately and averaged over 16 cardiac cycles. No data window was applied (the rectangular window was used, in effect), therefore the procedure used is similar to the Bartlett procedure. Individual systolic or diastolic segments could be of different durations; for the present illustration, all periodograms were computed with the same number of samples, which was taken to be the maximum RR interval in the ECG record of the subject. The averaged systolic and diastolic PSD estimates are shown in Figure 6.11. The averaging procedure provides a smoother estimate of the PSDs by removing beat-to-beat variations that are neither significant nor of interest. Spectral peaks may be clearly observed in the averaged periodograms, and may be considered to be more reliable estimates of resonance than the peaks found in individual periodograms.

Figure 6.12 illustrates a PCG signal cycle as well as the individual and averaged systolic and diastolic PSD estimates for a patient with systolic murmur, split S2, and opening snap of the mitral valve (see also Figures 4.28 and 5.7). It is unlikely that the patient held her breath during data acquisition. The presence of increased high-frequency power in the range $120 - 250\ Hz$ due to the systolic murmur is evident in the averaged systolic PSD. The diastolic PSDs are comparable to the corresponding normal diastolic PSDs in Figure 6.11.

6.5 MEASURES DERIVED FROM POWER SPECTRAL DENSITY FUNCTIONS

The Fourier spectrum or PSD provides us with a density function of signal amplitude, power, or energy versus frequency. We would typically have a large number of samples of the PSD over a wide frequency range, which may not lend itself to easy analysis. We may, of course, study the shape of the spectrum graphically, and observe its general characteristics. Such an approach is often referred to as *nonparametric* spectral analysis. The spectral models we shall study later in Section 7.4 are characterized by a small number of parameters, and are hence called *parametric* spectral analysis (or modeling) methods.

Problem: *Derive parameters or measures from a Fourier spectrum or PSD that can help in the characterization of the spectral variations or features contained therein.*

Solution: Since the PSD is a nonnegative function as well as a density function, we may readily treat it as a PDF, and compute statistics using moments. We may also detect peaks corresponding to resonance, measure their bandwidth or quality factor, and derive measures of concentration of power in specific frequency bands of

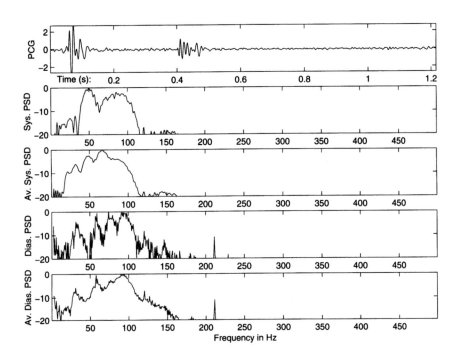

Figure 6.11 Top to bottom: A sample PCG signal over one cardiac cycle of a normal subject (male, 23 years; see also Figures 4.27 and 5.6); periodogram of the systolic portion of the signal (approximately $0 - 0.4$ *s*); averaged periodogram of the systolic parts of 16 cardiac cycles segmented as illustrated in Figure 4.27; periodogram of the diastolic portion of the signal shown in the first plot (approximately $0.4 - 1.2$ *s*); averaged periodogram of the diastolic parts of 16 cardiac cycles. The periodograms are on a log scale (*dB*).

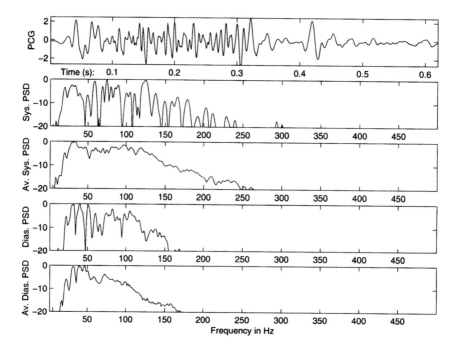

Figure 6.12 Top to bottom: A sample PCG signal over one cardiac cycle of a patient with systolic murmur, split S2, and opening snap of the mitral valve (female, 14 months; see also Figures 4.28 and 5.7); periodogram of the systolic portion of the signal (approximately 0 − 0.28 *s*); averaged periodogram of the systolic parts of 26 cardiac cycles segmented as illustrated in Figure 4.28; periodogram of the diastolic portion of the signal shown in the first plot (approximately 0.28 − 0.62 *s*); averaged periodogram of the diastolic parts of 26 cardiac cycles. The periodograms are on a log scale (*dB*).

interest or concern. Although the PSD itself is nonparametric, we may derive several parameters that, while not completely representing the entire PSD, may facilitate the identification of physiological and/or pathological phenomena. We shall investigate a few different approaches toward this end in the following subsections.

6.5.1 Moments of PSD functions

As the area under the PSD curve represents the total signal power or energy which need not be unity, we have to normalize all moments by the total power or energy of the signal E_x given by

$$E_x = \sum_{n=0}^{N-1} |x(n)|^2 = \frac{1}{N} \sum_{k=0}^{N-1} |X(k)|^2 \qquad (6.36)$$

$$= \frac{1}{2\pi} \int_0^{2\pi} |X(\omega)|^2 \, d\omega = \int_{f=0}^{1} |X(f)|^2 \, df.$$

Note that the frequency variables ω and f above are normalized. Assuming that the PSD has been obtained using one of the methods described in the preceding sections, we may replace $|X(.)|^2$ in the above expressions by $S_{xx}(.)$.

As a simple measure of the concentration of the signal power over its frequency range, we may compute the mean frequency $\bar{f}$ as the first-order moment

$$\bar{f} = f_s \, \frac{2}{E_x} \int_{f=0}^{0.5} f \, S_{xx}(f) \, df \qquad (6.37)$$

or as

$$\bar{f} = f_s \, \frac{2}{N E_x} \sum_{k=0}^{N/2} k \, S_{xx}(k), \qquad (6.38)$$

where N is the number of samples in the DFT-based representation of the PSD. The upper limit of integration of 0.5 represents integration from DC to the maximum frequency present in the signal, which is half the sampling frequency, the frequency variable having been normalized to the range $0 \leq f \leq 1$. Note that the integration or summation is performed over one-half period of the periodic function $S_{xx}(f)$ or $S_{xx}(k)$, which also possesses even symmetry about half the sampling frequency for real signals.

The median frequency f_{med} is defined as that frequency which splits the PSD in half:

$$f_{med} = \frac{m}{N} \, f_s \quad \text{with the largest } m \text{ such that} \qquad (6.39)$$

$$\frac{2}{N E_x} \sum_{k=0}^{m} S_{xx}(k) < \frac{1}{2}; \quad 0 \leq m \leq \frac{N}{2}.$$

We may also compute higher-order statistics such as

- variance f_{m2} as the second-order moment by using $(f - \bar{f})^2$ in place of f (the function of frequency that is multiplied with $S_{xx}(f)$) in Equation 6.37 or the equivalent expression in k in Equation 6.38; that is,

$$f_{m2} = f_s \frac{2}{E_x} \int_{f=0}^{0.5} (f - \bar{f})^2 \, S_{xx}(f) \, df \qquad (6.40)$$

or

$$f_{m2} = f_s \frac{2}{NE_x} \sum_{k=0}^{N/2} (k - \bar{k})^2 \, S_{xx}(k), \qquad (6.41)$$

where $\bar{k}$ is the frequency sample index corresponding to $\bar{f}$.

- skewness as

$$\text{skewness} = \frac{f_{m3}}{(f_{m2})^{3/2}}, \qquad (6.42)$$

where the third-order moment f_{m3} is computed with $(f - \bar{f})^3$ in place of f in Equation 6.37, that is,

$$f_{m3} = f_s \frac{2}{E_x} \int_{f=0}^{0.5} (f - \bar{f})^3 \, S_{xx}(f) \, df \qquad (6.43)$$

or

$$f_{m3} = f_s \frac{2}{NE_x} \sum_{k=0}^{N/2} (k - \bar{k})^3 \, S_{xx}(k). \qquad (6.44)$$

- kurtosis as

$$\text{kurtosis} = \frac{f_{m4}}{(f_{m2})^2}, \qquad (6.45)$$

where the fourth-order moment f_{m4} is computed with $(f - \bar{f})^4$ in place of f in Equation 6.37, that is,

$$f_{m4} = f_s \frac{2}{E_x} \int_{f=0}^{0.5} (f - \bar{f})^4 \, S_{xx}(f) \, df \qquad (6.46)$$

or

$$f_{m4} = f_s \frac{2}{NE_x} \sum_{k=0}^{N/2} (k - \bar{k})^4 \, S_{xx}(k). \qquad (6.47)$$

The mean frequency is a useful measure of the concentration of signal power, and could indicate the resonance frequency in the case of unimodal distributions. However, a nearly uniform PSD could lead to half the maximum frequency as the mean frequency, which by itself may not be a useful representation of the PSD. The presence of multiple resonance frequencies could also lead to a mean frequency that may not be a useful measure. Multimodal PSDs may be characterized better by a series of peak frequencies, along with measures of their relative levels and bandwidths or quality factors (to be described in the next subsection).

The square-root of f_{m2} provides a measure of spectral spread (standard deviation about the mean) and an indication of the bandwidth (but not at -3 dB) about the mean frequency. The skewness is zero if the density function is symmetric about the mean frequency; otherwise, it indicates the extent of asymmetry of the distribution. Kurtosis indicates if the PSD is a long-tailed function.

Moments of PSDs may be useful in characterizing the general trends in the distribution of the power of a signal over its bandwidth. The higher-order moments are sensitive to noise or spurious variations in the PSD estimate, and may not yield reliable measures if the PSD pattern is not simple or if the PSD estimate is poor (has a high variance). The reliability of moments may be improved by smoothing the PSD estimate, or by fitting a smooth parametric curve (Gaussian, Laplacian, spline, etc.) as a model of the PSD estimate and computing the moments of the model. Saltzberg and Burch [136] discuss the relationship between moments of PSDs and ZCR, and their application to EEG analysis.

6.5.2 Spectral power ratios

The moments described in the preceding subsection provide general statistical characterization of the PSD treated as a PDF. In the case of analysis of biomedical signals, it may be more advantageous to define specific measures based upon *a priori* information or empirical knowledge about the signals, systems, and the physiological or pathological processes of concern. For example, in the case of PCG analysis for the detection of murmurs, we could specifically investigate the presence of signal power in the frequency range beyond that of S1 and/or S2. If a specific type of pathology of interest is known to cause a shift in the frequency content within a certain band of frequencies, we may measure spectral power ratios over partitions of the band of interest. We have already seen in Sections 6.2.1 and 6.2.2 how such measures have been used for the analysis of ventricular elasticity, diagnosis of myocardial infarction, and detection of murmurs.

The fraction of signal power in a frequency band of interest $(f_1 : f_2)$ may be computed as

$$E_{(f_1:f_2)} = \frac{2}{E_x} \int_{f=f_1}^{f_2} |X(f)|^2 \, df = \frac{2}{NE_x} \sum_{k=k_1}^{k_2} |X(k)|^2, \qquad (6.48)$$

where k_1 and k_2 are the DFT indices corresponding to f_1 and f_2, respectively. Fractions of power as above may be computed for several bands of interest that may or may not span the entire signal bandwidth.

In a variation of the above fractional-power measure, Johnson et al. [163] compared the integral of the magnitude spectrum of the systolic murmurs due to aortic stenosis over the band $75 : 150$ Hz to that over the band $25 : 75$ Hz. They considered the higher-frequency band to represent the *predictive area* (PA) of the spectrum related to the aortic stenosis, and the lower-frequency band to represent a *constant area* (CA)

that would be common to all systolic PCG signal segments. The ratio of PA to CA was defined as

$$\frac{PA}{CA} = \frac{\int_{f=f_2}^{f_3} |X(f)|\, df}{\int_{f=f_1}^{f_2} |X(f)|\, df},\qquad(6.49)$$

with $f_1 = 25\ Hz$, $f_2 = 75\ Hz$, and $f_3 = 150\ Hz$. The $\frac{PA}{CA}$ ratio is provided for the PSDs of systolic murmurs of four patients with aortic stenosis in Figure 6.2; Johnson et al. showed that the ratio correlates well with the severity of aortic stenosis.

Binnie et al. [145, 146] describe the application of spectrum analysis to EEG for the detection of epilepsy. Their method was based upon partitioning or banding of the EEG spectrum into not only the traditional δ, θ, α, and β bands, but also into seven other nonuniform bands specified as $1 - 2$, $2 - 4$, $4 - 6$, $6 - 8$, $8 - 11$, $11 - 14$, and $> 14\ Hz$. Additional features related to form factor FF (see Section 5.6.4) were also used. In a study with 275 patients with suspected epilepsy, 90% of the signals of the patients with pathology were classified as abnormal by their methods; conversely, 86% of the patients whose EEGs were classified as abnormal had confirmed pathology.

When analyzing a spectral peak, we may also compute the $-3\ dB$ bandwidth of the peak, and furthermore, its quality factor as the ratio of the peak frequency to the bandwidth. Such measures may be computed for not only the dominant peak, but several peaks at progressively lower levels of signal power. Essentially, each potential resonance peak is treated and characterized as a bandpass filter. Durand et al. [167] used such measures to characterize the PSDs of sounds produced by prosthetic heart valves (to be discussed in Section 6.6).

6.6 APPLICATION: EVALUATION OF PROSTHETIC HEART VALVES

Efficient opening and closing actions of cardiac valves are of paramount importance for proper pumping of blood by the heart. When native valves fail, they may be replaced by mechanical prosthetic valves or by bioprosthetic valves extracted from pigs. Mechanical prosthetic valves are prone to sudden failure due to fracture of their components. Bioprosthetic valves fail gradually due to tissue degeneration and calcification, and have been observed to last $7 - 12$ years [167]. Follow-up of the health of patients with prosthetic valves requires periodic, noninvasive assessment of the functional integrity of the valves.

Problem: *Deposition of calcium causes the normally pliant and elastic bioprosthetic valve leaflets to become stiff. Propose a method to assess the functional integrity of bioprosthetic valves.*

Solution: Based on the theory that valve opening and closure contribute directly to heart sounds, analysis of PCG components offers a noninvasive and passive approach to evaluation of prosthetic valves. The increased stiffness is expected to lead to higher-frequency components in the opening or closing sounds of the valve. Durand et al. [167] studied the spectra of the entire S1 signal segment to evaluate the sounds contributed by the closure of porcine (pig) bioprosthetic valves implanted in the mitral position in humans. They demonstrated that, whereas normal S1 spectra were limited in bandwidth to about 100 Hz, degenerated bioprosthetic valves created significant spectral energy in the range $100 - 250$ Hz. Figure 6.13 shows the relative power spectra of S1 in the case of a normal bioprosthetic valve and a degenerated bioprosthetic valve.

Durand et al. derived several parameters from S1 spectra and used them to discriminate normal from degenerated bioprosthetic valves. Some of the parameters used by them are the first and second dominant peak frequencies; the bandwidth and quality factor of the dominant peak; integrated mean area above -20 dB; the highest frequency found at -3 dB; total area and RMS value of the spectrum; area and RMS value in the $20 - 100$ Hz, $100 - 200$ Hz, and $200 - 300$ Hz bands; and the median frequency. Normal versus degenerated valve classification accuracies as high as 98% were achieved.

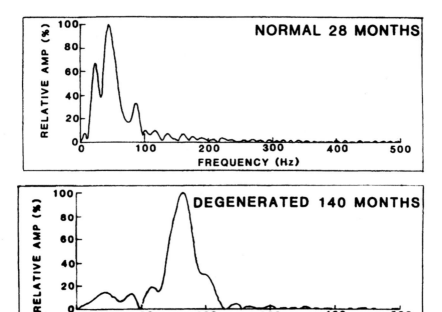

Figure 6.13   First heart sound spectra in the case of normal and degenerated porcine bioprosthetic valves implanted in the mitral position. Reproduced with permission from L.G. Durand, M. Blanchard, G. Cloutier, H.N. Sabbah, and P.D. Stein, Comparison of pattern recognition methods for computer-assisted classification of spectra of heart sounds in patients with a porcine bioprosthetic valve implanted in the mitral position, *IEEE Transactions on Biomedical Engineering*, 37(12):1121–1129, 1990 ©IEEE.

Durand et al. [121] also studied the sounds of bioprosthetic valves in the aortic position. They argued that the aortic and pulmonary components (A2 and P2, respectively) of S2, each lasting about 50 ms, are not temporally correlated during normal breathing. The two components of S2 are separated by $30 - 60 \ ms$ during inspiration, but get closer and could overlap during expiration. Furthermore, P2 is weaker than A2 if the PCG is recorded in the aortic area. Thus P2 may be suppressed and A2 strengthened by coherent detection and averaging of S2 over several cardiac and breath cycles; see Section 4.11. Durand et al. performed spectral analysis of A2 extracted as above for the purpose of evaluation of bioprosthetic valves in the aortic position. Among a selection of spectral analysis methods including the basic periodogram, Welch's averaged periodogram, all-pole modeling (see Section 7.5), and pole-zero modeling (see Section 7.6), they found the basic periodogram to provide the best compromise for estimating both the spectral distribution and the dominant frequency peaks of bioprosthetic valve sounds.

Cloutier et al. [168] studied the bias and variability of several diagnostic spectral parameters computed from simulated closing sounds of bioprosthetic valves in the mitral position. They found that the most-dominant spectral peak frequency and its quality factor were best estimated using an FFT-based PSD estimate with a rectangular window. However, the $-3 \ dB$ bandwidth of the most-dominant spectral peak, the frequency of the second-dominant peak, and a few other parameters were best estimated by the Steiglitz-McBride method of pole-zero modeling (see Section 7.6.2). Some other parameters were best estimated by all-pole modeling using the covariance method (see Section 7.5). It was concluded that a single method would not provide the best estimates of all possible spectral parameters of interest.

6.7   REMARKS

We have investigated the frequency-domain characteristics of a few biomedical signals and the corresponding physiological systems, with particular attention to the PCG and the cardiovascular system. Frequency-domain analysis via PSDs and parameters derived from PSDs can enable us to view the signal from a different perspective than the time domain. Certain signals such as the PCG and EEG may not lend themselves to easy interpretation in the time domain, and therefore may benefit from a move to the frequency domain.

PSDs and their parameters facilitate investigation of the behavior of physiological systems in terms of rhythms, resonance, and parameters that could be related to the physical characteristics of anatomical entities (for example, the loss of elasticity of the myocardial muscles due to ischemia or infarction, the extent of aortic valvular stenosis, or the extent of calcification and stiffness of bioprosthetic valves). Pathological states may also be derived or simulated by modifying the spectral parameters or representations of the corresponding normal physiological states and signals.

It is worthwhile to pause at this stage of our study, and recognize the importance of the topics presented in the preceding chapters. A good understanding of the physiological systems that produce the biomedical signals we deal with, as well as of the pathological processes that alter their characteristics, is of paramount importance before we may process the signals. Preprocessing the signals to remove artifacts and detect events is essential before we may derive parameters to facilitate their analysis in the time and/or frequency domains. Design of biomedical signal analysis techniques requires a thorough understanding of the characteristics and properties of the biomedical systems behind the signals, in addition to detailed knowledge of mathematical principles, computer techniques, and digital signal processing algorithms.

6.8 STUDY QUESTIONS AND PROBLEMS

1. The impulse response of a filter is specified by the series of sample values $\{3, 1, -1\}$.

   (a) What will be the response of the filter to the input whose sample values are $\{4, 4, 2, 1\}$?

   (b) Is the filter response obtained by linear convolution or circular convolution of the input with the impulse response?

   (c) What will be the response with the type of convolution other than the one you indicated as the answer to the questions above?

   (d) How would you implement convolution of the two signals listed above using the FFT? Which type of convolution will this procedure provide? How would you get the other type of convolution for the signals in this problem via the FFT-based procedure?

2. A conjugate symmetric (even) signal $x_e(n)$ is defined as a signal with the property $x_e(n) = x_e^*(-n)$. A conjugate antisymmetric (odd) signal $x_o(n)$ is defined as a signal with the property $x_o(n) = -x_o^*(-n)$. An arbitrary signal $x(n)$ may be expressed as the sum of its conjugate symmetric and conjugate antisymmetric parts as $x(n) = x_e(n) + x_o(n)$, where $x_e(n) = \frac{1}{2}[x(n) + x^*(-n)]$ and $x_o(n) = \frac{1}{2}[x(n) - x^*(-n)]$. Prove that

$$FT[x_e(n)] = \text{real}[X(\omega)],$$

   and

$$FT[x_o(n)] = j\text{imag}[X(\omega)],$$

   where $FT[x(n)] = X(\omega)$, and FT stands for the Fourier transform [86].

3. A signal $x(t)$ is transmitted through a channel. The received signal $y(t)$ is a scaled, shifted, and noisy version of $x(t)$ given as $y(t) = \alpha x(t - t_0) + \eta(t)$ where α is a scale factor, t_0 is the time delay, and $\eta(t)$ is noise. Assume that the noise process has zero mean and is statistically independent of the signal process, and that all processes are stationary.

   Derive expressions for the PSD of $y(t)$ in terms of the PSDs of x and η [105, 5].

4. Consider a continuous-time sinusoidal signal of frequency $10\ Hz$.

   (a) Derive an analytical expression for the ACF of the signal.

   (b) Draw a schematic plot of the ACF, including detailed labeling of the time axis.

   (c) State the relationship of the PSD to the ACF.

   (d) Derive the analytical expression for the PSD of the given signal.

   (e) Draw a schematic plot of the PSD, including detailed labeling of the frequency axis.

5. Two real signals $x_1(n)$ and $x_2(n)$ are combined to form a complex signal defined as $y(n) = x_1(n) + jx_2(n)$. Derive a procedure to extract the DFTs $X_1(k)$ and $X_2(k)$ of $x_1(n)$ and $x_2(n)$, respectively, from the DFT $Y(k)$ of $y(n)$.

6. Distinguish between ensemble averages and temporal (time) averages. Identify applications of first-order and second-order averages of both types in PCG analysis.

7. Propose, in point form, a procedure to process PCG signals to identify the possible presence of a murmur due to aortic stenosis.

8. Propose an algorithm to detect the presence of the alpha rhythm in an EEG signal.

   Propose an extension to the algorithm to detect the joint presence of the same rhythm in four simultaneously recorded EEG channels.

6.9 LABORATORY EXERCISES AND PROJECTS

Note: Data files related to the exercises are available at the site
ftp://ftp.ieee.org/uploads/press/rangayyan/

1. Using MATLAB, prepare a signal that contains the sum of two cosine waves of equal amplitude at 40 Hz and 45 Hz. Let the sampling rate be 1 kHz.

   (a) Compute the power spectrum of the signal with a rectangular window of duration 2 s.

   (b) Compute the power spectrum of the signal with a Hamming window of duration 2 s.

   (c) Compute the power spectrum of the signal with a rectangular window of duration 0.5 s.

   (d) Compute the power spectrum of the signal with a Hamming window of duration 0.5 s.

   To obtain the power spectrum, you may take the FFT and square the result. Compare the spectra obtained in parts (a) – (d) and comment upon their similarities and/or differences. In order to visualize the differences clearly, use 2,048-point FFTs and plot the logarithm of the magnitude-squared spectra with an expanded scale from 0 to 100 Hz only. Be sure to label the frequency axis in Hz!

   What should the ideal spectrum look like?

2. Two VAG signals are given in the files vag1.dat and vag2.dat (see also the file vag.m). The sampling rate is 2 kHz. Obtain and plot their power spectra (PSDs) using MATLAB. Label the frequency axis in Hz!

   Compute the mean frequency as the first moment of the PSD for each signal. Compute also the variance (second central moment) of each PSD. What are the units of these parameters?

   Compare the spectra and the parameters derived and give your evaluation of the frequency content of the signals.

3. The file safety.wav contains the speech signal for the word "safety" uttered by a male speaker, sampled at 8 kHz (see also the file safety.m). The signal has a significant amount of background noise (as it was recorded in a normal computer laboratory). Develop procedures to segment the signal into voiced, unvoiced, and silence (background noise) portions using short-time RMS, turns count, or ZCR measures. Compute the PSD for each segment that you obtain and study its characteristics.

4. The files pec1.dat, pec33.dat, and pec52.dat give three-channel recordings of the PCG, ECG, and carotid pulse signals (sampled at 1,000 Hz; you may read the signals using the program in the file plotpec.m). The signals in pec1.dat and pec52.dat are normal; the PCG signal in pecg33.dat has systolic murmur, and is of a patient suspected to have pulmonary stenosis, ventricular septal defect, and pulmonary hypertension.

Apply the Pan-Tompkins method for QRS detection to the ECG channel and the Lehner and Rangayyan method to detect the dicrotic notch in the carotid pulse channel. Extrapolate the timing information from the ECG and carotid pulse channels to segment the PCG signal into two parts: the systolic part from the onset of an S1 and to the onset of the following S2, and the diastolic part from the onset of an S2 to the onset of the following S1. Compute the PSD of each segment.

Extend the procedure to average the systolic and diastolic PSDs over several cardiac cycles. Compare the PSDs obtained for the three cases.

5. Compute the mean frequency and the ratio of the energy in the range $100 - 300 \, Hz$ to the total energy for each PSD derived in the previous problem. What can you infer from these measures?

6. Compute the PSDs of a few channels of the EEG in the file eeg1-xx.dat using Welch's procedure (see also the file eeg1.m). Study the changes in the PSDs derived with variations in the window width, the number of segments averaged, and the type of the window used. Compare the results with the PSDs computed using the entire signal in each channel. Discuss the results in terms of the effects of the procedures and parameters on spectral resolution and leakage.

7

Modeling Biomedical Signal-generating Processes and Systems

We have thus far concentrated on the processing and analysis of biomedical signals. The signals were treated in their own right as conveyors of diagnostic information. While it was emphasized that the design and application of signal analysis procedures require an understanding of the physiological and pathological processes and systems that generate the signals, no specific mathematical model was used to represent the genesis of the signals in the methods we have studied so far.

We shall now consider the modeling approach, where an explicit mathematical model is used to represent the process or the system that generates the signal of interest. The parameters of the model are then investigated for use in signal analysis, pattern recognition, and decision making. As we shall see, the model parameters may also be related to the physical or physiological aspects of the related systems. The *parametric modeling* approach often leads to succinct and efficient representation of signals and systems. Regardless of the emphasis on modeling, the final aim of the methods described in this chapter will be analysis of the signal of interest.

7.1 PROBLEM STATEMENT

Propose mathematical models to represent the generation of biomedical signals. Identify the possible relationships between the mathematical models and the physiological and pathological processes and systems that generate the signals. Explore the potential use of the model parameters in signal analysis, pattern recognition, and diagnostic decision making.

Given the diversity of the biomedical signals that we have already encountered and the many others that exist, a generic model cannot be expected to represent a large number of signals. Indeed, a very specific model is often required for each signal. Bioelectric signals such as the ECG and EMG may be modeled using the basic action potential or SMUAP as the building block. Sound and vibration signals such as the PCG and speech may be modeled using fluid-filled resonating chambers, turbulent flow across a baffle or through a constriction, vibrating pipes, and acoustic or vibrational excitation of a tract of variable shape. We shall investigate a few representative signals and models in the following sections, and then study a few modeling techniques that facilitate signal analysis based upon the parameters extracted.

7.2 ILLUSTRATION OF THE PROBLEM WITH CASE-STUDIES

7.2.1 Motor-unit firing patterns

We saw in Section 1.2.3 that the surface EMG of an active skeletal muscle is the spatio-temporal summation of the action potentials of a large number of motor units that have been recruited into action (see Figure 1.6). If we consider the EMG of a single motor unit, we have a train of SMUAPs; the same basic wave (spike, pulse, or wavelet) is repeated in a quasi-periodic sequence. For the sake of generality, we may represent the intervals between the SMUAPs by a random variable: although an overall periodicity exists and is represented by the firing rate in *pps*, the intervals between the pulses, known as the *inter-pulse interval* or IPI, may not precisely be the same from one SMUAP to another.

Agarwal and Gottlieb [169] modeled the single-motor-unit EMG as the convolution of a series of unit impulses or Dirac delta functions — known as a *point process* [170, 171, 172, 173] — with the basic SMUAP wave. The SMUAP train $y(t)$ may then be modeled as the output of a linear system whose impulse response $h(t)$ is the SMUAP, and the input is a point process $x(t)$:

$$y(t) = \int_0^t h(t - \tau)\, x(\tau)\, d\tau. \qquad (7.1)$$

Physiological conditions dictate that successive action potentials of the same motor unit cannot overlap: the interval between any two pulses should be greater than the SMUAP duration. In normal muscle activation, SMUAP durations are of the order of $3 - 20$ *ms* and motor unit firing rates are in the range $7 - 25$ *pps*; the IPI is therefore in the range $40 - 140$ *ms*, which is significantly higher than the SMUAP duration. An SMUAP train therefore consists of discrete (distinct and separated) events or waves.

The model as above permits independent analysis of SMUAP waveshape and firing pattern: the two are indeed physiologically separate entities. The SMUAP waveshape depends upon the spatial arrangement of the muscle fibers that constitute the motor unit, while the firing pattern is determined by the motor neuron that stimulates the muscle fibers. Statistics of the point process representing the IPI may be used to study the muscle activation process independently of the SMUAP waveshape. Details on point processes and their application to EMG modeling will be presented in Section 7.3.

7.2.2 Cardiac rhythm

The ECG is a quasi-periodic signal that is also cyclo-stationary in the normal case (see Section 1.2.4). Each beat is triggered by a pulse from the SA node. The P wave is the combined result of the action potentials of the atrial muscle units, while the QRS and T waves are formed by the spatio-temporal summation of the action potentials of the ventricular muscle units.

In rhythm analysis, one is more interested in the timing of the beats than in their individual waveshape (with the exception of PVCs). Diseases that affect the SA node could disturb the normal rhythm, and lead to abnormal variability in the RR intervals. Disregarding the details of atrial and ventricular ECG waves, an ECG rhythm may be modeled by a point process representing the firing pattern of the SA node. Sinus arrhythmia and HRV may then be investigated by studying the distribution and statistics of the RR interval.

Figure 7.1 illustrates the representation of ECG complexes in terms of the instantaneous heart rate values defined as the inverse of the RR interval of each beat, in terms of a series of RR interval values, and as a train of delta functions at the SA node firing instants [72]. A discrete-time signal may be derived by sampling the signal in Figure 7.1 (b) at equidistant points; the result, however, may not be continuous or differentiable [72]. The signal in Figure 7.1 (c), known as the interval series, has values $I_k = t_k - t_{k-1}$, where the instants t_k represent the time instants at which the QRS complexes occur in the ECG signal. The I_k series is defined as a function of interval number and not of time, and hence may pose difficulties regarding interpretation in the frequency domain. Finally, the signal in Figure 7.1 (d) is defined as a train of Dirac delta functions $s(t) = \sum \delta(t - t_k)$. The series of impulses represents a point process that may be analyzed and interpreted with relative ease, as will be seen in Section 7.3. The last two representations may be used to analyze cardiac rhythm and HRV, which will be described in Section 7.8 (see also Section 8.9).

7.2.3 Formants and pitch in speech

Speech signals are formed by exciting the vocal tract with either a pulse train or a random signal produced at the glottis, and possibly their combination as well (see Section 1.2.11). The shape of the vocal tract is varied according to the nature of the sound or phoneme to be produced; the system is therefore a time-variant system. We may model the output as the convolution of the (time-variant) impulse response of the vocal tract with the input glottal waveform. The input may be modeled by a random process for unvoiced speech and as a point process for voiced speech. Clearly, the speech signal is a nonstationary signal; however, the signal may be considered to be

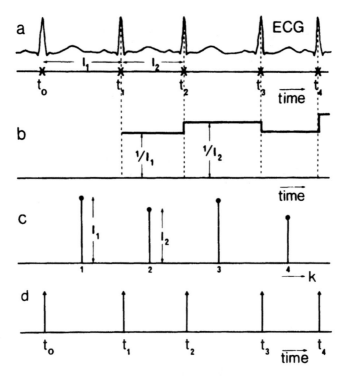

Figure 7.1   The train of ECG complexes in (a) is represented in terms of: (b) the instantaneous heart rate values defined as the inverse of the RR interval of each beat; (c) a series of RR interval values (known as the interval series); and (d) a train of delta functions at the SA node firing instants. Reproduced with permission from R.W. DeBoer, J.M. Karemaker, and J. Strackee, Comparing spectra of a series of point events particularly for heart rate variability studies, *IEEE Transactions on Biomedical Engineering,* 31(4): 384–387, 1984. ©IEEE.

quasi-stationary over short intervals of time during which the same phoneme is being produced.

Figure 7.2 illustrates the commonly used model for speech production [46]. The speech signal may be modeled using the same convolutional relationship as in Equation 7.1, with the limitation that the expression is valid over durations of time when the vocal-tract shape is held fixed and the same glottal excitation is applied. Then, $h(t)$ represents the impulse response of the vocal-tract system (filter) for the time interval considered, and $x(t)$ represents the glottal waveform that is input to the system. In the case of voiced speech, the IPI statistics of the point-process input, in particular its mean, are related to the pitch. Furthermore, the frequency response of the filter $H(\omega)$ representing the vocal tract determines the spectral content of the speech signal: the dominant frequencies or peaks are known as formants in the case of voiced speech.

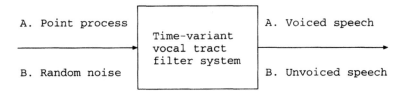

Figure 7.2   Model for production of speech, treating the vocal tract as a time-variant linear system. A point-process input generates quasi-periodic voiced speech, whereas a random-noise input generates unvoiced speech.

Point processes will be described in Section 7.3. Parametric spectral modeling and analysis techniques suitable for formant extraction will be described in Sections 7.4, 7.5, and 7.6.

7.2.4   Patello-femoral crepitus

Among the various types of VAG signals produced by the knee joint (see Section 1.2.13), the most common is a signal known as physiological patello-femoral crepitus (PPC) [174, 59, 175, 176, 177]. The PPC signal is a random sequence of vibrational pulses generated between the surfaces of the patella and the femur, typically observed during slow movement of the knee joint. The PPC signal may carry information on the state and lubrication of the knee joint. A mechanical model of the knee-joint surfaces that generate PPC, as proposed by Beverland et al. [176], will be described in Section 7.7.2.

Zhang et al. [174] proposed a model for generation of the PPC signal based on point processes, similar to that for the SMUAP train described in Section 7.2.1. The effects of the repetition rate (or IPI) and the basic patello-femoral pulse (PFP) waveform on the spectrum of the PPC signal were analyzed separately. It was suggested that the model could represent the relationships between physiological parameters such as the mean and standard deviation of the IPI as well as the PFP waveshape, and parameters that could be measured from the PPC signal such as its mean, RMS, and

PSD-based features. Illustrations related to this application will be provided at the end of Section 7.3.

7.3 POINT PROCESSES

Problem: *Formulate a mathematical model representing the generation of a train of SMUAPs, and derive an expression for the PSD of the signal.*

Solution: In the model for EMG generation proposed by Agarwal and Gottlieb [169], a point process is used to represent the motor neuron firing sequence, and the SMUAP train is modeled by the convolution integral as in Equation 7.1. The IPI is treated as a sequence of independent random variables with identical normal (Gaussian) PDFs.

Let the interval between the ith SMUAP and the preceding one be τ_i, and let the origin be set at the instant of appearance of the first SMUAP at $i = 0$ with $\tau_0 = 0$. The time of arrival of the ith SMUAP is then given by $t_i = \tau_1 + \tau_2 + \cdots + \tau_i$. The variable t_i is the sum of i independent random variables; note that $\tau_i > 0$. It is assumed that the mean μ and variance σ^2 of the random variable representing each IPI are the same. Then, the mean of t_i is $i\mu$, and its variance is $i\sigma^2$. Furthermore, t_i is also a random variable with the Gaussian PDF

$$p_{t_i}(t_i) = \frac{1}{\sqrt{2\pi i}\,\sigma} \exp\left[-\frac{(t_i - i\mu)^2}{2i\sigma^2}\right]. \tag{7.2}$$

If the SMUAP train has $N + 1$ SMUAPs labeled as $i = 0, 1, 2, \ldots, N$, the motor neuron firing sequence is represented by the point process

$$x(t) = \sum_{i=0}^{N} \delta(t - t_i). \tag{7.3}$$

The Fourier transform of the point process is

$$
\begin{aligned}
X(\omega) &= \int_{-\infty}^{\infty} \sum_{i=0}^{N} \delta(t - t_i)\,\exp(-j\omega t)\,dt \tag{7.4} \\
&= \sum_{i=0}^{N} \exp(-j\omega t_i).
\end{aligned}
$$

$X(\omega)$ is a function of the random variable t_i, which is, in turn, a function of i random variables $\tau_1, \tau_2, \ldots, \tau_i$. Therefore, $X(\omega)$ is random. The ensemble average of $X(\omega)$ may be obtained by computing its expectation, taking into account the PDF of t_i, as follows [169]:

$$\overline{X}(\omega) = E[X(\omega)] = \sum_{i=0}^{N} E[\exp(-j\omega t_i)]. \tag{7.5}$$

$$E[\exp(-j\omega t_i)] = \int_{-\infty}^{\infty} \exp(-j\omega t_i)\,p_{t_i}(t_i)\,dt_i. \tag{7.6}$$

Using the expression for $p_{t_i}(t_i)$ in Equation 7.2, we get

$$E[\exp(-j\omega t_i)] = \frac{1}{\sqrt{2\pi i}\sigma} \int_{-\infty}^{\infty} \exp(-j\omega t_i) \exp\left[-\frac{(t_i - i\mu)^2}{2i\sigma^2}\right] dt_i. \quad (7.7)$$

Substituting $t_i - i\mu = r$, where r is a temporary variable, we get

$$E[\exp(-j\omega t_i)] = \frac{1}{\sqrt{2\pi i}\sigma} \exp(-j\omega i\mu) \int_{-\infty}^{\infty} \exp\left[-\frac{r^2}{2i\sigma^2}\right] \exp(-j\omega r) \, dr. \quad (7.8)$$

Using the property that the Fourier transform of $\exp(-\frac{t^2}{2\sigma^2})$ is $\sigma\sqrt{2\pi}\exp(-\frac{\sigma^2\omega^2}{2})$ [1], we get

$$E[\exp(-j\omega t_i)] = \exp(-j\omega i\mu) \exp\left[-\frac{i\sigma^2\omega^2}{2}\right]. \quad (7.9)$$

Finally, we have

$$\overline{X}(\omega) = \sum_{i=0}^{N} \exp(-j\omega i\mu) \exp\left[-\frac{i\sigma^2\omega^2}{2}\right]. \quad (7.10)$$

The ensemble-averaged Fourier transform of the SMUAP train is given by

$$\overline{Y}(\omega) = \overline{X}(\omega)H(\omega), \quad (7.11)$$

where $H(\omega)$ is the Fourier transform of an individual SMUAP. The Fourier transform of an SMUAP train is, therefore, a multiplicative combination of the Fourier transform of the point process representing the motor neuron firing sequence and the Fourier transform of an individual SMUAP.

Illustration of application to EMG: Figure 7.3 illustrates EMG signals synthesized using the point-process model as above using $1, 20, 40,$ and 60 motor units, all with the same biphasic SMUAP of 8 ms duration and IPI statistics $\mu = 50$ ms and $\sigma = 6.27$ ms [169]. It is seen that the EMG signal complexity increases as more motor units are activated. The interference patterns obscure the shape of the SMUAP used to generate the signals, and were observed to closely resemble real EMG signals.

Figure 7.4 shows the magnitude spectra of synthesized EMG signals with one motor unit and 15 motor units, with biphasic SMUAP duration of 8 ms, $\mu = 20$ ms, and $\sigma = 4.36$ ms [169]. The smooth curve superimposed on the second spectrum in Figure 7.4 was derived from the mathematical model described in the preceding paragraphs. An important point to observe from the spectra is that the average magnitude spectrum of several identical motor units approaches the spectrum of a single MUAP. The spectral envelope of an SMUAP train or that of an interference pattern of several SMUAP trains with identical SMUAP waveshape is determined by the shape of an individual SMUAP.

Figure 7.5 shows the magnitude spectra of surface EMG signals recorded from the gastrocnemius-soleus muscle, averaged over $1, 5,$ and 15 signal records [169]. The

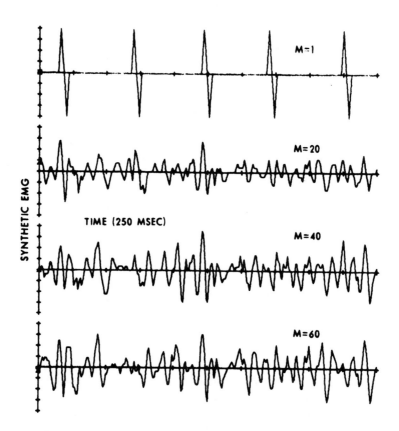

Figure 7.3   Synthesis of an SMUAP train and EMG interference pattern using the point-process model. Top to bottom: SMUAP train of a single motor unit, and interference patterns of the activities of 20, 40, and 60 motor units. SMUAP duration $= 8$ ms. IPI statistics $\mu = 50$ ms and $\sigma = 6.27$ ms. The duration of each signal is 250 ms. Reproduced with permission from G.C. Agarwal and G.L. Gottlieb, An analysis of the electromyogram by Fourier, simulation and experimental techniques, *IEEE Transactions on Biomedical Engineering*, 22(3): 225–229, 1975. ©IEEE.

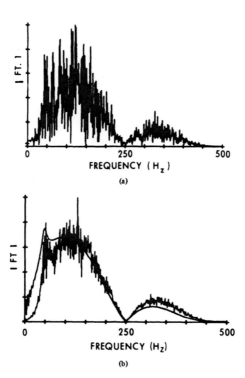

Figure 7.4 Magnitude spectra of synthesized EMG signals with (a) one motor unit, and (b) 15 motor units, with biphasic SMUAP duration of 8 ms, $\mu = 20$ ms, and $\sigma = 4.36$ ms. The smooth curve superimposed on the spectrum in (b) was derived from the point-process model with 10 SMUAPs. Reproduced with permission from G.C. Agarwal and G.L. Gottlieb, An analysis of the electromyogram by Fourier, simulation and experimental techniques, *IEEE Transactions on Biomedical Engineering*, 22(3): 225–229, 1975. ©IEEE.

spectra in Figures 7.4 and 7.5 demonstrate comparable features. If all of the motor units active in a composite EMG record were to have similar or identical MUAPs, the spectral envelope of the signal could provide information on the MUAP waveshape (via an IFT). As we have noted earlier in Section 1.2.3, MUAP shape could be useful in the diagnosis of neuromuscular diseases. In reality, however, many motor units of different MUAP shapes could be contributing to the EMG signal even at low levels of effort, and analysis as above may have limited applicability. Regardless, the point-process model provides an interesting approach to model EMG signals. The same model is applicable to the generation of voiced-speech signals, as illustrated in Figure 7.2. For other models to represent the characteristics of the EMG signal, refer to the papers by Parker et al. [178], Lindström and Magnusson [179], Zhang et al. [180], Parker and Scott [181], Shwedyk et al. [182], Person and Libkind [183], Person and Kudina [184], de Luca [185, 24], Lawrence and de Luca [150], and de Luca and van Dyk [186].

Illustration of application to PPC: Zhang et al. [174] proposed a point-process model to represent knee-joint PPC signals, which they called PFP trains or signals (see Section 7.2.4). Figure 7.6 illustrates the PSDs of two point processes simulated with mean repetition rate $\mu_r = 21$ *pps* and coefficient of variation $CV_r = \sigma_r/\mu_r = 0.1$ and 0.05, where σ_r is the standard deviation of the repetition rate. A Gaussian distribution was used to model the IPI statistics. The spectra clearly show the most-dominant peak at the mean repetition rate of the point process, followed by smaller peaks at its harmonics. The higher-order harmonics are better defined in the case with the lower CV_r; in the limit, the PSD will be a periodic impulse train with all impulses of equal strength when the point process is exactly periodic ($\sigma_r = 0, CV_r = 0$).

Zhang et al. [174] simulated PFP trains for different IPI statistics using a sample PFP waveform from a real VAG signal recorded at the patella of a normal subject using an accelerometer. The duration of the PFP waveform was 21 *ms*, and the IPI statistics μ_r and CV_r were limited such that the PFP trains synthesized would have non-overlapping PFP waveforms and resemble real PFP signals. Figures 7.7 and 7.8 illustrate the PSDs of synthesized PFP signals for different μ_r but with the same CV_r, and for the same μ_r but with different CV_r, respectively. The PSDs clearly illustrate the influence of IPI statistics on the spectral features of signals generated by point processes. Some important observations to be made are:

- The PSD envelope of the PFP train remains the same, regardless of the IPI statistics.

- The PSD envelope of the PFP train is determined by the PSD of an individual PFP waveform.

- The PSD envelope of the PFP train is modulated by a series of impulses with characteristics determined by the IPI statistics. The first impulse indicates the mean repetition rate.

- The point process has a highpass effect: low-frequency components of the PSD of the basic PFP are suppressed due to multiplication with the PSD of the point process.

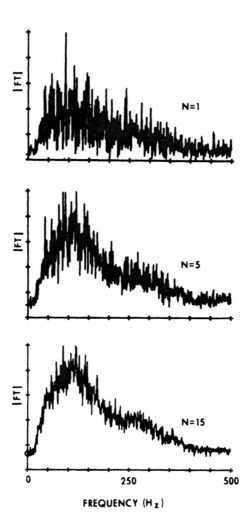

Figure 7.5 Magnitude spectra of surface EMG signals recorded from the gastrocnemius-soleus muscle, averaged over 1, 5, and 15 signal records. Reproduced with permission from G.C. Agarwal and G.L. Gottlieb, An analysis of the electromyogram by Fourier, simulation and experimental techniques, *IEEE Transactions on Biomedical Engineering*, 22(3): 225–229, 1975. ©IEEE.

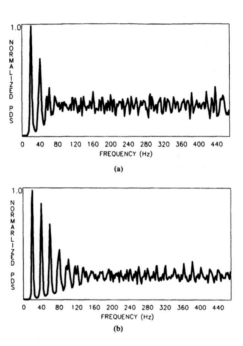

Figure 7.6   Normalized PSDs of synthesized point processes with (a) μ_r = 21 *pps* and CV_r = 0.1, and (b) μ_r = 21 *pps* and CV_r = 0.05. *Note:* PDS = power density spectrum = PSD. Reproduced with permission from Y.T. Zhang, C.B. Frank, R.M. Rangayyan, and G.D. Bell, Mathematical modelling and spectrum analysis of the physiological patello-femoral pulse train produced by slow knee movement, *IEEE Transactions on Biomedical Engineering,* 39(9):971–979, 1992. ©IEEE.

- Physiological signals rarely exhibit precise periodicity. The CV_r value will be reasonably large, thereby limiting the effect of repetition to low frequencies in the PSD of the PFP train.

The observations made above are valid for all signals generated by point processes, including SMUAP trains and voiced-speech signals.

Zhang et al. [174] verified the point-process model for PFP signals by computing the IPI statistics and PSDs of real PFP signals recorded from normal subjects. Figure 7.9 shows the IPI histograms computed from the PFP signals of two normal subjects. The IPI statistics computed for the two cases were $\mu_r = 25.2$ *pps* and $CV_r = 0.07$ for the first, and $\mu_r = 16.1$ *pps* and $CV_r = 0.25$ for the second signal. While the IPI histogram for the first signal appears to be close to a Gaussian distribution, the second is not. The PSDs of the two signals are shown in Figure 7.10. The PSDs of the real signals demonstrate features that are comparable to those observed from the PSDs of the synthesized signals, and agree with the observations listed above. The envelopes of the two PSDs demonstrate minor variations: the basic PFP waveform in the two cases were not identical.

7.4 PARAMETRIC SYSTEM MODELING

The importance of spectral analysis of biomedical signals was established in Chapter 6. However, the methods described were based on the computation and use of the Fourier spectrum; while this approach is, to begin with, nonparametric, we saw how a few parameters could be computed from Fourier spectra. The limitations of such an approach were also discussed in Chapter 6. We shall now study methods for parametric modeling and analysis that, although based on time-domain data and models at the outset, can facilitate parametric characterization of the spectral properties of signals and systems.

Problem: *Explore the possibility of parametric modeling of signal characteristics using the general linear system model.*

Solution: The difference equation that gives the output of a general linear, shift-invariant (or time-invariant), discrete-time system is

$$y(n) = -\sum_{k=1}^{P} a_k \, y(n-k) + G \sum_{l=0}^{Q} b_l \, x(n-l), \qquad (7.12)$$

with $b_0 = 1$. (*Note:* The advantage of the negative sign before the summation with a_k will become apparent later in this section; some model formulations use a positive sign, which does not make any significant difference in the rest of the derivation.) The input to the system is $x(n)$; the output is $y(n)$; the parameters $b_l, l = 0, 1, 2, \ldots, Q$, indicate how the present and Q past samples of the input are combined, in a linear manner, to generate the present output sample; the parameters $a_k, k = 1, 2, \ldots, P$, indicate how the past P samples of the output are linearly combined (in a feedback loop) to produce the current output; G is a gain factor; and P and Q determine

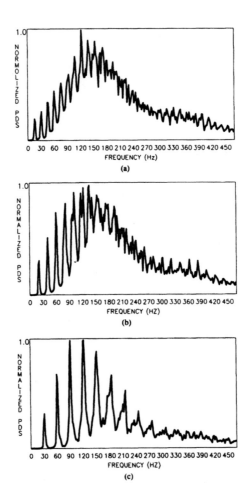

Figure 7.7 Normalized PSDs of synthesized PFP trains using a real PFP waveform with a duration of 21 ms, $CV_r = 0.05$, and (a) $\mu_r = 16$ pps, (b) $\mu_r = 21$ pps, and (c) $\mu_r = 31$ pps. *Note:* PDS = power density spectrum = PSD. Reproduced with permission from Y.T. Zhang, C.B. Frank, R.M. Rangayyan, and G.D. Bell, Mathematical modelling and spectrum analysis of the physiological patello-femoral pulse train produced by slow knee movement, *IEEE Transactions on Biomedical Engineering,* 39(9):971–979, 1992. ©IEEE.

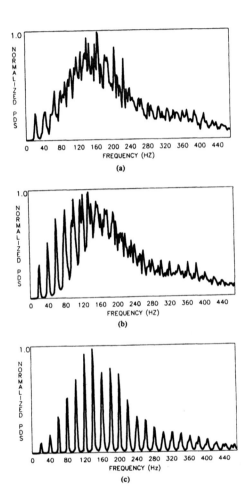

Figure 7.8   Normalized PSDs of synthesized PFP trains using a real PFP waveform with a duration of 21 ms, $\mu_r = 21$ pps, and (a) $CV_r = 0.1$, (b) $CV_r = 0.05$, and (c) $CV_r = 0.01$. *Note:* PDS = power density spectrum = PSD. Reproduced with permission from Y.T. Zhang, C.B. Frank, R.M. Rangayyan, and G.D. Bell, Mathematical modelling and spectrum analysis of the physiological patello-femoral pulse train produced by slow knee movement, *IEEE Transactions on Biomedical Engineering*, 39(9):971–979, 1992. ©IEEE.

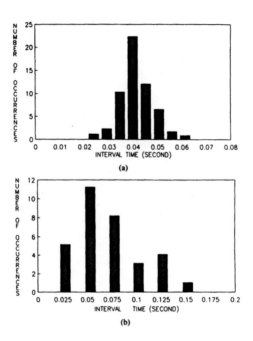

Figure 7.9   IPI histograms computed from real PFP trains recorded from two normal subjects. The statistics computed were (a) $\mu_r = 25.2$ *pps* and $CV_r = 0.07$, and (b) $\mu_r = 16.1$ *pps* and $CV_r = 0.25$. Reproduced with permission from Y.T. Zhang, C.B. Frank, R.M. Rangayyan, and G.D. Bell, Mathematical modelling and spectrum analysis of the physiological patello-femoral pulse train produced by slow knee movement, *IEEE Transactions on Biomedical Engineering*, 39(9):971–979, 1992. ©IEEE.

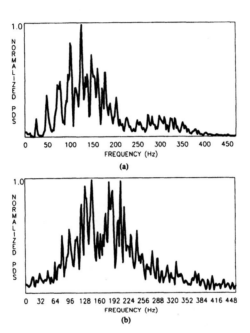

Figure 7.10   Normalized PSDs of the real PFP trains recorded from two normal subjects whose IPI histograms are shown in Figure 7.9.   The IPI statistics of the two cases are (a) $\mu_r = 25.2$ *pps* and $CV_r = 0.07$, and (b) $\mu_r = 16.1$ *pps* and $CV_r = 0.25$.   *Note:* PDS = power density spectrum = PSD. Reproduced with permission from Y.T. Zhang, C.B. Frank, R.M. Rangayyan, and G.D. Bell, Mathematical modelling and spectrum analysis of the physiological patello-femoral pulse train produced by slow knee movement, *IEEE Transactions on Biomedical Engineering,* 39(9):971–979, 1992. ©IEEE.

the order of the system. The summation over x represents the *moving-average* or MA part of the system; the summation over y represents the *autoregressive* or AR part of the system; the entire system may be viewed as a combined *autoregressive, moving-average* or ARMA system. The feedback part typically makes the impulse response of the system infinitely long; the system may then be viewed as an IIR filter (see Figures 3.29 and 3.30).

Equation 7.12 indicates that the output of the system is simply a linear combination of the present input sample and a few past input samples, and a few past output samples. The use of the past input and output samples in computing the present output sample represents the memory of the system. The model also indicates that the present output sample may be *predicted* as a *linear combination* of the present and a few past input samples, and a few past output samples. For this reason, the model is also known as the *linear prediction* or LP model [187, 46, 77].

Applying the z-transform to Equation 7.12, we can obtain the transfer function of the system as

$$H(z) = \frac{Y(z)}{X(z)} = G \, \frac{1 + \sum_{l=1}^{Q} b_l \, z^{-l}}{1 + \sum_{k=1}^{P} a_k \, z^{-k}} \, . \tag{7.13}$$

(The advantage of the negative sign before the summation with a_k in Equation 7.12 is now apparent in the numerator – denominator symmetry of Equation 7.13.) The system is completely characterized by the parameters $a_k, k = 1, 2, \ldots, P$; $b_l, l = 1, 2, \ldots, Q$; and G. In most applications the gain factor G is not important; the system is therefore completely characterized by the a and b parameters, with the exception of a gain factor. Furthermore, we may factorize the numerator and denominator polynomials in Equation 7.13 and express the transfer function as

$$H(z) = G \, \frac{\prod_{l=1}^{Q} (1 - z_l \, z^{-1})}{\prod_{k=1}^{P} (1 - p_k \, z^{-1})} \, , \tag{7.14}$$

where $z_l, l = 1, 2, \ldots, Q$, are the zeros of the system and $p_k, k = 1, 2, \ldots, P$, are the poles of the system. The model may now be referred to as a *pole-zero model*. It is evident from Equation 7.14 that the system is completely characterized by its poles and zeros but for a gain factor.

Equations 7.12, 7.13, and 7.14 demonstrate the applicability of the same conceptual model in the time and frequency domains. The a and b parameters are directly applicable in both the time and the frequency domains in expressing the input – output relationship or the system transfer function. The poles and zeros are more specific to the frequency domain, although the contribution of each pole or zero to the time-domain impulse response of the system may be derived directly from its coordinates in the z-plane [1].

Given a particular input signal $x(n)$ and the corresponding output of the system $y(n)$, we could derive their z-transforms $X(z)$ and $Y(z)$ and hence obtain the system transfer function $H(z)$ in some form. Difficulties arise at values of z for which $X(z) = 0$; as the system is linear, and $Y(z) = H(z)X(z)$, we have $Y(z) = 0$ at such points as well. Then, $H(z)$ cannot be determined at the corresponding values of z. [The simplest test signal is the unit-impulse function $x(n) = \delta(n)$, for which $X(z) = 1$ for all z: the response of a linear shift-invariant system to an impulse completely characterizes the system with the corresponding $y(n) = h(n)$ or its z-domain equivalent $H(z)$.] Methods to determine an AR or ARMA model for a given signal for which the corresponding input to the system is not known (or is assumed to be a point process or a random process) will be described in the following sections.

7.5 AUTOREGRESSIVE OR ALL-POLE MODELING

Problem: *How can we obtain an AR (or LP) model when the input to the system that caused the given signal as its output is unknown?*

Solution: In the AR or all-pole model [187, 46], the output is modeled as a linear combination of P past values of the output and the present input sample as

$$y(n) = -\sum_{k=1}^{P} a_k \, y(n-k) + G \, x(n). \tag{7.15}$$

(The discussion on AR modeling here closely follows that in Makhoul [187], with permission.) Some model formulations use a positive sign in place of the negative sign before the summation in the above equation. It should be noted that the model as in Equation 7.15 does not account for the presence of noise.

The all-pole transfer function corresponding to Equation 7.15 is

$$H(z) = \frac{G}{1 + \sum_{k=1}^{P} a_k \, z^{-k}}. \tag{7.16}$$

In the case of biomedical signals such as the EEG or the PCG, the input to the system is totally unknown. Then, we can only approximately predict the current sample of the output signal using its past values as

$$\tilde{y}(n) = -\sum_{k=1}^{P} a_k \, y(n-k), \tag{7.17}$$

where the $\tilde{\ }$ indicates that the predicted value is only approximate. The error in the predicted value (also known as the residual) is

$$e(n) = y(n) - \tilde{y}(n) = y(n) + \sum_{k=1}^{P} a_k \, y(n-k). \tag{7.18}$$

The general signal-flow diagram of the AR model viewed as a prediction or error filter is illustrated in Figure 7.11.

The least-squares method: In the least-squares method, the parameters a_k are obtained by minimizing the MSE with respect to all of the parameters. The procedure is similar to that used to derive the optimal Wiener filter (see Section 3.5 and Haykin [77]).

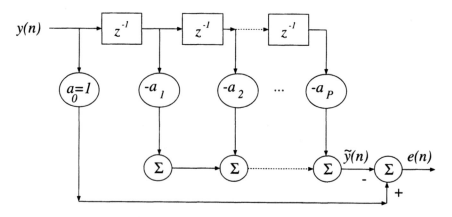

Figure 7.11 Signal-flow diagram of the AR model.

Given an observed signal $y(n)$, the following procedure is applicable for minimization of the MSE [187]. The total squared error (TSE) ε is given by

$$\varepsilon = \sum_n e^2(n) = \sum_n \left(y(n) + \sum_{k=1}^{P} a_k \, y(n-k) \right)^2 . \qquad (7.19)$$

(*Note:* The TSE is the same as the MSE except for a scale factor.) Although the range of the summation in Equation 7.19 is important, we may minimize ε without specifying the range for the time being. Minimization of ε is performed by applying the conditions

$$\frac{\partial \varepsilon}{\partial a_k} = 0, \;\; 1 \le k \le P \qquad (7.20)$$

to Equation 7.19, which yields

$$\sum_{k=1}^{P} a_k \sum_n y(n-k) \, y(n-i) = -\sum_n y(n) \, y(n-i), \;\; 1 \le i \le P. \qquad (7.21)$$

For a given signal $y(n)$, Equation 7.21 provides a set of P equations in the P unknowns $a_k, k = 1, 2, \ldots, P$, known as the *normal equations*; the similarities between the normal equations here and those in the case of the Wiener filter (see Section 3.5) will become apparent later.

By expanding Equation 7.19 and using the relationship in Equation 7.21, the minimum TSE ε_P for the model of order P is obtained as

$$\varepsilon_P = \sum_n y^2(n) + \sum_{k=1}^{P} a_k \sum_n y(n)y(n-k). \qquad (7.22)$$

The expression for TSE will be simplified in the following sections.

The autocorrelation method: If the range of summation in Equations 7.19 and 7.21 is specified to be $-\infty < n < \infty$, the error is minimized over an infinite duration, and we have

$$\phi_y(i) = \sum_{n=-\infty}^{\infty} y(n) \, y(n-i), \tag{7.23}$$

where $\phi_y(i)$ is the ACF of $y(n)$. In practice, the signal $y(n)$ will be available only over a finite interval, say $0 \leq n \leq N - 1$; the given signal may then be assumed to be zero outside this range and treated as a windowed version of the true signal, as we have already seen in Section 6.4. Then, the ACF may be expressed as

$$\phi_y(i) = \sum_{n=i}^{N-1-i} y(n) \, y(n-i), \quad i \geq 0, \tag{7.24}$$

where the scale factor $\frac{1}{N}$ is omitted. (It will become apparent later that the scale factor is immaterial in the derivation of the model coefficients.) The normal equations then become

$$\sum_{k=1}^{P} a_k \, \phi_y(i-k) = -\phi_y(i), \quad 1 \leq i \leq P. \tag{7.25}$$

We now see that an AR model may be derived for a signal with the knowledge of only its ACF; the signal samples themselves are not required. It is seen now that the scale factor $\frac{1}{N}$ that was omitted in defining the ACF is of no consequence in deriving the model coefficients. It may be advantageous to use the normalized ACF values given as $\overline{\phi}_y(i) = \phi_y(i)/\phi_y(0)$, which have the property that $|\overline{\phi}_y(i)| \leq 1$.

The minimum TSE is given by

$$\varepsilon_P = \phi_y(0) + \sum_{k=1}^{P} a_k \, \phi_y(k). \tag{7.26}$$

Application to random signals: If the signal $y(n)$ is a sample of a random process, the error $e(n)$ is also a sample of a random process. We then have to use the expectation operation to obtain the MSE as follows:

$$\varepsilon = E[e^2(n)] = E\left[\left(y(n) + \sum_{k=1}^{P} a_k \, y(n-k) \right)^2 \right]. \tag{7.27}$$

Applying the condition for minimum error as in Equation 7.20, we get the normal equations as

$$\sum_{k=1}^{P} a_k \, E[y(n-k) \, y(n-i)] = -E[y(n) \, y(n-i)], \quad 1 \leq i \leq P. \tag{7.28}$$

The minimum MSE is

$$\varepsilon_P = E[y^2(n)] + \sum_{k=1}^{P} a_k \, E[y(n) \, y(n-k)].$$ (7.29)

If the signal is a sample of a stationary random process, we have $E[y(n-k) \, y(n-i)] = \phi_y(i-k)$. This leads to the same normal equations as in Equation 7.25. If the process is ergodic, the ACF may be computed as a time average as in Equation 7.24.

If the signal is a sample of a nonstationary random process, $E[y(n-k) \, y(n-i)] = \phi_y(n-k, n-i)$; the ACF is a function of not only the shift but also time. We would then have to compute the model parameters for every instant of time n; we then have a time-variant model. Modeling and analysis of nonstationary signals will be postponed to Chapter 8.

Computation of the gain factor G: Since we assumed earlier that the input to the system being modeled is unknown, the gain parameter G is not important. Regardless, the derivation of G demonstrates a few important points. Equation 7.18 may be rewritten as

$$y(n) = -\sum_{k=1}^{P} a_k \, y(n-k) + e(n).$$ (7.30)

Comparing this with Equation 7.15, we see that the only input signal $x(n)$ which can result in $y(n)$ at the output is given by the condition $Gx(n) = e(n)$. This condition indicates that the input signal is proportional to the error of prediction when the estimated model parameters are equal to the real system parameters. Regardless of the input, a condition that could be applied is that the energy of the output be equal to that of the signal $y(n)$ being modeled. Since the transfer function $H(z)$ is fixed, we then have the condition that the total energy of the input signal be equal to the total energy of the error ε_P.

As illustrated in the model for speech generation in Figure 7.2, two types of input that are of interest are the impulse function and a random process that is stationary white noise. In the case when $x(n) = \delta(n)$, we have the impulse response $h(n)$ at the output, and

$$h(n) = -\sum_{k=1}^{P} a_k \, h(n-k) + G \, \delta(n).$$ (7.31)

Multiplying both sides of the expression above with $h(n-i)$ and summing over all n, we get expressions in terms of the ACF $\phi_h(i)$ of $h(n)$ as

$$\phi_h(i) = -\sum_{k=1}^{P} a_k \, \phi_h(i-k), \quad 1 \le |i| \le \infty$$ (7.32)

and

$$\phi_h(0) = -\sum_{k=1}^{P} a_k \, \phi_h(k) + G^2.$$ (7.33)

Due to the condition that the energy of the output of the system be equal to that of $y(n)$, the condition $\phi_h(0) = \phi_y(0)$ must be satisfied. Comparing Equations 7.32 and 7.25, we then have

$$\phi_h(i) = \phi_y(i), \quad 0 \le i \le P. \tag{7.34}$$

Therefore, for a model of order P, the first $(P+1)$ ACF terms of the impulse response $h(n)$ must be equal to the corresponding ACF terms of the signal $y(n)$ being modeled. It follows from Equations 7.33, 7.34, and 7.26 that

$$G^2 = \varepsilon_P = \phi_y(0) + \sum_{k=1}^{P} a_k \, \phi_y(k). \tag{7.35}$$

In the case when the input is a sequence of uncorrelated samples of a random process (white noise) with zero mean and unit variance, we could use the same procedure as for the impulse-input case, with the difference being that expectations are taken instead of summing over all n. (The conditions to be noted in this case are $E[x(n)] = 0$ and $E[x(n)\,x(n-i)] = \delta(i)$.) The same relations as above for the impulse-input case are obtained. The identical nature of the results for the two cases follows from the fact that the two types of input have identical ACFs and PSDs. These characteristics are relevant in the speech model shown in Figure 7.2.

Computation of the model parameters: For low orders of the model, Equation 7.25 may be solved directly. However, direct methods may not be feasible when P is large.

The normal equations in Equation 7.25 may be written in matrix form as

$$\begin{bmatrix} \phi_y(0) & \phi_y(1) & \cdots & \phi_y(P-1) \\ \phi_y(1) & \phi_y(0) & \cdots & \phi_y(P-2) \\ \vdots & \vdots & \ddots & \vdots \\ \phi_y(P-1) & \phi_y(P-2) & \cdots & \phi_y(0) \end{bmatrix} \begin{bmatrix} a_1 \\ a_2 \\ \vdots \\ a_P \end{bmatrix} = - \begin{bmatrix} \phi_y(1) \\ \phi_y(2) \\ \vdots \\ \phi_y(P) \end{bmatrix}. \tag{7.36}$$

For real signals, the $P \times P$ ACF matrix is symmetric and the elements along any diagonal are identical, that is, it is a Toeplitz matrix.

It is worth noting the following similarities and differences between the normal equations in the case of the Wiener filter as given in Equation 3.85 and those above in Equation 7.36:

- The matrix on the left-hand side is the ACF of the input to the filter in the case of the Wiener filter, whereas it is the ACF of the output of the prediction filter in the present case.

- The filter vector on the left-hand side contains the coefficients of the filter being designed in both cases — the optimal Wiener filter or the optimal prediction filter.

- The vector on the right-hand side is the CCF between the input and the desired response in the case of the Wiener filter, whereas it is the ACF of the output of the prediction filter in the present case.

Haykin [77] provides a more detailed correspondence between the AR model and the Wiener filter.

A procedure known as Durbin's method [188, 189] or the Levinson-Durbin algorithm (see Makhoul [187], Rabiner and Schafer [46], or Haykin [77]) provides a recursive method to solve the normal equations in Equation 7.36. The procedure starts with a model order of 1; computes the model parameters, the error, and a secondary set of parameters known as the reflection coefficients; updates the model order and the parameters; and repeats the procedure until the model of the desired order is obtained. The Levinson-Durbin algorithm is summarized below.

Initialize model order $i = 0$ and error $\varepsilon_0 = \phi_y(0)$. Perform the following steps recursively for $i = 1, 2, \ldots, P$.

1. Increment model order i and compute the ith reflection coefficient γ_i as

$$\gamma_i = -\frac{1}{\varepsilon_{i-1}} \left[\phi_y(i) + \sum_{j=1}^{i-1} a_{i-1,j}\, \phi_y(i-j) \right], \tag{7.37}$$

where $a_{i-1,j}$ denotes the jth model coefficient at iteration $(i-1)$; the iteration index is also the recursively updated model order.

2. Let $a_{i,i} = \gamma_i$.

3. Update the predictor coefficients as

$$a_{i,j} = a_{i-1,j} + \gamma_i\, a_{i-1,i-j}, \quad 1 \le j \le i - 1. \tag{7.38}$$

4. Compute the error value as

$$\varepsilon_i = (1 - \gamma_i^2)\, \varepsilon_{i-1}. \tag{7.39}$$

The final model parameters are given as $a_k = a_{P,k}$, $1 \le k \le P$. The Levinson-Durbin algorithm computes the model parameters for all orders up to the desired order P. As the order of the model is increased, the TSE reduces, and hence we have $0 \le \varepsilon_i \le \varepsilon_{i-1}$. The reflection coefficients may also be used to test the stability of the model (filter) being designed: $|\gamma_i| < 1, i = 1, 2, \ldots, P$, is the required condition for stability of the model of order P.

The covariance method: In deriving the autocorrelation method, the range of summation of the prediction error in Equations 7.19 and 7.21 was specified to be $-\infty < n < \infty$. If, instead, we specify the range of summation to be a finite interval, say, $0 \le n \le N - 1$, we get

$$\sum_{k=1}^{P} a_k\, C(k, i) = -C(0, i), \quad 1 \le i \le P \tag{7.40}$$

instead of Equation 7.25 based upon the ACF, and the minimum TSE is given by

$$\varepsilon_P = C(0,0) + \sum_{k=1}^{P} a_k\, C(0, k) \tag{7.41}$$

instead of Equation 7.26, where

$$C(i, k) = \sum_{n=0}^{N-1} y(n - i)y(n - k) \qquad (7.42)$$

is the covariance of the signal $y(n)$ in the specified interval. The matrix formed by the covariance function is symmetric as $C(i, k) = C(k, i)$, similar to the ACF matrix in Equation 7.36; however, the elements along each diagonal will not be equal, as $C(i + 1, k + 1) = C(i, k) + y(-i - 1)y(-k - 1) - y(N - 1 - i)y(N - 1 - k)$. Computation of the covariance coefficients also requires $y(n)$ to be known for $-P \le n \le N - 1$. The distinctions disappear as the specified interval of summation (error minimization) tends to infinity.

7.5.1 Spectral matching and parameterization

The AR model was derived in the preceding section based upon time-domain formulations in the autocorrelation and covariance methods. We shall now see that equivalent formulations can be derived in the frequency domain, which can lead to a different interpretation of the model. Applying the z-transform to Equation 7.18, we get

$$E(z) = \left[1 + \sum_{k=1}^{P} a_k z^{-k}\right] Y(z) = A(z)Y(z), \qquad (7.43)$$

and

$$H(z) = \frac{G}{A(z)}, \qquad (7.44)$$

where

$$A(z) = 1 + \sum_{k=1}^{P} a_k z^{-k}, \qquad (7.45)$$

and $E(z)$ is the z-transform of $e(n)$. We can now view the error $e(n)$ as the result of passing the signal being modeled $y(n)$ through the filter $A(z)$, which may be considered to be an *inverse filter*. In the case of $y(n)$ being a deterministic signal, applying Parseval's theorem, the TSE to be minimized may be written as

$$\varepsilon = \sum_{n=-\infty}^{\infty} e^2(n) = \frac{1}{2\pi} \int_{-\pi}^{\pi} |E(\omega)|^2 \, d\omega, \qquad (7.46)$$

where $E(\omega)$ is obtained by evaluating $E(z)$ on the unit circle in the z-plane. Using $S_y(\omega)$ to represent the PSD of $y(n)$, we have

$$\varepsilon = \frac{1}{2\pi} \int_{-\pi}^{\pi} S_y(\omega)|A(\omega)|^2 \, d\omega, \qquad (7.47)$$

where $A(\omega)$ is the frequency response of the inverse filter, and is given by evaluating $A(z)$ on the unit circle in the z-plane.

From Equations 7.15, 7.16, and 7.44, we get

$$\tilde{S}_y(\omega) = |H(\omega)|^2 = \frac{G^2}{|A(\omega)|^2} = \frac{G^2}{\left|1 + \sum_{k=1}^{P} a_k \exp(-jk\omega)\right|^2}. \tag{7.48}$$

Here, $\tilde{S}_y(\omega)$ represents the PSD of the modeled signal $\tilde{y}(n)$ that is an approximation of $y(n)$ as in Equation 7.17. From Equation 7.43 we have

$$S_y(\omega) = \frac{|E(\omega)|^2}{|A(\omega)|^2}. \tag{7.49}$$

Now, $\tilde{S}_y(\omega)$ is the model's approximation of $S_y(\omega)$. Comparing Equations 7.48 and 7.49, we see that the error PSD $|E(\omega)|^2$ is modeled by a uniform (or "flat" or "white") PSD equal to G^2. For this reason, the filter $A(z)$ is also known as a "whitening" filter.

From Equations 7.46, 7.48, and 7.49, we get the TSE as

$$\varepsilon = \frac{G^2}{2\pi} \int_{-\pi}^{\pi} \frac{S_y(\omega)}{\tilde{S}_y(\omega)} d\omega. \tag{7.50}$$

As the model is derived by minimizing the TSE ε, we see now that the model is effectively minimizing the integrated ratio of the signal PSD $S_y(\omega)$ to its approximation $\tilde{S}_y(\omega)$. Makhoul [187] describes the equivalence of the model in the following terms:

- As the model order $P \to \infty$, the TSE is minimized, that is, $\varepsilon_P \to 0$.

- For a model of order P, the first $(P + 1)$ ACF values of its impulse response are equal to those of the signal being modeled. Increasing P increases the range of the delay parameter (time) over which the model ACF is equal to the signal ACF.

- Given that the PSD and the ACF are Fourier-transform pairs, the preceding point leads to the frequency-domain statement that increasing P leads to a better fit of $\tilde{S}_y(\omega)$ to $S_y(\omega)$. As $P \to \infty$, the model ACF and PSD become identical to the signal ACF and PSD, respectively. Thus *any spectrum may be approximated by an all-pole model of an appropriate order* (see Section 7.5.2 for a discussion on the optimal model order).

Noting from Equation 7.35 that $G^2 = \varepsilon_P$, Equation 7.50 yields another important property of the model as

$$\frac{1}{2\pi} \int_{-\pi}^{\pi} \frac{S_y(\omega)}{\tilde{S}_y(\omega)} d\omega = 1. \tag{7.51}$$

Equations 7.50 and 7.51 lead to the following spectral-matching properties of the AR model [187]:

- Due to the fact that the TSE is determined by the ratio of the true PSD to the model PSD, the spectral-matching process performs uniformly over the entire frequency range irrespective of the spectral shape. (Had the error measure been dependent on the difference between the true PSD and the model PSD, the spectral match would have been better at higher-energy frequency coordinates than at lower-energy frequency coordinates.)

- $S_y(\omega)$ will be greater than $\tilde{S}_y(\omega)$ at some frequencies and lesser at others, while satisfying Equation 7.51 on the whole; the contribution to the TSE is more significant when $S_y(\omega) > \tilde{S}_y(\omega)$ than in the opposite case. Thus, when the error is minimized, the fitting of $\tilde{S}_y(\omega)$ to $S_y(\omega)$ is better at frequencies where $S_y(\omega) > \tilde{S}_y(\omega)$. Thus the model PSD fits better at the *peaks* of the signal PSD.

- The preceding point leads to another interpretation: the AR-model spectrum $\tilde{S}_y(\omega)$ is a good estimate of the *spectral envelope* of the signal PSD. This is particularly useful when modeling quasi-periodic signals such as voiced speech, PCG, and other signals that have strong peaks in their spectra representing harmonics, formants, or resonance. By following the envelope, the effects of repetition, that is, the effects of the point-process excitation function (see Section 7.3), are removed.

Since the model PSD is entirely specified by the model parameters (as in Equation 7.48), we now have a *parametric representation* of the PSD of the given signal (subject to the error in the model). The TSE may be related to the signal PSD as follows [187].

$$\hat{\tilde{y}}(0) = \frac{1}{2\pi} \int_{-\pi}^{\pi} \log[\tilde{S}_y(\omega)] \, d\omega \qquad (7.52)$$

represents the zeroth coefficient of the (power or real) cepstrum (see Sections 4.8.3 and 5.4.2) of $\tilde{y}(n)$. Using the relationship in Equation 7.48, we get

$$\hat{\tilde{y}}(0) = \log G^2 - \frac{1}{2\pi} \int_{-\pi}^{\pi} \log|A(\omega)|^2 \, d\omega. \qquad (7.53)$$

As all the roots (zeros) of $A(z)$ are inside the unit circle in the z-plane (for the AR model to be stable), the integral in the above equation is zero [187]. We also have $G^2 = \varepsilon_P$. Therefore,

$$\varepsilon_P = \exp[\hat{\tilde{y}}(0)]. \qquad (7.54)$$

The minimum of ε_P is reached as $P \to \infty$, and is given by

$$\varepsilon_{\min} = \varepsilon_\infty = \exp[\hat{y}(0)]. \qquad (7.55)$$

This relationship means that the TSE ε_P is the geometric mean of the model PSD $\tilde{S}_y(\omega)$, which is always positive for a positive-definite PSD. The quantity ε_P represents that portion of the signal's information content that is not predictable by a model of order P.

7.5.2 Optimal model order

Given that the AR model performs better and better as the order P is increased, where do we stop? Makhoul [187] shows that if the given signal is the output of a P-pole system, then an AR model of order P would be the optimal model with the minimum error. But how would one find in practice if the given signal was indeed produced by a P-pole system?

One possibility to determine the optimal order for modeling a given signal is to follow the trend in the TSE as the model order P is increased. This is feasible in a recursive procedure such as the Levinson-Durbin algorithm, where models of all lower orders are computed in deriving a model of order P, and the error at each order is readily available. The procedure could be stopped when there is no significant reduction in the error as the model order is incremented.

Makhoul [187] describes the use of a normalized error measure $\bar{\varepsilon}_P$ defined as

$$\bar{\varepsilon}_P = \frac{\varepsilon_P}{\phi_y(0)} = \frac{\exp[\hat{\bar{y}}(0)]}{\phi_y(0)}. \tag{7.56}$$

As the model order $P \to \infty$,

$$\bar{\varepsilon}_{\min} = \bar{\varepsilon}_\infty = \frac{\exp[\hat{y}(0)]}{\phi_y(0)}. \tag{7.57}$$

$\bar{\varepsilon}_{\min}$ is a monotonically decreasing function of P, with $\bar{\varepsilon}_0 = 1$ and $\bar{\varepsilon}_\infty = \bar{\varepsilon}_{\min}$; furthermore, it can be expressed as a function of the model PSD as

$$\bar{\varepsilon}_P = \frac{\exp\left[\frac{1}{2\pi}\int_{-\pi}^{\pi} \log \tilde{S}_y(\omega)\, d\omega\right]}{\frac{1}{2\pi}\int_{-\pi}^{\pi} \tilde{S}_y(\omega)\, d\omega}. \tag{7.58}$$

It is evident that $\bar{\varepsilon}_P$ depends only upon the shape of the model PSD, and that $\bar{\varepsilon}_{\min}$ is determined solely by the signal PSD. The quantity $\bar{\varepsilon}_P$ may be viewed as the ratio of the geometric mean of the model PSD to its arithmetic mean, which is a measure of the spread of the PSD: the smaller the spread, the closer is the ratio to unity; the larger the spread, the closer is the ratio to zero. If the signal is the result of an all-pole system with P_0 poles, $\bar{\varepsilon}_P = \bar{\varepsilon}_{P_0}$ for $P \geq P_0$, that is, the curve remains flat. In practice, the incremental reduction in the normalized error may be checked with a condition such as

$$1 - \frac{\bar{\varepsilon}_{P+1}}{\bar{\varepsilon}_P} < \Delta, \tag{7.59}$$

where Δ is a small threshold. The optimal order may be considered to have been reached if the condition is satisfied for several consecutive model orders.

Another measure based upon an information-theoretic criterion proposed by Akaike [190] may be expressed as [187]

$$I(P) = \log \bar{\varepsilon}_P + \frac{2P}{N_e}, \tag{7.60}$$

where N_e is the effective number of data points in the signal taking into account windowing (for example, $N_e = 0.4N$ for a Hamming window, where N is the number of data samples). The first term in the equation above decreases while the second term increases as P is increased. Akaike's measure $I(P)$ may be computed up to the maximum order P of interest or the maximum that is feasible, and then the model of the order for which $I(P)$ is at its minimum could be taken as the optimal model.

Model parameters: The AR (all-pole) model $H(z)$ and its inverse $A(z)$ are uniquely characterized by any one set of the following sets of parameters [187]:

- The model parameters $a_k, k = 1, 2, \ldots, P$. The series of a_k parameters is also equal to the impulse response of the inverse filter.

- The impulse response $h(n)$ of the AR model.

- The poles of $H(z)$, which are also the roots (zeros) of $A(z)$.

- The reflection coefficients $\gamma_i, i = 1, 2, \ldots, P$.

- The ACF (or PSD) of the a_k coefficients.

- The ACF (or PSD) of $h(n)$.

- The cepstrum of a_k or $h(n)$.

With the inclusion of the gain factor G as required, all of the above sets have a total of $(P + 1)$ values, and are equivalent in the sense that one set may be derived from another. Any particular set of parameters may be used, depending upon its relevance, interpretability, or relationship to the real-world system being modeled.

Illustration of application to EEG signals: Identification of the existence of rhythms of specific frequencies is an important aspect of EEG analysis. The direct relationship between the poles of an AR model and resonance frequencies makes this technique an attractive tool for the analysis of EEG signals.

Figure 7.12 shows the FFT spectrum and AR-model spectra with $P = 6$ and $P = 10$ for the o1 channel of the EEG signal shown in Figure 1.22. The FFT spectrum in the lowest trace of Figure 7.12 includes many spurious variations which make its interpretation difficult. On the other hand, the AR spectra indicate distinct peaks at about 10 Hz corresponding to an alpha rhythm; a peak at 10 Hz is clearly evident even with a low model order of $P = 6$ (the middle trace in Figure 7.12).

The poles of the AR model with order $P = 10$ are plotted in Figure 7.13. The dominant pole (closest to the unit circle in the z-plane) appears at 9.9 Hz, corresponding to the peak observed in the spectrum in the top-most plot in Figure 7.12. The radius of the dominant pole is $|z| = 0.95$; the other complex-conjugate pole pairs have $|z| \leq 0.76$. The model with $P = 6$ resulted in two complex-conjugate pole pairs and two real poles, with the dominant pair at 10.5 Hz with $|z| = 0.91$; the magnitude of the other pole pair was 0.74. A simple search for the dominant (complex) pole can thus provide an indication of the prevalent EEG rhythm with fairly low AR model orders.

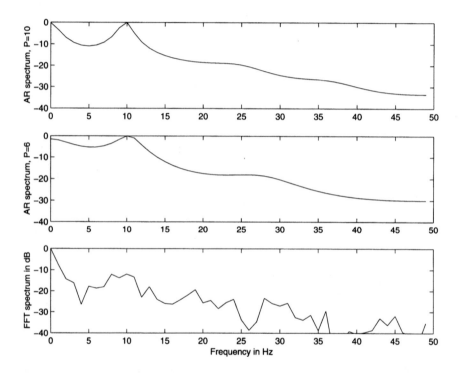

Figure 7.12   From bottom to top: FFT-based and AR-model spectra with $P = 6$ and $P = 10$ for the o1 channel of the EEG signal shown in Figure 1.22.

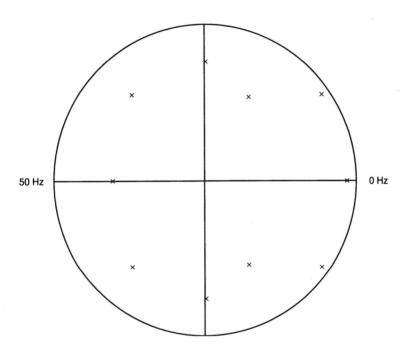

Figure 7.13 Poles of the AR model with $P = 10$ for the o1 channel of the EEG signal shown in Figure 1.22. See also Figure 7.12.

Illustration of application to PCG signals: Application of AR modeling is an attractive possibility for the analysis of PCG signals due to the need to identify significant frequencies of resonance in the presence of multiple components, artifacts, and noise. Although the model coefficients themselves do not carry any physical correlates or significance, the poles may be related directly to the physical or physiological characteristics of hearts sounds and murmurs.

Figures 7.14 and 7.15 illustrate the FFT-based spectrum of a segment containing one S1 and the subsequent systolic portion of the PCG signal of a normal subject, the AR-model spectra for order $P = 10$ and $P = 20$, and the poles of the model of order $P = 20$ (see also Figures 4.27, 5.6, and 6.11). Figures 7.16 and 7.17 illustrate the same items for a segment containing one S2 and the subsequent diastolic portion of the same subject. It is evident that the AR spectra follow the dominant peaks in the spectra of the original signals. The spectra for the models of order $P = 20$ provide closer fits than those for $P = 10$; peaks in the $P = 10$ spectra gloss over multiple peaks in the original spectra. Observe the presence of poles close to the unit circle in the z-plane at frequencies corresponding to the peaks in the spectra of the signals. The AR-model spectra are smoother and easier to interpret than the periodogram-based spectra illustrated in Figure 6.11 for the same subject. The spectra for the diastolic portion indicate more medium-frequency energy than those for the systolic portion, as expected. The model coefficients or poles provide a compact parametric representation of the signals and their spectra.

Figures 7.18 and 7.19 illustrate the FFT-based spectrum of a segment containing one S1 and the subsequent systolic portion of the PCG signal of a subject with systolic murmur, split S2, and opening snap of the mitral valve (see also Figures 4.28, 5.7, and 6.12); the AR-model spectra for order $P = 10$ and $P = 20$; and the poles of the model of order $P = 20$. Figures 7.20 and 7.21 illustrate the same items for a segment containing one S2 and the subsequent diastolic portion of the same subject. The systolic murmur has given rise to more medium-frequency components than in the case of the normal subject in Figure 7.14. The AR-model spectra clearly indicate additional and stronger peaks at $150\ Hz$ and $250\ Hz$, which are confirmed by poles close to the unit circle at the corresponding frequencies in Figure 7.19.

7.5.3 Relationship between AR and cepstral coefficients

If the poles of $H(z)$ are inside the unit circle in the complex z-plane, from the theory of complex variables, $\ln H(z)$ can be expanded into a Laurent series as

$$\ln H(z) = \sum_{n=1}^{\infty} \hat{h}(n)\, z^{-n}. \tag{7.61}$$

Given the definition of the complex cepstrum as the inverse z-transform of the logarithm of the z-transform of the signal, and the fact that the left-hand side of the equation above represents the z-transform of $\hat{h}(n)$, it is clear that the coefficients of the series $\hat{h}(n)$ are the cepstral coefficients of $h(n)$. If $H(z)$ has been approximated

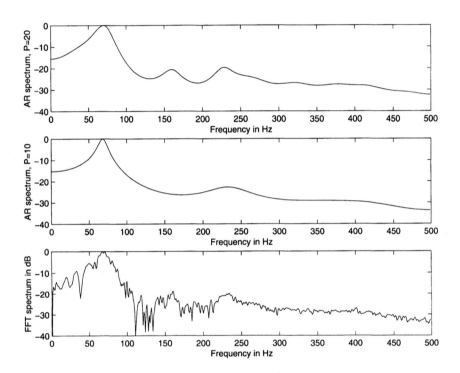

Figure 7.14   Bottom to top: FFT-based spectrum of the systolic portion of the PCG of a normal subject (male, 23 years); AR-model spectrum with order $P = 10$; AR-model spectrum with order $P = 20$. (See also Figures 4.27, 5.6, and 6.11.)

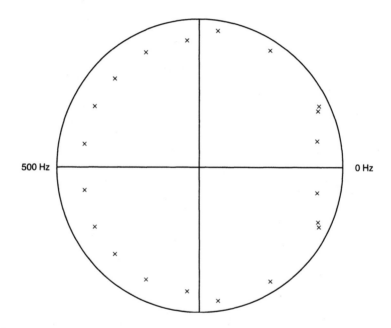

Figure 7.15 Poles of the AR model with order $P = 20$ of the systolic portion of the PCG of a normal subject. (See also Figure 7.14.)

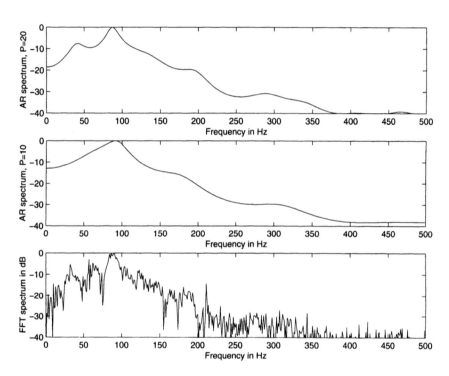

Figure 7.16 Bottom to top: FFT-based spectrum of the diastolic portion of the PCG of a normal subject (male, 23 years); AR-model spectrum with order $P = 10$; AR-model spectrum with order $P = 20$. (See also Figures 4.27, 5.6, and 6.11.)

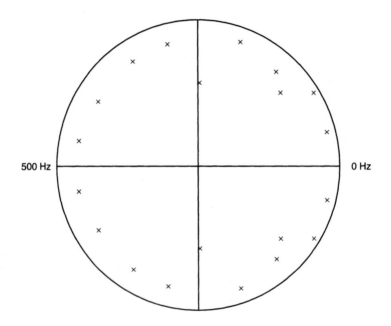

Figure 7.17   Poles of the AR model with order $P = 20$ of the diastolic portion of the PCG of a normal subject. (See also Figure 7.16.)

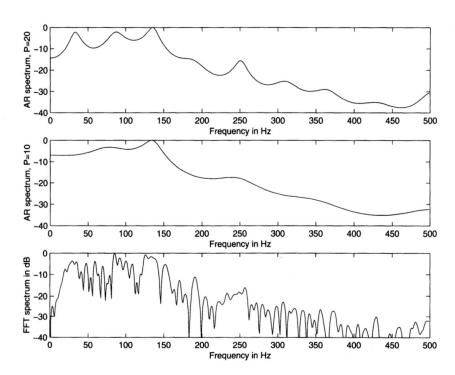

Figure 7.18 Bottom to top: FFT-based spectrum of the systolic portion of the PCG of a subject with systolic murmur, split S2, and opening snap of the mitral valve (female, 14 months); AR-model spectrum with order $P = 10$; AR-model spectrum with order $P = 20$. (See also Figures 4.28, 5.7, and 6.12.)

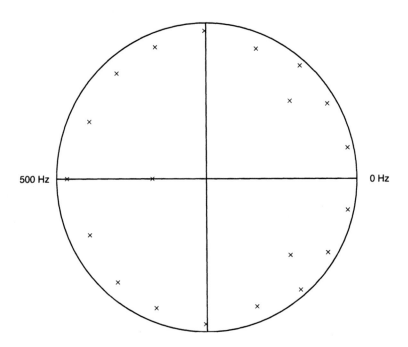

Figure 7.19   Poles of the AR model with order $P = 20$ of the systolic portion of the PCG of a subject with systolic murmur, split S2, and opening snap of the mitral valve. (See also Figure 7.18.)

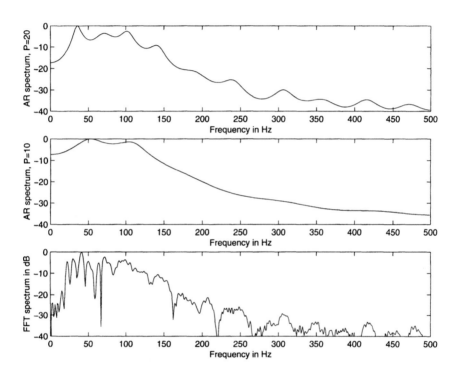

Figure 7.20 Bottom to top: FFT-based spectrum of the diastolic portion of the PCG of a subject with systolic murmur, split S2, and opening snap of the mitral valve (female, 14 months); AR-model spectrum with order $P = 10$; AR-model spectrum with order $P = 20$. (See also Figures 4.28, 5.7, and 6.12.)

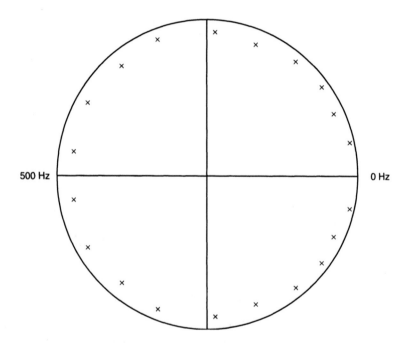

Figure 7.21   Poles of the AR model with order $P = 20$ of the diastolic portion of the PCG of a subject with systolic murmur, split S2, and opening snap of the mitral valve. (See also Figure 7.20.)

by an AR model with coefficients a_k, $1 \leq k \leq P$, we have

$$\ln\left(\frac{1}{1 + \sum_{k=1}^{P} a_k z^{-k}}\right) = \sum_{n=1}^{\infty} \hat{h}(n) z^{-n}. \tag{7.62}$$

Differentiating both sides of the equation above with respect to z^{-1}, we get

$$\frac{-\left(\sum_{k=1}^{P} k a_k z^{-k+1}\right)}{1 + \sum_{k=1}^{P} a_k z^{-k}} = \sum_{n=1}^{\infty} n \hat{h}(n) z^{-n+1}, \tag{7.63}$$

or

$$-\sum_{k=1}^{P} k a_k z^{-k+1} = \left(1 + \sum_{k=1}^{P} a_k z^{-k}\right) \sum_{n=1}^{\infty} n \hat{h}(n) z^{-n+1}. \tag{7.64}$$

By equating the constant term and the like powers of z^{-1} on both sides, the following relationship can be obtained [191]:

$$\hat{h}(1) = -a_1, \tag{7.65}$$

$$\hat{h}(n) = -a_n - \sum_{j=1}^{n-1} \left(1 - \frac{j}{n}\right) a_j \hat{h}(n-j), \quad 1 < n \leq P.$$

As we saw in Section 4.8.3, phase unwrapping is a major issue in estimating the cepstral coefficients using the inverse Fourier transform of the logarithm of the Fourier transform of a given signal [115]. Estimation of the cepstral coefficients using the AR coefficients has the advantage that it does not require phase unwrapping. Although the cepstral coefficients are deduced from the AR coefficients, it is expected that the nonlinear characteristics of the transformation could lead to an improvement in signal classification using the former than the latter set of coefficients. Cepstral coefficients have provided better classification than AR coefficients in speech [191], EMG [192], and VAG [58] signal analysis.

7.6   POLE-ZERO MODELING

Although AR or all-pole modeling can provide good spectral models for any kind of spectra with appropriately high model orders, it has a few limitations. The AR model essentially follows the peaks in the PSD of the signal being modeled, and thus resonance characteristics are represented well. However, if the signal has spectral nulls or valleys (anti-resonance), the AR-model spectrum will not provide a good fit in such spectral segments. Spectral zeros are important in modeling certain signals, such as nasal speech signals [193]. Furthermore, an all-pole model assumes the signal to be a minimum-phase signal or a maximum-phase signal, and does not allow mixed-phase signals [128].

The main conceptual difficulty posed by pole-zero modeling is that it is inherently non-unique, because a zero can be approximated by a large number of poles, and vice-versa [187]. However, if the system being modeled has a number of influential zeros, the number of poles required for an all-pole model can become very large. For these reasons, ARMA or pole-zero modeling [187, 193, 128, 80, 194, 195] is important in certain applications.

The ARMA normal equations: From the ARMA model represented by Equation 7.13, we can write the model PSD as [187]

$$\tilde{S}_y(\omega) = |H(\omega)|^2 = G^2 \frac{|B(\omega)|^2}{|A(\omega)|^2} = G^2 \frac{S_b(\omega)}{S_a(\omega)}, \tag{7.66}$$

where

$$S_a(\omega) = \left| 1 + \sum_{k=1}^{P} a_k \, \exp(-jk\omega) \right|^2 \tag{7.67}$$

and

$$S_b(\omega) = \left| 1 + \sum_{l=1}^{Q} b_l \, \exp(-jl\omega) \right|^2. \tag{7.68}$$

The total spectral-matching error is given by

$$\varepsilon = \frac{1}{2\pi} \int_{-\pi}^{\pi} S_y(\omega) \frac{S_a(\omega)}{S_b(\omega)} \, d\omega, \tag{7.69}$$

which may be viewed as the residual energy after passing the modeled signal through the inverse filter $\frac{A(z)}{B(z)}$. In order to obtain the optimal pole-zero model, we need to determine the coefficients a_k and b_l such that the error ε is minimized.

Before taking the derivatives of ε with respect to a_k and b_l, the following relationships are worth noting. Taking the partial derivative of $S_a(\omega)$ in Equation 7.67 with respect to a_i, we get

$$\frac{\partial S_a(\omega)}{\partial a_i} = 2 \sum_{k=0}^{P} a_k \, \cos[(i-k)\omega], \tag{7.70}$$

with $a_0 = 1$. Similarly, from Equation 7.68 we get

$$\frac{\partial S_b(\omega)}{\partial b_i} = 2 \sum_{l=0}^{Q} b_l \, \cos[(i-l)\omega]. \tag{7.71}$$

Let

$$\phi_{ya\beta}(i) = \frac{1}{2\pi} \int_{-\pi}^{\pi} S_y(\omega) \frac{[S_a(\omega)]^\beta}{[S_b(\omega)]^\alpha} \, \cos(i\omega) \, d\omega. \tag{7.72}$$

$\phi_{y00}(i)$ is the inverse Fourier transform of $S_y(\omega)$, and hence simply $\phi_y(i)$.

Now we can take the partial derivative of ε in Equation 7.69 with respect to a_i as

$$\frac{\partial \varepsilon}{\partial a_i} = \frac{1}{2\pi} \int_{-\pi}^{\pi} \frac{S_y(\omega)}{S_b(\omega)} \frac{\partial}{\partial a_i} S_a(\omega) \, d\omega \tag{7.73}$$

$$= \frac{1}{2\pi} \int_{-\pi}^{\pi} \frac{S_y(\omega)}{S_b(\omega)} \, 2 \sum_{k=0}^{P} a_k \, \cos[(i-k)\omega] \, d\omega$$

$$= 2 \sum_{k=0}^{P} a_k \, \phi_{y10}(i-k), \quad 1 \le i \le P.$$

In the same manner, we can obtain

$$\frac{\partial \varepsilon}{\partial b_i} = -2 \sum_{l=0}^{Q} b_l \, \phi_{y21}(i-l), \quad 1 \le i \le Q. \tag{7.74}$$

Because $\phi_{y10}(i-k)$ in Equation 7.73 is not a function of a_k, we obtain a set of linear equations by setting the final expression in Equation 7.73 to zero, which could be solved to obtain the a_k coefficients in a manner similar to the procedures used in AR modeling. However, $\phi_{y21}(i-l)$ in Equation 7.74 is a function of the b_l coefficients, which leads to a set of nonlinear equations that must be solved to obtain the b_l coefficients; the zeros of the model may then be derived from the b_l coefficients. Obtaining the ARMA model therefore requires solving P linear equations and Q nonlinear equations.

Iterative solution of the ARMA normal equations: Makhoul [187] describes the following iterative procedure to solve the $(P + Q)$ ARMA model normal equations based on the Newton-Raphson procedure:

Let $\mathbf{a} = [a_1, a_2, \ldots, a_P]^T$, $\mathbf{b} = [b_1, b_2, \ldots, b_Q]^T$, and $\mathbf{c} = [a_1, a_2, \ldots, a_P, b_1, b_2, \ldots, b_Q]^T$ represent the model coefficients to be derived in vector form. The vector at iteration $(m + 1)$ is derived from that at iteration m as

$$\mathbf{c}_{m+1} = \mathbf{c}_m - \mathbf{J}^{-1} \left. \frac{\partial \varepsilon}{\partial \mathbf{c}} \right|_{\mathbf{c} = \mathbf{c}_m}, \tag{7.75}$$

where $\mathbf{J}$ is the $(P + Q) \times (P + Q)$ symmetric Hessian matrix defined as $\mathbf{J} = \frac{\partial^2 \varepsilon}{\partial \mathbf{c} \, \partial \mathbf{c}^T}$. The vector $\mathbf{c}$ may be partitioned as $\mathbf{c}^T = [\mathbf{a}^T, \mathbf{b}^T]$, and the iterative procedure may be expressed as

$$\begin{bmatrix} \mathbf{a}_{m+1} \\ \mathbf{b}_{m+1} \end{bmatrix} = \begin{bmatrix} \mathbf{a}_m \\ \mathbf{b}_m \end{bmatrix} \tag{7.76}$$

$$- \begin{bmatrix} \frac{\partial^2 \varepsilon}{\partial \mathbf{a} \, \partial \mathbf{a}^T} & \frac{\partial^2 \varepsilon}{\partial \mathbf{a} \, \partial \mathbf{b}^T} \\ \frac{\partial^2 \varepsilon}{\partial \mathbf{b} \, \partial \mathbf{a}^T} & \frac{\partial^2 \varepsilon}{\partial \mathbf{b} \, \partial \mathbf{b}^T} \end{bmatrix}_{\substack{\mathbf{a} = \mathbf{a}_m \\ \mathbf{b} = \mathbf{b}_m}}^{-1} \begin{bmatrix} \frac{\partial \varepsilon}{\partial \mathbf{a}} \\ \frac{\partial \varepsilon}{\partial \mathbf{b}} \end{bmatrix}_{\substack{\mathbf{a} = \mathbf{a}_m \\ \mathbf{b} = \mathbf{b}_m}}.$$

Equations 7.73 and 7.74 give the first-order partial derivatives required above. The second-order partial derivatives are given as follows [187]:

$$\frac{\partial^2 \varepsilon}{\partial a_i \, \partial a_j} = 2\phi_{y10}(i-j), \tag{7.77}$$

$$\frac{\partial^2 \varepsilon}{\partial a_i \, \partial b_j} = -2 \sum_{k=0}^{P} \sum_{l=0}^{Q} a_k \, b_l \, [\phi_{y20}(j+i-l-k) + \phi_{y20}(j-i-l+k)], \quad (7.78)$$

and

$$\frac{\partial^2 \varepsilon}{\partial b_i \, \partial b_j} = -2\phi_{y21}(i-j) + 4 \sum_{k=0}^{P} \sum_{l=0}^{Q} b_k \, b_l \, [\phi_{y31}(j+i-l-k) + \phi_{y31}(j-i-l+k)].$$

$$(7.79)$$

The iterative procedure works well if the initial estimate is close to the optimal model; otherwise, one of the noniterative methods described in the following sections may be considered.

7.6.1 Sequential estimation of poles and zeros

Given the difficulties with the nonlinear nature of direct pole-zero modeling, a few methods have been proposed to split the problem into two parts: identify the poles first by AR modeling, and then treat the residual error in some manner to estimate the zeros [187, 193, 128, 80, 194, 195]. (*Note:* Several notational differences exist between the various references cited here. The following derivations use notations consistent with those used so far in the present chapter.)

Shanks' method: Let us consider a slightly modified version of Equation 7.13 as

$$H(z) = \frac{Y(z)}{X(z)} = \frac{B(z)}{A(z)} = \frac{1 + \sum_{l=1}^{Q} b_l \, z^{-l}}{1 + \sum_{k=1}^{P} a_k \, z^{-k}}, \quad (7.80)$$

where the gain factor G has been set to be unity: $G = 1$. The difference equation relating the output to the input is given by a small change to Equation 7.12 as

$$y(n) = -\sum_{k=1}^{P} a_k \, y(n-k) + \sum_{l=0}^{Q} b_l \, x(n-l). \quad (7.81)$$

The effect of the numerator and denominator polynomials in Equation 7.80 may be separated by considering $Y(z) = V(z)B(z)$, where $V(z) = \frac{X(z)}{A(z)}$. This leads to the all-zero or MA part of the system

$$y(n) = \sum_{l=0}^{Q} b_l \, v(n-l), \quad (7.82)$$

with $v(n)$ given by the all-pole or AR part of the model as

$$v(n) = -\sum_{k=1}^{P} a_k \, v(n-k) + x(n). \quad (7.83)$$

Let us consider the case of determining the a_k and b_l coefficients (equivalently, the poles and zeros) of the system $H(z)$ given its impulse response. Recollect that $y(n) = h(n)$ when $x(n) = \delta(n)$; consequently, we have $X(z) = 1$, and $Y(z) = H(z)$. The impulse response of the system is given by

$$h(n) = -\sum_{k=1}^{P} a_k \, h(n-k) + \sum_{l=0}^{Q} b_l \, \delta(n-l), \qquad (7.84)$$

which simplifies to

$$h(n) = -\sum_{k=1}^{P} a_k \, h(n-k), \ n > Q. \qquad (7.85)$$

The effect of the impulse input does not last beyond the number of zeros in the system: the system output is then perfectly predictable from the preceding P samples, and hence an AR or all-pole model is adequate to model $h(n)$ for $n > Q$. As a consequence, Equation 7.32 is modified to

$$\phi_h(i) = -\sum_{k=1}^{P} a_k \, \phi_h(i-k), \ \ i > Q. \qquad (7.86)$$

This system of equations may be solved by considering P equations with $Q + 1 \leq i \leq Q + P$. Thus the a_k coefficients and hence the poles of the system may be computed independently of the b_l coefficients or the zeros by restricting the AR error analysis to $n > Q$.

In a practical application, the error of prediction needs to be considered, as the model order P will not be known or some noise will be present in the estimation. Kopec et al. [193] recommend that the covariance method described in Section 7.5 be used to derive the AR model by considering the error of prediction as

$$e(n) = h(n) + \sum_{k=1}^{P} a_k \, h(n-k), \qquad (7.87)$$

and minimizing the TSE defined as

$$\varepsilon = \sum_{n=Q+1}^{\infty} |e(n)|^2. \qquad (7.88)$$

The first Q points are left out as they are not predictable with an all-pole model.

Let us assume that the AR modeling part has been successfully performed by the procedure described above. Let

$$\tilde{A}(z) = 1 + \sum_{k=1}^{P} \tilde{a}_k \, z^{-k} \qquad (7.89)$$

represent the denominator polynomial of the system that has been estimated. The TSE in modeling $h(n)$ is given by

$$\varepsilon = \sum_{n=0}^{\infty} \left| h(n) - \sum_{l=0}^{Q} b_l \, \tilde{v}(n-l) \right|^2, \qquad (7.90)$$

where $\tilde{v}(n)$ is the impulse response of the AR model derived, with $\tilde{V}(z) = \frac{1}{\tilde{A}(z)}$. Minimization of ε above leads to the set of linear equations

$$\sum_{l=0}^{Q} b_l \, \phi_{\tilde{v}\tilde{v}}(l, j) = \phi_{h\tilde{v}}(0, j), \ \ 0 \leq j \leq Q, \tag{7.91}$$

where

$$\phi_{h\tilde{v}}(l, j) = \sum_{n=0}^{\infty} h(n-l) \, \tilde{v}(n-j) \tag{7.92}$$

is the CCF between $h(n)$ and $\tilde{v}(n)$, and $\phi_{\tilde{v}\tilde{v}}$ is the ACF of $\tilde{v}(n)$.

The frequency-domain equivalents of the steps above may be analyzed as follows. The TSE is

$$\begin{aligned}
\varepsilon &= \frac{1}{2\pi} \int_{-\pi}^{\pi} \left| H(\omega)\tilde{A}(\omega) - \sum_{l=0}^{Q} b_l \, \exp(-jl\omega) \right|^2 |\tilde{V}(\omega)|^2 \, d\omega \quad (7.93) \\
&= \frac{1}{2\pi} \int_{-\pi}^{\pi} |E_h(\omega) - B(\omega)|^2 \, |\tilde{V}(\omega)|^2 \, d\omega,
\end{aligned}$$

where $E_h(z) = H(z)\tilde{A}(z)$ is the AR model error in the z-domain. [Recall that the Fourier spectrum of a signal is obtained by evaluating the corresponding function of z with $z = \exp(j\omega)$.] The method described above, which is originally attributed to Shanks [80] and has been described as above by Kopec et al. [193], therefore estimates the numerator polynomial of the model by fitting a polynomial (spectral function) to the z-transform of the AR or all-pole model error.

Makhoul [187] and Kopec et al. [193] suggest another method labeled as *inverse LP* modeling, where the inverse of the AR model error $e_h^{-1}(n)$ given as the inverse z-transform of $E_h^{-1}(z)$ is subjected to all-pole modeling. The poles so obtained are the zeros of the original system being modeled.

7.6.2   Iterative system identification

Problem: *Given a noisy observation of the output of a linear system in response to a certain input, develop a method to estimate the numerator and denominator polynomials of a rational z-domain model of the system.*

Solution: In consideration of the difficulty in solving the nonlinear problem inherent in ARMA modeling or pole-zero estimation, Steiglitz and McBride [194] proposed an iterative procedure based upon an initial estimate of the denominator (AR) polynomial. Since their approach to system identification is slightly different from the LP approach we have been using so far in this chapter, it is appropriate to restate the problem.

The Steiglitz-McBride method: Figure 7.22 shows a block diagram illustrating the problem of system identification. The system is represented by its transfer function $H(z)$, input $x(n)$, output $y(n) = h(n) * x(n)$, and the noisy observation

$w(n) = y(n) + \eta(n)$, where $\eta(n)$ is a noise process that is statistically independent of the signals being considered. $H(z)$ is represented as a rational function of z, as

$$H(z) = \frac{Y(z)}{X(z)} = \frac{B(z)}{A(z)} = \frac{\sum_{l=0}^{Q} b_l z^{-l}}{\sum_{k=0}^{P} a_k z^{-k}}. \tag{7.94}$$

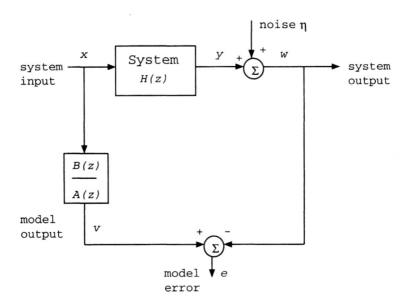

Figure 7.22   Schematic representation of system identification. Adapted from Steiglitz and McBride [194].

The error to be minimized may be written as [194]

$$\sum_{n=0}^{N-1} e^2(n) = \frac{1}{2\pi j} \oint \left| X(z)\frac{B(z)}{A(z)} - W(z) \right|^2 \frac{dz}{z}, \tag{7.95}$$

where the right-hand side represents the inverse z-transform of the function of z involved, and N is the number of data samples available. The functions of z within the integral essentially compare the predicted model output with the observed output of the physical system.

As seen earlier, this approach leads to a nonlinear problem. The problem may be simplified (linearized) by taking the approach of separate identification of the numerator and denominator polynomials: the estimation problem illustrated in Figure 7.23 treats $A(z)$ and $B(z)$ as separate systems. The error to be minimized may then be written as [194]

$$\sum_{n=0}^{N-1} e^2(n) = \frac{1}{2\pi j} \oint |X(z)B(z) - W(z)A(z)|^2 \frac{dz}{z}. \tag{7.96}$$

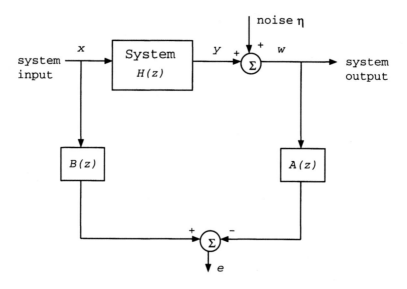

Figure 7.23 Schematic representation of system identification with separate estimation of $A(z)$ and $B(z)$. Adapted from Steiglitz and McBride [194].

(This approach was originally proposed by Kalman [196].)

The sample model error is given by

$$e(n) = \sum_{l=0}^{Q} b_l\, x(n-l) - \sum_{k=1}^{P} a_k\, w(n-k) - w(n). \qquad (7.97)$$

The model coefficients and the input – output data samples may be written in vector form as

$$\mathbf{c} = [b_0, b_1, \ldots, b_Q, -a_1, -a_2, \ldots, -a_P]^T, \qquad (7.98)$$

and

$$\mathbf{d}(n) = [x(n), x(n-1), \ldots, x(n-Q), w(n-1), w(n-2), \ldots, w(n-P)]^T, \qquad (7.99)$$

with the vectors being of size $(P + Q + 1)$. Then, the error is given by

$$e(n) = \mathbf{d}^T(n)\mathbf{c} - w(n). \qquad (7.100)$$

The condition for minimum TSE is given by

$$\frac{\partial}{\partial \mathbf{c}} \sum_{n=0}^{N-1} e^2(n) = 2 \sum_{n=0}^{N-1} \frac{\partial e(n)}{\partial \mathbf{c}} e(n) = 2 \sum_{n=0}^{N-1} \mathbf{d}(n)e(n) = 0. \qquad (7.101)$$

Substitution of the expression for the error in Equation 7.100 in the above condition gives

$$\left(\sum_{n=0}^{N-1} \mathbf{d}(n)\mathbf{d}^T(n) \right) \mathbf{c} = \sum_{n=0}^{N-1} w(n)\mathbf{d}(n). \qquad (7.102)$$

If we let

$$\mathbf{\Phi} = \sum_{n=0}^{N-1} \mathbf{d}(n)\mathbf{d}^T(n) \qquad (7.103)$$

represent the $(P + Q + 1) \times (P + Q + 1)$ correlation matrix of the combined string of input – output data samples $\mathbf{d}(n)$, and let

$$\mathbf{\Theta} = \sum_{n=0}^{N-1} w(n)\mathbf{d}(n) \qquad (7.104)$$

represent the correlation between the signal $w(n)$ and the data vector $\mathbf{d}(n)$ of size $(P + Q + 1)$, we get the solution to the estimation problem as

$$\mathbf{c} = \mathbf{\Phi}^{-1}\,\mathbf{\Theta}. \qquad (7.105)$$

Although the vectors and matrices related to the filter coefficients and the signal correlation functions are defined in a different manner, the solution above is comparable to that of the optimal Wiener filter (see Section 3.5 and Equation 3.84).

The limitation of the approach above is that the error used has no physical significance. The separation of the numerator and denominator functions as in Figure 7.23, while simplifying the estimation problem, has led to a situation that is far from reality.

To improve upon the match between the real physical situation and the estimation problem, Steiglitz and McBride [194] proposed an iterative procedure which is schematically illustrated in Figure 7.24. The basic approach is to treat the system identified using the simplified procedure described above as an initial estimate, labeled as $A_1(z)$ and $B_1(z)$; filter the original signals $x(n)$ and $w(n)$ with the system $\frac{1}{A_1(z)}$; use the filtered signals to obtain new estimates $A_2(z)$ and $B_2(z)$; and iterate the procedure until convergence is achieved.

The error to be minimized may be written as [194]

$$
\begin{aligned}
\sum_{n=0}^{N-1} e^2(n) &= \frac{1}{2\pi j} \oint \left| X(z)\frac{B_i(z)}{A_{i-1}(z)} - W(z)\frac{A_i(z)}{A_{i-1}(z)} \right|^2 \frac{dz}{z} \quad (7.106) \\
&= \frac{1}{2\pi j} \oint \left| X(z)\frac{B_i(z)}{A_i(z)} - W(z) \right|^2 \left| \frac{A_i(z)}{A_{i-1}(z)} \right|^2 \frac{dz}{z},
\end{aligned}
$$

with $A_0(z) = 1$. It is obvious that upon convergence, when $A_i(z) = A_{i-1}(z)$, the minimization problem above reduces to the ideal (albeit nonlinear) situation expressed in Equation 7.95 and illustrated in Figure 7.22.

Steiglitz and McBride [194] suggest a modified iterative procedure to further improve the estimate, by imposing the condition that the partial derivatives of the true error criterion with respect to the model coefficients be equal to zero at convergence. The true (ideal) model error is given in the z-domain as (refer to Figure 7.22)

$$E(z) = X(z)\frac{B(z)}{A(z)} - W(z) = V(z) - W(z). \qquad (7.107)$$

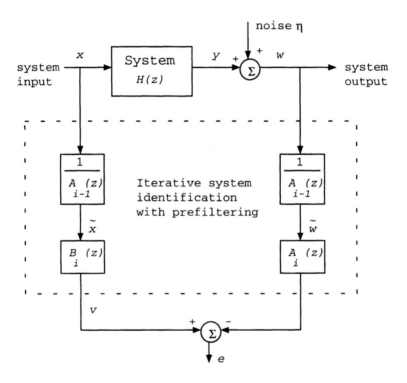

Figure 7.24 Schematic representation of system identification via iterative prefiltering. Adapted from Steiglitz and McBride [194].

$V(z)$ is the output of the model for the input $x(n)$. The derivatives of $E(z)$ with respect to the model coefficients are given by

$$\frac{\partial \varepsilon}{\partial a_i} = -\frac{X(z)B(z)}{A^2(z)} z^{-i} = -\frac{V(z)}{A(z)} z^{-i} = -\tilde{V}(z)z^{-i} \qquad (7.108)$$

and

$$\frac{\partial \varepsilon}{\partial b_i} = \frac{X(z)}{A(z)} z^{-i} = \tilde{X}(z)z^{-i}, \qquad (7.109)$$

where the superscript ˜ represents a filtered version of the corresponding signal, the filter transfer function being $\frac{1}{A(z)}$. A new data vector is defined as

$$\mathbf{d}_1(n) = [\tilde{x}(n), \tilde{x}(n-1), \dots, \tilde{x}(n-Q), \tilde{v}(n-1), \tilde{v}(n-2), \dots, \tilde{v}(n-P)]^T. \quad (7.110)$$

The error gradient in Equation 7.101 is modified to

$$\frac{\partial}{\partial \mathbf{c}} \sum_{n=0}^{N-1} e^2(n) = 2 \sum_{n=0}^{N-1} \frac{\partial e(n)}{\partial \mathbf{c}} e(n) = 2 \sum_{n=0}^{N-1} \mathbf{d}_1(n)e(n) \qquad (7.111)$$

$$= 2 \sum_{n=0}^{N-1} [\mathbf{d}_1(n)\mathbf{d}_1^T(n)\mathbf{c} - w(n)\mathbf{d}_1(n)],$$

where the last equality is true only at convergence. The rest of the procedure remains the same as before, but with the correlation functions defined as

$$\mathbf{\Phi_1} = \sum_{n=0}^{N-1} \mathbf{d}_1(n)\mathbf{d}_1^T(n) \qquad (7.112)$$

and

$$\mathbf{\Theta_1} = \sum_{n=0}^{N-1} w(n)\mathbf{d}_1(n). \qquad (7.113)$$

Once the a_k and b_l coefficients are obtained, the related polynomials may be solved to obtain the poles and zeros of the system being modeled. Furthermore, the polynomials may be used to derive spectral models of the system or the signal of interest. Note that the procedures given above are applicable to the special case of system identification when the impulse response $h(n)$ is given: we just need to change $x(n) = \delta(n)$ and $X(z) = 1$. Steiglitz and McBride [194] did not provide any proof of convergence of their methods; however, it was indicated that the method performed successfully in many practical applications.

The Steiglitz-McBride method was applied to the modeling and classification of PCG signals by Joo et al. [197]. The first and second peak frequencies were detected from the model spectra and used to analyze porcine prosthetic valve function. Murthy and Prasad [198] applied the Steiglitz-McBride method to ECG signals. Pole-zero models derived from ECG strips including a few cardiac cycles were were used to reconstruct and identify the ECG waveform over a single cycle, and also to reconstruct separately (that is, to segment) the P, QRS, and T waves.

7.6.3 Homomorphic prediction and modeling

Problem: *Given the relative ease of all-pole modeling, is it possible to convert the zeros of a system to poles?*

Solution: As mentioned earlier, an all-pole model assumes the signal being modeled to be a minimum-phase signal or a maximum-phase signal, and does not allow mixed-phase signals [128]. We have seen in Sections 4.8.3 and 5.4.2 that homomorphic filtering can facilitate the separation of the minimum-phase and maximum-phase components of a mixed-phase signal, and further facilitate the derivation of a minimum-phase version or correspondent (MPC) of a mixed-phase signal. Makhoul [187], Oppenheim et al. [128], and Kopec et al. [193] suggest methods to combine homomorphic filtering and LP into a procedure that has been labeled *homomorphic prediction* or *cepstral prediction*.

An intriguing property that arises in homomorphic prediction is that if a signal $x(n)$ has a rational z-transform, then $n\hat{x}(n)$ [where $\hat{x}(n)$ is the complex cepstrum of $x(n)$] has a rational z-transform whose poles correspond to the poles and zeros of $x(n)$. The basic property of the z-transform we need to recollect here is that if $X(z)$ is the z-transform of $x(n)$, then the z-transform of $nx(n)$ is $-z \frac{dX(z)}{dz}$. Now, the complex cepstrum $\hat{x}(n)$ of $x(n)$ is defined as the inverse z-transform of $\hat{X}(z) = \log X(z)$. Therefore, we have

$$ZT[n\hat{x}(n)] = -z \frac{d\hat{X}(z)}{dz} = -z \frac{1}{X(z)} \frac{dX(z)}{dz}, \qquad (7.114)$$

where $ZT[\ ]$ represents the z-transform operator. If $X(z) = \frac{B(z)}{A(z)}$, we get

$$ZT[n\hat{x}(n)] = -z \frac{A(z)B'(z) - B(z)A'(z)}{A(z)B(z)}, \qquad (7.115)$$

where the prime $'$ denotes the derivative of the associated function with respect to z. A general representation of a rational function of z (which represents an exponential signal in the z-domain) in terms of its poles and zeros is given by [128]

$$X(z) = A\, z^r \frac{\prod_{l=1}^{Q_i} (1 - z_{il}\, z^{-1}) \prod_{n=1}^{Q_o} (1 - z_{on}\, z)}{\prod_{k=1}^{P_i} (1 - p_{ik}\, z^{-1}) \prod_{m=1}^{P_o} (1 - p_{om}\, z)}, \qquad (7.116)$$

with the magnitudes of all of the z_i, z_o, p_i, and p_o coefficients being less than unity. The p_i and z_i values above give the P_i poles and Q_i zeros, respectively, of the system that are inside the unit circle in the z-plane; $\frac{1}{p_o}$ and $\frac{1}{z_o}$ give the P_o poles and Q_o zeros, respectively, that lie outside the unit circle. Of course, a causal and stable system will not have any poles outside the unit circle; the general representation above will permit the analysis and modeling of maximum-phase signals that are anti-causal. Computation of the complex cepstrum requires the removal of any linear phase component that may be present, and hence we could impose the condition $r = 0$. We then have

$$\hat{X}(z) = \log X(z) = \log A \qquad (7.117)$$

$$+ \sum_{l=1}^{Q_i} \log(1 - z_{il} z^{-1}) + \sum_{n=1}^{Q_o} \log(1 - z_{on} z)$$

$$- \sum_{k=1}^{P_i} \log(1 - p_{ik} z^{-1}) - \sum_{m=1}^{P_o} \log(1 - p_{om} z),$$

and furthermore,

$$-z \frac{d\hat{X}(z)}{dz} = - \sum_{l=1}^{Q_i} \frac{z_{il} z^{-1}}{(1 - z_{il} z^{-1})} + \sum_{n=1}^{Q_o} \frac{z_{on} z}{(1 - z_{on} z)} \qquad (7.118)$$

$$+ \sum_{k=1}^{P_i} \frac{p_{ik} z^{-1}}{(1 - p_{ik} z^{-1})} - \sum_{m=1}^{P_o} \frac{p_{om} z}{(1 - p_{om} z)}.$$

From the expression above, it is evident that $n\hat{x}(n)$ has simple (first-order) poles at every pole as well as every zero of $x(n)$. Therefore, we could apply an all-pole modeling procedure to $n\hat{x}(n)$, and then separate the poles so obtained into the desired poles and zeros of $x(n)$. An initial all-pole model of $x(n)$ can assist in the task of separating the poles from the zeros. Oppenheim et al. [128] show further that even if $X(z)$ is irrational, $n\hat{x}(n)$ has a rational z-transform with first-order poles corresponding to each irrational factor in $X(z)$.

Illustration of application to a synthetic speech signal: Figures 7.25 and 7.26 show examples of the application of several pole-zero and all-pole modeling techniques to a synthetic speech signal [128]. The impulse response of the synthetic system with two poles at $292\ Hz$ and $3,500\ Hz$ with bandwidth $79\ Hz$ and $100\ Hz$, and one zero at $2,000\ Hz$ with bandwidth $200\ Hz$, is shown in Figure 7.25 (a), with its log-magnitude spectrum in Figure 7.26 (a). The formant or resonance structure of the signal is evident in the spectral peaks. (The sampling rate is $12\ kHz$.) Excitation of the system with a pulse train with repetition rate $120\ Hz$ resulted in the signal in Figure 7.25 (b), whose spectrum is shown in Figure 7.26 (b); the spectrum clearly shows the effect of repetition of the basic wavelet in the series of waves that are superimposed on the basic spectrum of the wavelet. Application of homomorphic filtering to the signal in Figure 7.25 (b) provided an estimate of the basic wavelet as shown in Figure 7.25 (c), with the corresponding spectrum in Figure 7.26 (c).

The pole-zero modeling method of Shanks was applied to the result of homomorphic filtering in Figure 7.25 (c) with four poles and two zeros. The impulse response of the model and the corresponding spectrum are shown in Figure 7.25 (d) and Figure 7.26 (d), respectively. It is seen that the two peaks and the valley in the original spectrum are faithfully reproduced in the modeled spectrum. The frequencies of the poles (and their bandwidths) given by the model were $291\ Hz$ ($118\ Hz$) and $3,498\ Hz$ ($128\ Hz$), and those of the zero were $2,004\ Hz$ ($242\ Hz$), which compare well with those of the synthesized system listed in the preceding paragraph.

Application of the autocorrelation method of LP modeling with six poles to the original signal in Figure 7.25 (a) resulted in the model impulse response and spectrum illustrated in Figures 7.25 (e) and 7.26 (e). While the all-pole model spectrum has

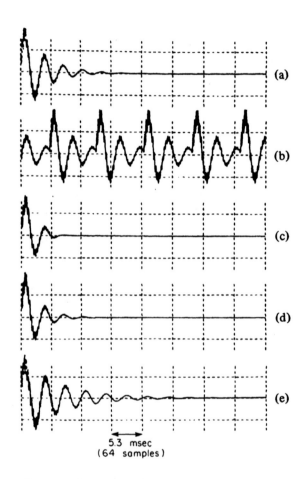

5.3 msec
(64 samples)

Figure 7.25 Time-domain signals: (a) impulse response of a 4-pole, 2-zero synthetic system; (b) synthesized voiced-speech signal obtained by triggering the system with an impulse train; (c) result of basic wavelet extraction via application of homomorphic filtering to the signal in (b); (d) impulse response of a 4-pole, 2-zero model of the signal in (c) obtained by Shanks' method; (e) impulse response of a 6-pole AR model. Reproduced with permission from A.V. Oppenheim, G.E. Kopec, and J.M. Tribolet, Signal analysis by homomorphic prediction, *IEEE Transactions on Acoustics, Speech, and Signal Processing,* 24(4):327–332, 1976. ©IEEE.

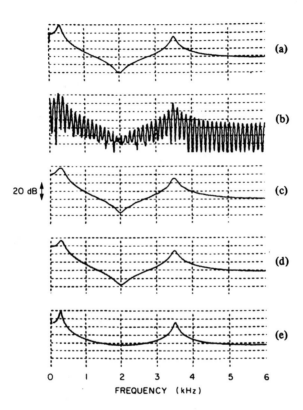

20 dB

FREQUENCY (kHz)

Figure 7.26 Log-magnitude spectra of the time-domain signals in Figure 7.25: (a) actual spectral response of the 4-pole, 2-zero synthetic system; (b) spectrum of the synthesized voiced-speech signal obtained by triggering the system with an impulse train; (c) spectrum of the basic wavelet extracted via application of homomorphic filtering to the signal corresponding to (b); (d) spectral response of a 4-pole, 2-zero model of the signal in (c) obtained by Shanks' method; (e) spectral response of a 6-pole AR model. Reproduced with permission from A.V. Oppenheim, G.E. Kopec, and J.M. Tribolet, Signal analysis by homomorphic prediction, *IEEE Transactions on Acoustics, Speech, and Signal Processing,* 24(4):327–332, 1976. ©IEEE.

followed the spectral peaks well, it has failed to represent the valley or null related to the zero.

Illustration of application to a real speech signal: Figure 7.27 (a) shows the log-magnitude spectrum of a real speech signal (pre-emphasized) of the nasalized vowel /U/ in the word "moon" [193]. Part (b) of the same figure shows the spectrum after homomorphic filtering to remove the effects of repetition of the basic wavelet. Parts (c) and (d) show 10-pole, 6-zero model spectra obtained using Shanks' method and inverse LP modeling, respectively. The spectra of the models have successfully followed the peaks and valleys in the signal spectrum.

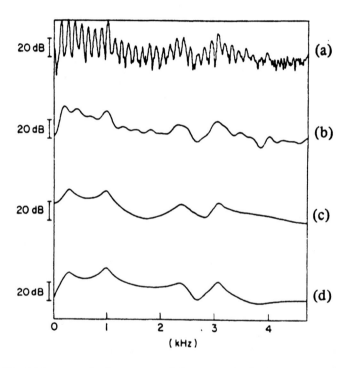

Figure 7.27   (a) Log-magnitude spectrum of the pre-emphasized, real speech signal of the nasalized vowel /U/ in the word "moon"; (b) spectrum after homomorphic filtering to remove the effects of repetition of the basic wavelet; (c) spectral response of a 10-pole, 6-zero model obtained by Shanks' method; (d) spectral response of a 10-pole, 6-zero model obtained by inverse LP modeling. Reproduced with permission from G.E. Kopec, A.V. Oppenheim, and J.M. Tribolet, Speech analysis by homomorphic prediction, *IEEE Transactions on Acoustics, Speech, and Signal Processing,* 25(1):40–49, 1977. ©IEEE.

Shanks' method was applied to the minimum-phase and maximum-phase components of ECG signals obtained via homomorphic filtering by Murthy et al. [199]. Akay et al. [200] used ARMA techniques to model diastolic heart sounds for the detection of coronary heart disease; however, only the dominant poles of the model were used in pattern analysis (see Section 7.10 for details of this application).

7.7 ELECTROMECHANICAL MODELS OF SIGNAL GENERATION

While purely mathematical models of signal generation such as point processes and linear system models provide the advantage of theoretical elegance and convenience, they may not be able to represent certain physical and physiological aspects of the systems that generate the signals. For example, the models we have seen in the preceding sections cannot directly accommodate the physical dimensions of blood vessels or valves, the loss in the compliance of a valve leaflet due to stenosis, or the lubrication (or the lack thereof) or friction between joint surfaces.

Sikarskie et al. [201] proposed a model to characterize aortic valve vibration for the analysis of its contribution to S2; in addition to mathematical relationships, they included physical factors such as the valve forcing function, valve mass, and valve stiffness. It was shown that the amplitude and frequency of A2 depend strongly on the valve forcing function and valve stiffness. Valve mass was shown to have little effect on the amplitude and frequency of A2; blood density was shown to have no effect on the same parameters. We shall now study two representative applications of electromechanical modeling, where mechanical models and their electrical counterparts are used to represent the generation and altered characteristics of sounds in arteries and knee joints.

7.7.1 Sound generation in coronary arteries

Problem: *Propose an electromechanical model to characterize the sounds produced due to blood flow in stenosed arteries.*

Solution: Blood vessels are normally flexible, elastic, and pliant, with smooth internal surfaces. When a segment of a blood vessel is hardened due to the deposition of calcium and other minerals, the segment becomes rigid. Furthermore, the development of plaque inside the vessel causes narrowing or constriction of the vessel, which impedes the flow of blood. The result is a turbulent flow of blood, with accompanying high-frequency sounds.

Wang et al. [202, 203] proposed a sound-source model combining an incremental-network model of the left coronary-artery tree with a transfer-function model describing the resonance characteristics of arterial chambers. The network model, illustrated in Figure 7.28, predicts flow in normal and stenosed arteries. It was noted that stenotic branches may require division into multiple segments in the model due to greater geometric variations. Furthermore, it was observed that a stenotic segment may exhibit post-stenotic dilation as illustrated in Figure 7.29, due to increased pressure fluctuations caused by turbulence at the point of stenosis.

The resonance frequency of a segment depends upon the length and diameter of the segment, as well as upon the distal (away from the heart) hydraulic pressure loading the segment. The physical parameters required for the model were obtained from arteriograms of the patient being examined. The terminal resistances, labeled Z in Figure 7.28, represent loading of the resistive arteriolar beds, assumed to be directly related to the areas that the terminal branches serve.

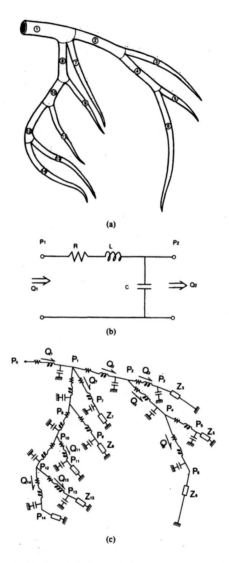

Figure 7.28   Electromechanical model of a coronary artery tree. (a) The left coronary-artery tree is divided into 14 branches. (b) Circuit model of a segment. (c) Circuit model of the artery tree. Reproduced with permission from J.Z. Wang, B. Tie, W. Welkowitz, J.L. Semmlow, and J.B. Kostis, Modeling sound generation in stenosed coronary arteries, *IEEE Transactions on Biomedical Engineering,* 37(11):1087–1094, 1990, ©IEEE; and J.Z. Wang, B. Tie, W. Welkowitz, J. Kostis, and J. Semmlow, Incremental network analogue model of the coronary artery, *Medical & Biological Engineering & Computing,* 27:416–422, 1989. ©IFMBE.

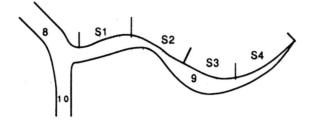

S1: Proximal to Stenosis Segment
S2: Stenotic Segment
S3: Poststenotic Dilation Segment
S4: Distal to Dilation Segment

Figure 7.29 Hypothetical example of stenosis in coronary artery branch 9. Reproduced with permission from J.Z. Wang, B. Tie, W. Welkowitz, J.L. Semmlow, and J.B. Kostis, Modeling sound generation in stenosed coronary arteries, *IEEE Transactions on Biomedical Engineering,* 37(11):1087–1094, 1990. ©IEEE.

Wang et al. related the network elements (resistance R, inertance or inductance L, and capacitance C) to the physical parameters of the artery segments as

$$R = 8\pi\nu \, \frac{l}{A^2}, \tag{7.119}$$

$$L = \rho \, \frac{l}{A},$$

$$C = Alh \, \frac{D}{E},$$

where $\nu = 0.04 \ g \, cm^{-1} \, s$ is the viscosity of blood, $\rho = 1.0 \ g \, cm^{-3}$ is the density of blood, $E = 2 \times 10^6 \ g \, cm^{-1} \, s^2$ is the Young's modulus of the blood vessel, D is the diameter of the segment, $A = \pi\frac{D^2}{4}$ is the cross-sectional area of the segment, $h \approx 0.08D$ is the wall thickness of the segment, and l is the length of the segment. Wang et al. [203] remarked that while the network elements may be assumed to be approximately constant during diastole, the assumption would not be valid during systole due to variations in the parameters of the segments.

In analyzing the artery – network model, voltage is analogous to pressure (P), and current is analogous to blood flow (Q). State-variable differential equations were used by Wang et al. [203] to derive the flow through the artery tree model for various pressure waveforms. It was hypothesized that turbulence at the point of stenosis would provide the excitation power, and that the stenotic segment and the dilated segment distal to the point of stenosis (see Figure 7.29) would act as resonance chambers.

Wang et al. [202] used the following relationships to compute the RMS pressure fluctuation (see also Fredberg [204]):

$$\langle P^2 \rangle_{\max} = 10^{-4} \rho u^2 f(x), \tag{7.120}$$

$$f(x) = 25.1 - 37.1x + 15.5x^2 - 0.08x^3 - 0.89x^4 + 0.12x^5,$$

$$x = 10^{-3} \frac{ud}{\nu} \left(\frac{D}{d} \right)^{0.75},$$

where u is the blood velocity in the stenotic segment, and d is the diameter of the stenotic segment. The incremental network model was used to estimate the blood velocity in each segment.

The wide-band spectrum of the sound associated with turbulent flow was modeled as (see also Fredberg [204]):

$$S(f) = \frac{0.7 \frac{d}{U} \langle P^2 \rangle_{\max}}{1 + 0.5 \left[f \frac{d}{U} \right]^{\frac{10}{3}}}, \tag{7.121}$$

where U is the velocity of blood in a normal segment and f is frequency in Hz. Wang et al. used the function $S(f)$ as above as the source of excitation power to derive the response of their network model. It was observed that the model spectra indicated two resonance frequencies, the magnitude and frequency of which depended upon the geometry and loading of the segments. Wang et al. cautioned that the results of the model are sensitive to errors in the estimation of the required parameters from arteriograms or other sources.

Figure 7.30 illustrates the model spectra for segment 12 of the artery tree model in Figure 7.28 with no stenosis and with stenosis of two grades. Narrowing of the segment with increasing stenosis is seen to shift the second peak in the spectrum to higher frequencies, while the magnitude and frequency of the first peak are both reduced. The results were confirmed by comparing the model spectra with spectra of signals recorded from a few patients with stenosed coronary arteries. Examples of spectral analysis of signals recorded from patients before and after angioplasty to correct for stenosis will be presented in Section 7.10.

7.7.2 Sound generation in knee joints

Problem: *Develop a mechanical analog of the knee joint to model the generation of the pulse train related to physiological patello-femoral crepitus.*

Solution: Beverland et al. [176] studied the PPC signals produced during very slow movement of the leg (at about $4^\circ/s$). The signals were recorded by taping accelerometers to the skin above the upper pole and/or the lower pole of the patella. Reproducible series of bursts of vibration were recorded in their experiments. Figure 7.31 illustrates two channels of simultaneously recorded PPC signals from the upper and lower poles of the patella during extension and flexion of the leg. The signals display reversed similarity when extension versus flexion or upper-pole versus lower-pole recordings are compared.

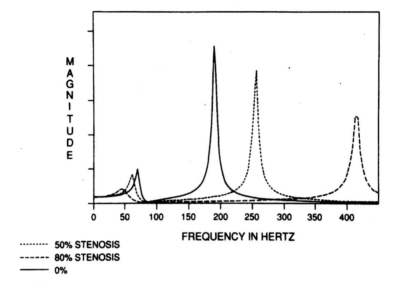

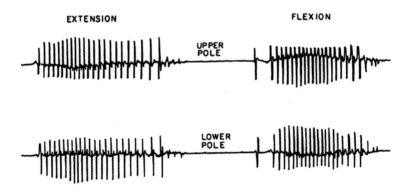

Figure 7.30 Shift in frequency components predicted by the transfer-function model for the case of stenosis in element 12 in the model of the coronary artery in Figure 7.28. Reproduced with permission from J.Z. Wang, B. Tie, W. Welkowitz, J.L. Semmlow, and J.B. Kostis, Modeling sound generation in stenosed coronary arteries, *IEEE Transactions on Biomedical Engineering,* 37(11):1087–1094, 1990. ©IEEE.

Figure 7.31 Simultaneously recorded PPC signals from the upper and lower poles of the patella during extension and flexion of the leg. The duration of the signal was not specified. Reproduced with permission from D.E. Beverland, W.G. Kernohan, G.F. McCoy, and R.A.B. Mollan, What is physiological patellofemoral crepitus?, *Proceedings of XIV International Conference on Medical and Biological Engineering and VII International Conference on Medical Physics,* pp 1249–1250, Espoo, Finland, 1985. ©IFMBE

Beverland et al. proposed a mechanical model to explain the generation of the PPC signals. The patella was considered to behave like a see-saw in the model, which was supported by the observation that a pivot point exists at the mid-point of the patella. The apparatus constructed, as illustrated in Figure 7.32, included a rubber wheel to represent the trochlear surface of the femur, on top of which was tensioned a rectangular piece of hardboard to represent the patella.

It was argued that as the wheel in the model is slowly rotated clockwise (representing extension), it would initially stick to the overlying patella (hardboard) due to static friction. This would tend to impart an anticlockwise rotatory motion, as a rotating cogwheel would impart an opposite rotation to a cog in contact with it (as illustrated in the upper right-hand corner of Figure 7.32). The upper end of the patella would then move toward the wheel. A point would be reached where the static friction would be overcome, when the patella would slip and the rotation is suddenly reversed, with the upper pole jerking outward and the lower pole jerking inward. The actions would be the opposite to those described above in the case of flexion. The mechanical model was shown to generate signals similar to those recorded from subjects, thereby confirming the *stick-slip* frictional model for the generation of PPC signals.

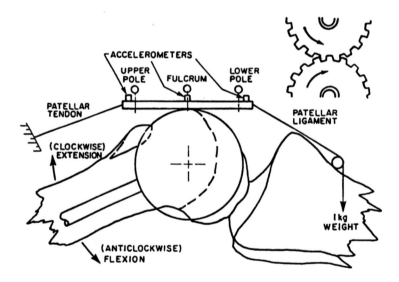

Figure 7.32 Apparatus to mimic the generation of PPC signals via a stick-slip frictional model. Reproduced with permission from D.E. Beverland, W.G. Kernohan, G.F. McCoy, and R.A.B. Mollan, What is physiological patellofemoral crepitus?, *Proceedings of XIV International Conference on Medical and Biological Engineering and VII International Conference on Medical Physics,* pp 1249–1250, Espoo, Finland, 1985. ©IFMBE

7.8 APPLICATION: ANALYSIS OF HEART-RATE VARIABILITY

Problem: *Explore the applicability of Fourier spectral analysis methods to study heart-rate data.*

Solution: DeBoer et al. [72] applied Fourier analysis techniques to two types of data derived from heart-rate data. (See also Akselrod et al. [205].) They noted that the standard Fourier analysis methods cannot be applied directly to a series of point events. Therefore, they derived three types of signals from trains of ECG beats as illustrated in Figure 7.1.

The *interval spectrum* was derived by computing the Fourier spectrum of the interval series, normalized as $\tilde{I}_k = (I_k - \bar{I}) / \bar{I}$, where $\bar{I}$ is the mean interval length. The frequency axis was scaled by considering the time-domain data to be spaced at distances equal to the mean interval length $\bar{I}$, that is, the effective sampling frequency is $1 / \bar{I}$.

The *spectrum of counts* was derived by taking the Fourier transform of the impulse-train representation, derived from RR interval data as shown in Figure 7.1. The signal was normalized and scaled as $\tilde{s}(t) = \sum [\bar{I} \, \delta(t - t_k)] - N$, where N is the number of data samples, and the Fourier transform was computed. The spectra computed were smoothed with a 27-point rectangular window. DeBoer et al. demonstrated that the two spectra exhibit similar characteristics under certain conditions of slow or slight modulation of the data about the mean heart rate.

The RR interval data of a subject breathing freely and the two spectra derived from the data are shown in Figure 7.33. Three peaks are seen in both the spectra, which were explained as follows [72]:

- the effect of respiration at about $0.3 \, Hz$;

- the peak at $0.1 \, Hz$ related to $10 \, s$ waves seen in the blood pressure signal; and

- a peak at a frequency lower than $0.1 \, Hz$ caused by the thermo-regulatory system.

Figure 7.34 shows the RR interval data and spectra for a subject breathing at a fixed rate of $0.16 \, Hz$. The spectra display well-defined peaks at both the average heart rate ($1.06 \, Hz$) and at the breathing rate, as well as their harmonics. The spectra clearly illustrate the effect of respiration on the heart rate, and may be used to analyze the coupling between the cardiovascular and respiratory systems.

Note that direct Fourier analysis of a stream of ECG signals will not provide the same information as above. The reduced representation (model) of the RR interval data, as illustrated in Figure 7.1, has permitted Fourier analysis of the heart rate and its relationship with respiration. The methods have application in studies on HRV [69, 70, 71, 73, 74].

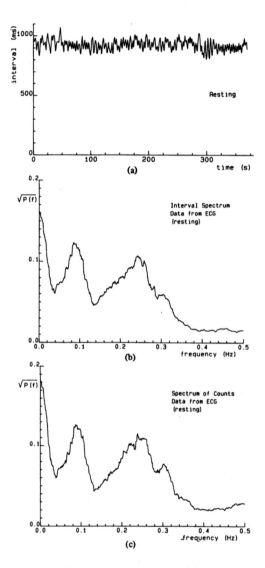

Figure 7.33 (a) 400 RR interval values from a healthy subject breathing freely. (b) Interval spectrum computed from a total of 940 intervals, including the 400 shown in (a) at the beginning. (c) Spectrum of counts. The spectra are shown for the range $0 - 0.5 \ Hz$ only. Reproduced with permission from R.W. DeBoer, J.M. Karemaker, and J. Strackee, Comparing spectra of a series of point events particularly for heart rate variability studies, *IEEE Transactions on Biomedical Engineering*, 31(4): 384–387, 1984. ©IEEE.

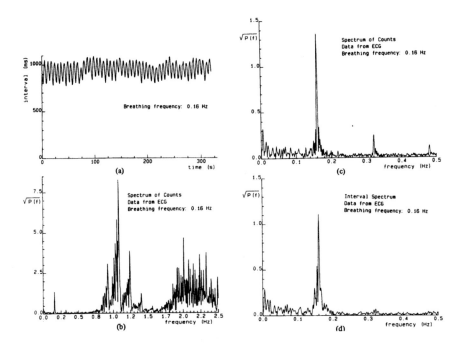

Figure 7.34 (a) 340 RR interval values from a healthy subject breathing at a fixed rate of 0.16 Hz. (b) Spectrum of counts for the range $0 - 2.5$ Hz. (c) Spectrum of counts for the range $0 - 0.5$ Hz. (d) Interval spectrum. Reproduced with permission from R.W. DeBoer, J.M. Karemaker, and J. Strackee, Comparing spectra of a series of point events particularly for heart rate variability studies, *IEEE Transactions on Biomedical Engineering*, 31(4): 384–387, 1984. ©IEEE.

7.9 APPLICATION: SPECTRAL MODELING AND ANALYSIS OF PCG SIGNALS

Iwata et al. [206, 207] applied AR modeling and parametric spectral analysis techniques to PCG signals for the detection of murmurs as well as the detection of the onset of S1 and S2. Their techniques included AR modeling, extraction of the dominant poles for pattern classification, and spectral tracking, which are explained in the following paragraphs.

Dominant poles: After the $a_k, k = 1, 2, \ldots, P$, coefficients of an all-pole or AR model of order P have been computed, the polynomial $A(z)$ may be factorized and solved to obtain the locations of the poles $p_k, k = 1, 2, \ldots, P$, of the system. The closer a pole is to the unit circle in the z-plane, the narrower is its bandwidth, and the stronger is its contribution to the impulse response of the system. Poles that are close to the unit circle may be related to the resonance frequencies of the system, and used in system identification and pattern recognition.

In view of the nonstationary nature of the signal, Iwata et al. [206] computed a new model with order $P = 8$ for every window or frame of duration 25 ms, allowing an overlap of 12.5 ms between adjacent frames (with the sampling rate $f_s = 2 \ kHz$). The frequency of a pole p_k was calculated as

$$f_k = \frac{\angle p_k}{2\pi} f_s, \tag{7.122}$$

and its bandwidth as

$$bw_k = \frac{\log |p_k|}{\pi} f_s. \tag{7.123}$$

Conditions based upon the difference in the spectral power estimate of the model from one frame to the next, and the existence of poles with $f_k < 300 \ Hz$ with the minimal bandwidth for the model considered, were used to segment each PCG signal into four phases: S1, a systolic phase spanning the S1 – S2 interval, S2, and a diastolic phase spanning the interval from one S2 to the following S1. (See also Section 4.10.)

Figures 7.35 and 7.36 show the PCG signals, spectral contours, the spectral power estimate, and the dominant poles for a normal subject and a patient with murmur due to patent ductus arteriosus (PDA). Most of the dominant poles of the model for the normal subject are below 300 Hz; the model for the patient with PDA indicates many dominant poles above 300 Hz.

The mean and standard deviation of the poles with $bw_k < 80 \ Hz$ of the model of each PCG phase were computed and used for pattern classification. The five coefficients of a fourth-order polynomial fitted to the series of spectral power estimates of the models for each phase were also used as features. Twenty-six out of 29 design samples and 14 out of 19 test samples were correctly classified. However, the number of cases was low compared to the number of features used in most of the six categories.

Spectral tracking: In another application of AR modeling for the analysis of PCG signals, Iwata et al. [207] proposed a spectral-tracking procedure based upon AR modeling to detect S1 and S2. PCG signals were recorded at the apex with a

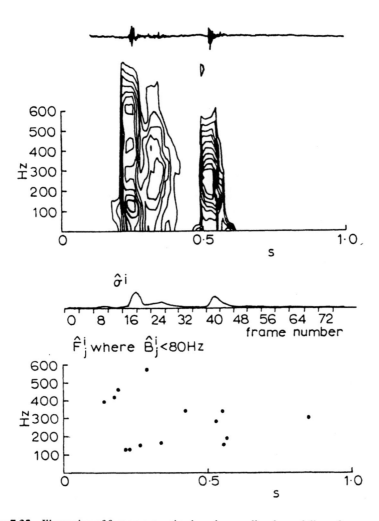

Figure 7.35 Illustration of feature extraction based upon all-pole modeling of a normal PCG signal. From top to bottom: PCG signal; model spectrum in the form of iso-intensity contours; model spectral power estimate $\hat{\sigma}^i$, where i refers to the frame number; the frequencies $\hat{F}^i_j$ of the dominant poles with bandwidth $\hat{B}^i_j < 80\ Hz$. Reproduced with permission from A. Iwata, N. Suzumara, and K. Ikegaya, Pattern classification of the phonocardiogram using linear prediction analysis, *Medical & Biological Engineering & Computing*, 15:407–412, 1977 ©IFMBE.

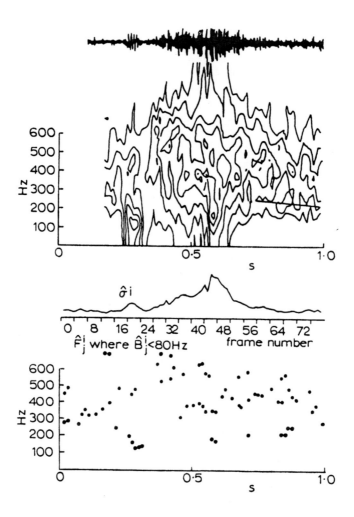

Figure 7.36   Illustration of feature extraction based upon all-pole modeling of the PCG signal of a patient with murmur due to patent ductus arteriosus. From top to bottom: PCG signal; model spectrum in the form of iso-intensity contours; model spectral power estimate $\hat{\sigma}^i$, where i refers to the frame number; the frequencies $\hat{F}_j^i$ of the dominant poles with bandwidth $\hat{B}_j^i < 80\ Hz$. Reproduced with permission from A. Iwata, N. Suzumara, and K. Ikegaya, Pattern classification of the phonocardiogram using linear prediction analysis, *Medical & Biological Engineering & Computing*, 15:407–412 ©IFMBE.

highpass filter that, at $100\ Hz$, had a gain $-40\ dB$ below the peak gain at $300\ Hz$ (labeled Ap-H). The signals were lowpass filtered with a gain of $-20\ dB$ at $1,000\ Hz$ and sampled at $2\ kHz$. The AR model was computed with order $P = 8$ for frames of length $25\ ms$; the frame-advance interval was only $5\ ms$. The model PSD was computed as

$$\tilde{S}(\omega) = \frac{\sigma_r^2}{2\pi} \frac{1}{\sum_{k=0}^{P} \phi_a(k)\ \cos(\omega T)}, \qquad (7.124)$$

where

$$\phi_a(k) = \sum_{j=0}^{P-k} a_j\ a_{j+k}, \qquad (7.125)$$

with the a_k being the AR model coefficients, $P = 8$, $T = 0.5\ ms$, σ_r^2 being the model residual energy (error), and $a_0 = 1$.

Based upon a study of the spectra of 69 normal and abnormal PCG signals, Iwata et al. [207] found the mean peak frequency of S1 to be $127\ Hz$ and that of S2 to be $170\ Hz$; it should be noted that the PCG signals were highpass filtered (as described in the preceding paragraph) at the time of data acquisition. The model spectral power at $100\ Hz$ was used as the tracking function to detect S1: the peak in the tracking function after the location t_R of the R wave in the ECG was taken to be the position of S1. The tracking function to detect S2 was based upon the spectral power at $150\ Hz$; the peak in the interval $t_R + 0.25RR \le t \le t_R + 0.6RR$, where RR is the inter-beat interval, was treated as the position of S2. The use of a normalized spectral density function based upon the AR model coefficients but without the σ_r^2 factor in Equation 7.124 was recommended, in order to overcome problems due to the occurrence of murmurs close to S2.

Figure 7.37 illustrates the performance of the tracking procedure with a normal PCG signal. The peaks in the $100\ Hz$ and $150\ Hz$ spectral-tracking functions (lowest traces) coincide well with the timing instants of S1 and S2, respectively. Figure 7.38 illustrates the application of the tracking procedure to the PCG signal of a patient with mitral insufficiency. The systolic murmur completely fills the interval between S1 and S2, and no separation is seen between the sounds and the murmur. Whereas the $150\ Hz$ spectral-tracking function labeled (b) in the figure does not demonstrate a clear peak related to S2, the normalized spectral-tracking function labeled (c) shows a clear peak corresponding to S2. The two additional PCG traces shown at the bottom of the figure (labeled Ap-L for the apex channel including more low-frequency components with a gain of $-20\ dB$ at $40\ Hz$, and 3L-H for a channel recorded at the third left-intercostal space with the same bandwidth as the Ap-H signal) illustrate S2 more distinctively than the Ap-H signal, confirming the peak location of the spectral-tracking function labeled (c) in the figure.

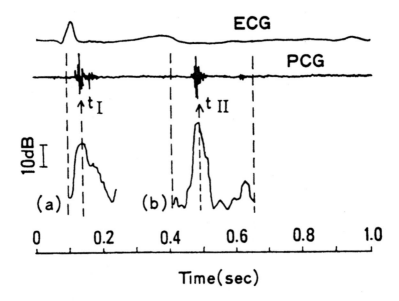

Figure 7.37 Illustration of the detection of S1 and S2 via spectral tracking based upon all-pole modeling of a normal PCG signal. From top to bottom: ECG signal; PCG signal; spectral-tracking functions at 100 Hz for S1 and 150 Hz for S2. The S1 and S2 locations detected are marked as t_I and t_{II}, respectively. Reproduced with permission from A. Iwata, N. Ishii, N. Suzumara, and K. Ikegaya, Algorithm for detecting the first and the second heart sounds by spectral tracking, *Medical & Biological Engineering & Computing*, 18:19–26, 1980 ©IFMBE.

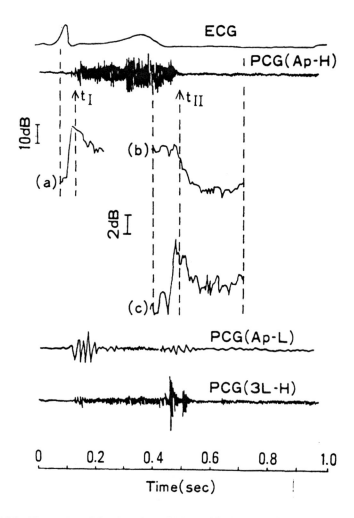

Figure 7.38 Illustration of the detection of S1 and S2 via spectral tracking based upon all-pole modeling of a PCG signal with systolic murmur due to mitral insufficiency. From top to bottom: ECG signal; PCG (Ap-H) signal; spectral-tracking functions at 100 Hz for S1 and 150 Hz for S2; normalized spectral-tracking function at 150 Hz for S2; PCG (Ap-L) signal from the apex with more low-frequency components included; PCG (3L-H) signal from the third left-intercostal space with the same filters as for Ap-H. The S1 and S2 locations detected are marked as t_I and t_{II}, respectively. Reproduced with permission from A. Iwata, N. Ishii, N. Suzumara, and K. Ikegaya, Algorithm for detecting the first and the second heart sounds by spectral tracking, *Medical & Biological Engineering & Computing*, 18:19–26, 1980 ©IFMBE.

7.10   APPLICATION: DETECTION OF CORONARY ARTERY DISEASE

The diastolic segment of a normal PCG signal after S2 is typically silent; in particular, the central portion of the diastolic segment after the possible occurrence of atrio-ventricular valve-opening snaps is silent. Akay et al. [65] conjectured that blood flow in the coronary arteries is maximum during mid-diastole, and further that coronary artery disease (occlusion, stenosis, etc.) could present high-frequency sounds in this period due to turbulent blood flow (see Section 7.7.1).

Akay et al. [65] studied the spectra of mid-diastolic segments of the PCGs, averaged over $20 - 30$ beats, of normal subjects and patients with coronary artery disease confirmed by angiography. It was found that the PCG signals in the case of coronary artery disease exhibited greater portions of their energy above $300\ Hz$ than the normal signals.

Figure 7.39 illustrates the AR-model spectra of two normal subjects and two patients with coronary artery disease. The signals related to coronary artery disease are seen to possess a high-frequency peak in the range $400 - 600\ Hz$ that is not evident in the normal cases.

Akay et al. [208] further found that the high relative-power levels of resonance frequencies in the range of $400 - 600\ Hz$ that were evident in patients with coronary artery disease were reduced after angioplasty. Figure 7.40 shows the spectra of the diastolic heart sounds of a patient before and after coronary artery occlusion was corrected by angioplasty. It may be readily observed that the high-frequency components that were present before surgery ("preang.") are not present after the treatment ("postang."). (The minimum-norm method of PSD estimation used by Akay et al. [208] — labeled as "MINORM" in the figure — is not discussed in this book.)

7.11   REMARKS

We have studied in this chapter how mathematical models may be derived to represent physiological processes that generate biomedical signals, and furthermore, how the models may be related to changes in signal characteristics due to functional and pathological processes. The important point to note in the modeling approach is that the models provide a small number of *parameters* that characterize the signal and/or system of interest; the modeling approach is therefore useful in *parametric analysis* of signals and systems. As the number of parameters derived is usually much smaller than the number of signal samples, the modeling approach could also assist in data compression and compact representation of signals and related information.

Pole-zero models could be used to view physiological systems as control systems. Pathological states may be derived or simulated by modifying the parameters of the related models. Models of signals and systems are also useful in the design and control of prostheses.

A combination of mathematical modeling with electromechanical modeling can allow the inclusion of physical parameters such as the diameter of a blood vessel,

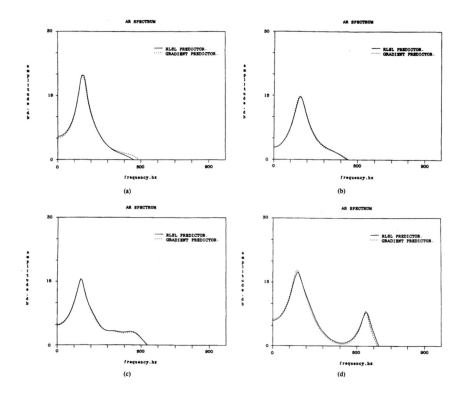

Figure 7.39   Diastolic heart sound spectra of (a, b) two normal subjects and (c, d) two patients with coronary artery disease. The method of estimating AR models identified in the figure as "RLSL" will be described in Section 8.6.2; the gradient predictor method is not discussed in this book. Reproduced with permission from A.M. Akay, J.L. Semmlow, W. Welkowitz, M.D. Bauer, and J.B. Kostis, Detection of coronary occlusions using autoregressive modeling of diastolic heart sounds, *IEEE Transactions on Biomedical Engineering,* 37(4):366–373, 1990. ©IEEE.

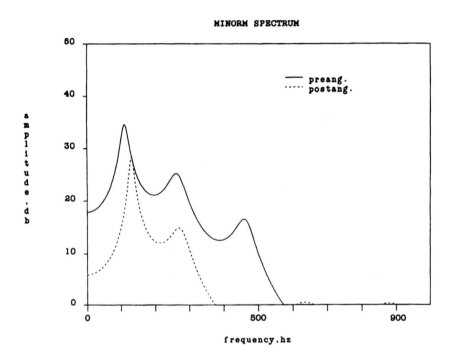

Figure 7.40   Diastolic heart sound spectra before (preang.) and after angioplasty (postang.) of a patient for whom coronary artery occlusion was corrected. (The minimum-norm method of PSD estimation used by Akay et al. [208] — labeled as "MINORM" in the figure — is not discussed in this book.) Reproduced with permission from A.M. Akay, J.L. Semmlow, W. Welkowitz, M.D. Bauer, and J.B. Kostis, Noninvasive detection of coronary stenoses before and after angioplasty using eigenvector methods, *IEEE Transactions on Biomedical Engineering,* 37(11):1095–1104, 1990. ⓒIEEE.

constriction due to plaque buildup, stiffness due to stenosis, and friction coefficient. Although accurate estimation of such parameters for human subjects may not always be possible, the models could lead to better understanding of the related biomedical signals and systems.

7.12   STUDY QUESTIONS AND PROBLEMS

1. Consider the simple linear prediction model given by $\tilde{y}(n) = ay(n - 1)$. Define the prediction error, and derive the optimal value for a by minimizing the total squared error.

2. The autoregressive model coefficients of a signal are $a_0 = 1, a_1 = 1, a_2 = 0.5$. What is the transfer function of the model? Draw the pole-zero diagram of the model. What are the resonance frequencies of the system?

3. The autoregressive model coefficient vectors of a number of signals are made available to you. Propose two measures to compare the signals for (a) similarity, and (b) dissimilarity.

4. In autoregressive modeling of signals, show why setting the derivative of the total squared error with respect to any coefficient to zero will always lead to the minimum error (and not the maximum).

5. What type of a filter can convert the autocorrelation matrix of a signal to a diagonal matrix?

6. A biomedical signal is sampled at $500\ Hz$ and subjected to AR modeling. The poles of the model are determined to be at $0.4 \pm j0.5$ and $-0.7 \pm j0.6$.

   (a) Derive the transfer function of the model.

   (b) Derive the difference equation in the time domain.

   (c) What are the resonance frequencies of the system that is producing the signal?

7. A model is described by the relationship

$$y(n) = x(n) + 0.5x(n - 1) + 0.25x(n - 2),$$

where $x(n)$ is the input and $y(n)$ is the output.
What is the type of this system among AR, MA, and ARMA systems?
What is the model order?
What is its transfer function?
Draw the pole-zero diagram of the system.
Comment upon the stability of the system.

8. A model is described by the relationship

$$y(n) = -0.5y(n - 1) - y(n - 2) + x(n) + 0.5x(n - 1) - x(n - 2),$$

where $x(n)$ is the input and $y(n)$ is the output.
What is the type of this system among AR, MA, and ARMA systems?
What is the model order?
What is its transfer function?

Draw the pole-zero diagram of the system.

Comment upon the stability of the system.

7.13 LABORATORY EXERCISES AND PROJECTS

Note: Data files related to the exercises are available at the site
ftp://ftp.ieee.org/uploads/press/rangayyan/

1. The file safety.wav contains the speech signal for the word "safety" uttered by a male speaker, sampled at 8 kHz (see the file safety.m). The signal has a significant amount of background noise (as it was recorded in a normal computer laboratory). Develop procedures to segment the signal into voiced, unvoiced, and silence (background noise) portions using short-time RMS, turns count, or ZCR measures.

   Apply the AR modeling procedure to each segment using the command *lpc* in MATLAB. Compute the AR-model PSD for each segment. Compare the model PSD with the FFT-based PSD for each segment. What are the advantages and disadvantages of the model-based PSD in the case of voiced and unvoiced sounds?

2. Derive the poles of the models you obtained in the preceding problem. Express each pole in terms of not only its z-plane coordinates but also its frequency and bandwidth. Study the variations in the pole positions as the type of the sound varies from one segment to the next over the duration of the signal.

3. The files pec1.dat, pec33.dat, and pec52.dat give three-channel recordings of the PCG, ECG, and carotid pulse signals (sampled at $1,000$ Hz; you may read the signals using the program in the file plotpec.m). The signals in pec1.dat and pec52.dat are normal; the PCG signal in pecg33.dat has systolic murmur, and is of a patient suspected to have pulmonary stenosis, ventricular septal defect, and pulmonary hypertension.

   Segment each signal into its systolic and diastolic parts. Apply the AR modeling procedure to each segment and derive the model PSD. Compare the result with the corresponding PSDs obtained using Welch's procedure.

4. Derive the poles of the models you obtained in the preceding problem. Express each pole in terms of not only its z-plane coordinates but also its frequency and bandwidth. Study the variations in the pole positions from the systolic part to the diastolic part of each signal. What are the major differences between the pole plots for the normal cases and the case with murmur?

5. The files ECG3, ECG4, ECG5, and ECG6 contain ECG signals sampled at the rate of 200 Hz (see the file ECGS.m). Apply the Pan-Tompkins method for QRS detection to each signal. Create impulse sequences including a delta function at every QRS location for the four signals. Create also the interval series for each signal as illustrated in Figure 7.1. Compute the spectra corresponding to the two representations of cardiac rhythm and study their relationship to the heart rate and its variability in each case.

8

Analysis of Nonstationary Signals

A stationary (or homogeneous) signal is one that possesses the same statistical measures for all time, or at least over the duration of observation. We have seen in the preceding chapters that most biomedical signals, being manifestations of dynamic systems and patho-physiological processes, are *nonstationary* (or heterogeneous): Figure 3.3 shows that the variance of the speech signal used as an example varies with time; Figure 3.4 shows that the spectrum or frequency content of the speech signal also varies considerably over its duration. Figures 6.11 and 6.12 show that the spectrum of a heart sound signal or PCG varies from systole to diastole, and could vary in between the two events as well.

When the characteristics of a signal being studied vary considerably over the duration of interest, measures and transforms computed over the entire duration do not carry useful information: they gloss over the dynamics of the signal. A single PSD computed from a long EMG, PCG, VAG, or speech record is of no practical value. The PSD does not provide information on time localization of the frequency components of the signal. We addressed this concern in PCG signal analysis in Section 6.4.5 by segmenting the PCG into its systolic and diastolic parts by using the ECG and carotid pulse signals as timing references. But how would we be able to handle the situation when murmurs are present in systole and diastole, and we need to analyze the spectra of the murmurs without the contributions of S1 and S2?

Furthermore, the EEG signal changes its nature in terms of rhythms, waves, transients, and spindles for which no independent references are available. In fact, the EEG is a conglomeration of a number of mental and physiological processes going on in the brain at any given instant of time. The VAG signal has nonstationary characteristics related to the cartilage surfaces that come into contact depending upon the activity performed, and no other source of information can assist in identifying time instants when the signal properties change. Indeed, a VAG signal contains no specific events that may be identified as such, but is a concatenation of nonspecific vibrations (with, perhaps, the exception of clicks). Would we able to extend the application of the well-established signal analysis techniques that we have studied so far to such nonstationary signals?

8.1   PROBLEM STATEMENT

Develop methods to study the dynamic characteristics of nonstationary biomedical signals. Propose schemes to apply the well-established Fourier and autoregressive modeling techniques to analyze and parameterize nonstationary signals.

In order to limit the scope of the present chapter, we shall consider the extension of only Fourier spectral analysis and AR modeling to nonstationary signals. The case-studies presented in the following section will provide the motivation for the study from the perspective of a few representative biomedical signals. Approaches to solving the stated problem will be presented in the sections to follow.

This chapter concentrates on segmentation-based analysis of nonstationary signals. Topics such as the Kalman filter, time-frequency distributions, and wavelets are not considered.

8.2   ILLUSTRATION OF THE PROBLEM WITH CASE-STUDIES

8.2.1   Heart sounds and murmurs

We noted in Section 6.4.5 that the spectral contents of S1 and S2 are different due to the different states of contraction or relaxation of the ventricular muscles and the differences in their blood content during the corresponding cardiac phases. In the normal case, the QRS in the ECG signal and the dicrotic notch in the carotid pulse signal may be used to split the PCG into S1 and S2, and separate PSDs may be obtained for the signal parts as illustrated in Section 6.4.5. However, when a PCG signal contains murmurs in systole and/or diastole and possibly valve opening snaps (see Figure 6.12), it may be desirable to split the signal further.

Iwata et al. [206] applied AR modeling to PCG signals by breaking the signal into fixed segments of 25 *ms* duration (see Section 7.9). While this approach may be satisfactory, it raises questions on optimality. What should be the window duration? Is it necessary to break the intervals between S1 and S2 into multiple segments? Would it not be more efficient to compute a single AR model for the entire durations of each of S1, S2, systolic murmur, and diastolic murmur – that is, a total of only four models? It is conceivable that each model would be more accurate as all available signal samples would be used to estimate the required ACF if the signal were to be segmented adaptively as mentioned above.

8.2.2 EEG rhythms and waves

The scalp EEG represents a combination of the multifarious activities of many small zones of the cortical surface beneath each electrode. The signal changes its characteristics in relation to mental tasks, external stimuli, and physiological processes. As we have noted in Section 1.2.5 and observed in Figure 1.21, a visual stimulus blocks the alpha rhythm; slower waves become prominent as the subject goes to deeper stages of sleep; and patients with epilepsy may exhibit sharp spikes and trains of spike-and-wave complexes. Description of an EEG record, as outlined in Sections 1.2.5 and 4.2.4, requires the identification of several types of waves and rhythms. This suggests that the signal may first have to be broken into segments, each possessing certain properties that remain the same for the duration of the segment. Each segment may then be described in terms of its characteristic features.

8.2.3 Articular cartilage damage and knee-joint vibrations

Movement of the knee joint consists of coupled translation and rotation. The configuration of the patella is such that some portion of the articular surface is in contact with the femur throughout knee flexion and to almost full extension (see Section 1.2.13 and Figure 1.31). Goodfellow et al. [209] demonstrated that initial patello-femoral engagement occurs at approximately $20°$ of flexion involving both the medial and lateral facets. Figure 8.1 shows the patellar contact areas at different joint angles. As the knee is flexed, the patello-femoral contact area moves progressively upward, involving both the medial and lateral facets. At $90°$ of flexion, the band of contact engages the upper pole of the patella. The odd facet does not articulate with the lateral margin of the medial femoral condyle until about $120° - 135°$ of knee flexion.

Articular cartilage is composed of a solid matrix and synovial fluid [210]; it has no nerves, blood vessels, or lymphatics, and is nourished by the synovial fluid covering its free surface. During articulation, friction between the bones is reduced as a result of the lubrication provided by the viscous synovial fluid [49, 52]. The material properties of articular cartilage and cartilage thickness are variable not only from joint to joint but also within the same joint. In case of abnormal cartilage alterations in the matrix structure such as increased hydration, disruption of the collagen fibrillar network and dis-aggregation or loss of proteoglycans occur. As the compositional and biomechanical properties of abnormal articular cartilage continue to deteriorate, substance loss eventually occurs. This may be focal or diffuse, restricted to superficial fraying and fibrillation, or partial-thickness loss to full-thickness loss. In some cases, focal swelling or blistering of the cartilage may be seen before there is fraying of the articular surface [211].

Chondromalacia patella (soft cartilage of the patella) is a condition in which there is degeneration of patellar cartilage, often associated with anterior knee pain. Exposed subchondral bone and surface fibrillation of the articular cartilage are evident on the posterior patellar surface in chondromalacia patella [212]. Chondromalacia patella is usually graded in terms of the severity of the lesions [213, 214] as follows:

- *Grade I:* Softening, cracking, and blistering, but no loss of articular cartilage.

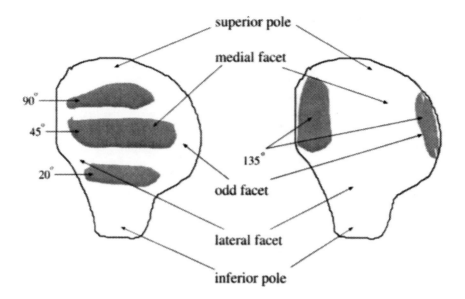

Figure 8.1   Contact areas of the patella with the femur during patello-femoral articulation. Adapted, with permission, from S. Krishnan, Adaptive Signal Processing Techniques for Analysis of Knee Joint Vibroarthrographic Signals, Ph.D. Thesis, University of Calgary, 1999.

- *Grade II:* Damage is moderate and there is some loss of cartilage.

- *Grade III:* Severe damage of fibrocartilage has occurred but bone is not exposed.

- *Grade IV:* The cartilage is eroded and the subchondral bone is exposed.

Osteoarthritis is a degenerative joint disease that involves specific changes to bone in addition to cartilage. In the late stages of osteoarthritis, there is full-thickness articular cartilage degeneration and exposed bone. Other structural changes include fibrous changes to the synovium, joint capsule thickening, and further alterations to the bone such as osteophyte formation [215]. Higher-grade chondromalacia may be categorized as osteoarthritis.

The menisci are subject to vertical compression, horizontal distraction, and rotary and shearing forces of varying degrees in the course of normal activities [216]. Advance of the aging process in both articular cartilage and fibrocartilage causes progressive liability to horizontal cleavage lesion [216].

The semi-invasive procedure of *arthroscopy* (fiber-optic inspection of joint surfaces, usually under general anesthesia) is often used for diagnosis of cartilage pathology. Through an arthroscope, the surgeon can usually see the patello-femoral joint, the femoral condyles, the tibial plateau (menisci), the anterior cruciate ligament, and the medial and lateral synovial spaces. Arthroscopy has emerged as the "gold standard" for relatively low-risk assessment of joint surfaces in order to determine the prognosis and treatment for a variety of conditions. Figure 8.2 shows the different stages of chondromalacia patella as viewed during arthroscopy.

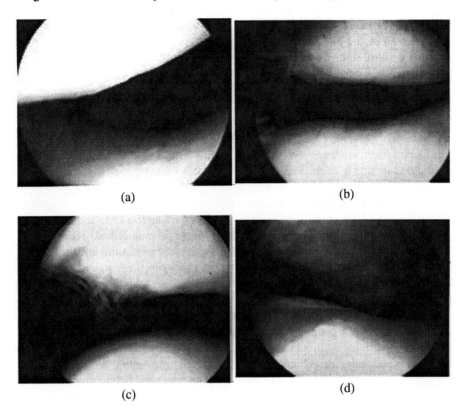

(a)

(b)

(c)

(d)

Figure 8.2 Arthroscopic views of the patello-femoral joint. (a) Normal cartilage surfaces. (b) Chondromalacia Grade II at the patella. (c) Chondromalacia Grade III at the patella. (d) Chondromalacia Grade IV at the patella and the femur; the bones are exposed. The under-surface of patella is at the top and the femoral condyle is at the bottom. Figure courtesy: G.D. Bell, Sport Medicine Centre, University of Calgary.

Abnormal structures and surfaces in the knee joint are more likely to generate sound during extension and flexion movements than normal structures. Softened articular cartilage in chondromalacia patella, and cracks, fissures, or thickened areas in osteoarthritis almost certainly increase the friction between the articular surfaces, and are therefore likely to increase the sounds emitted during normal joint movement [217, 54]. Injury to the menisci in the form of tearing causes irregularity in shape and disruption to normal joint movement, and may produce sharp clicking sounds during normal knee movement [218, 59, 54].

It is obvious from this discussion that the VAG signal is nonstationary. Different aspects of the articulating surfaces come into contact at different joint angles; their quality in terms of lubrication and functional integrity could vary from one position to another. Inspection of the VAG signals and their spectrograms illustrated in Sections 3.6.3 and 3.10 reveals that the nature of a VAG signal changes significantly over the duration of the signal. As no prior or independent information is available about changes in the knee-joint structures that could lead to vibrations, adaptive segmentation of the VAG signal is required before it may be analyzed, using the methods we have studied so far in this book. Illustration of adaptive segmentation of VAG signals will be provided in Sections 8.6.1 and 8.6.2.

8.3   TIME-VARIANT SYSTEMS

The linear system model represented by Equation 7.1 is a time-invariant system: the coefficients a_k and b_l of the system do not change with time, and consequently, the poles and zeros of the system stay fixed for all time. A nonstationary (or dynamic) system will possess coefficients that do vary with time: we saw in Sections 3.6.2 and 3.6.3 that the coefficient (tap-weight) vectors of the adaptive LMS and RLS filters are expressed as functions of time. (*Note:* The Wiener filter described in Section 3.5, once optimized for a given set of signal and noise statistics, is a time-invariant filter.) Since the coefficients of an LMS or RLS filter vary with time, so do the transfer function and the frequency response of the filter. It follows that the impulse response of such a system also varies with time.

Let us consider an all-pole filter for the sake of simplicity; the filter characteristics are determined by the positions of the poles to within a gain factor. If the poles are expressed in terms of their polar coordinates, their angles correspond to (resonance) frequencies and their radii are related to the associated bandwidths. We may therefore characterize time-variant or nonstationary systems and signals by describing their pole positions in the complex z-plane — or, equivalently, the related frequencies and bandwidths — as functions of time. A description of the variation or the modulation of the pole parameters over time can thus capture the nonstationary or dynamic nature of a time-variant system or signal. Variations in the gain factor also lead to nonstationarities in the signal produced by the system. Appel and v. Brandt [219, 220] describe the simulation of different types of nonstationary behavior of signals and systems.

In the general case of a nonstationary system that is an AR process, we may modify Equation 7.17 to indicate that the model coefficients are functions of time:

$$\tilde{y}(n) = -\sum_{k=1}^{P} a_k(n)y(n-k). \tag{8.1}$$

Methods related to the Kalman filter or the least-squares approach may be used to analyze such a system [77, 221, 222, 223] (not considered in this book). Time-varying AR and ARMA modeling techniques have been applied to analyze EEG [224], EGG [38], and HRV [225] signals; the application to HRV signals will be discussed in Section 8.9.

8.3.1 Characterization of nonstationary signals and dynamic systems

The output of a time-variant or dynamic system will be a nonstationary signal. The system may be characterized in terms of its time-variant model coefficients, transfer function, or related parameters derived thereof. Various short-time statistical measures computed over moving windows may be used to characterize a nonstationary signal; the measures may also be used to test for stationarity, or lack thereof, of a signal.

- **Mean:** The short-time mean represents the average or DC level of the signal in the analysis window. Variation of the mean from one window to another is usually an indication of the presence of a wandering base-line or low-frequency artifact, as in the case of the ECG signal in Figure 3.6. Clearly, the signal in Figure 3.6 is nonstationary in the mean. However, the mean is not an important measure in most signals, and is typically blocked at the data-acquisition stage via capacitive coupling and/or a highpass filter. Furthermore, since a DC level carries no sound or vibration information, its removal is of no consequence in the analysis of signals such as heart sounds, speech, VAG, and the VMG.

- **Variance:** Figure 3.3 illustrates the short-time variance for a speech signal. It is evident that the variance is high in regions of high signal variability (swings or excursions) about the mean, as in the case of the vowels in the signal. The variance is low in the regions related to the fricatives in the signal where the amplitude swing is small, in spite of their high-frequency nature. Since the mean of the signal is zero, the variance is equal to the MS value, and represents the average power level in the corresponding signal windows. Although variations in the power level of speech signals may be useful in making voiced/ unvoiced/ silence decision, the parameter does not bear much information, and provides a limited representation of the general statistical variability of signal characteristics. Regardless of the interpretation of the parameter, it is seen that the speech signal in Figure 3.3 is nonstationary in its variance (and the related measures of SD, MS, and RMS). From the discussion in Section 1.2.11, it is also clear that the vocal-tract system producing the speech signal is a dynamic system with time-varying configuration and filtering characteristics.

- **Measures of activity:** We have studied several measures of activity that indicate the "busy-ness" of the given signal, such as turning points, ZCR, and turns count (in Chapters 3 and 5). The short-time count of turning points is plotted in Figure 3.1 for a speech signal: it is evident that the signal is more active or busy in the periods related to the fricatives than those related to the vowels (a trend that is the opposite of that in the short-time variance of the same signal shown in Figure 3.3). The short-time turns count plot of the EMG signal in Figure 5.8 indicates the rising level of complexity of the signal with the level of breathing. Although turning points, ZCR, and turns count are not among the traditional statistical measures derived from PDFs, they characterize signal variability and complexity in different ways. Both the examples cited above illustrate variation of the parameters measured over the duration of the corresponding signals: the signals are therefore nonstationary in terms of the number of turning points or the turns count.

- **ACF:** The ACF was defined in Section 3.1.1 in general as $\phi_{xx}(t_1, t_1 + \tau) = E[x(t_1)x(t_1 + \tau)]$. In Section 3.1.2, one of the conditions for (wide-sense or second-order) stationarity was defined as the ACF being independent of time, that is, $\phi_{xx}(t_1, t_1 + \tau) = \phi_{xx}(\tau)$. A nonstationary signal will not satisfy this condition, and will have an ACF that varies with time. Since the ACF is based on the expectation of pairs of signal samples separated by a certain time difference or lag, it is a more general measure of signal variability than the variance and related measures. Note that the ACF for zero lag is the MS value of the signal.

  One faces limitations in computing the ACF of short-time segments of a signal to investigate (non)stationarity: the shorter the analysis window, the shorter the maximum lag up to which the ACF may be estimated reliably. Regardless, the short-time ACF may be used to track nonstationarities in a signal. If the signal is the result of a dynamic AR system, the system parameters may be derived from the ACF (see Section 7.5).

- **PSD:** The PSD and ACF of a signal are inter-related by the Fourier transform. Therefore, a signal that is (non)stationary in its ACF is also (non)stationary in its PSD. However, the PSD is easier to interpret than the ACF, as we have seen in Chapter 6. The spectrogram of the speech signal in Figure 3.4 indicates significant variations in the short-time PSD of the signal: the speech signal is clearly nonstationary in its PSD (and ACF). The spectrograms of VAG signals in Sections 3.6.3 and 3.10 illustrate the nonstationary nature of VAG signals.

- **Higher-order statistics:** A major limitation of signal analysis using the ACF (or equivalently the PSD) is that the phase information is lost. The importance of phase in signals is discussed by Oppenheim and Lim [226]. Various conditions under which a signal may be reconstructed from its magnitude spectrum only or from its phase spectrum only are described by Hayes and Oppenheim [227] and Oppenheim and Lim [226]. Analysis based only upon the ACF cannot be applied to signals that are of mixed phase (that is, not minimum phase), that are the result of nonlinear systems, or that follow a PDF other than a Gaussian [228].

  The general n^{th}-order moment of a random signal $x(t)$ at the instant t_1 is defined as [228, 229, 77]

$$m_x^n(t_1, t_1 + \tau_1, t_1 + \tau_2, \ldots, t_1 + \tau_{n-1}) = \qquad (8.2)$$
$$E[x(t_1)x(t_1 + \tau_1)x(t_1 + \tau_2) \cdots x(t_1 + \tau_{n-1})],$$

where $\tau_1, \tau_2, \ldots, \tau_{n-1}$ are various delays or lags. It is evident that the ACF is a special case of the above with $n = 2$, that is, the ACF is the second-order moment.

A set of parameters known as cumulants may be related to the moments as follows: The second-order and third-order cumulants are equal to the corresponding moments. The fourth-order cumulant is related to the fourth-order moment as [77, 228, 229]

$$
\begin{aligned}
c_x^4(t_1, t_1 + \tau_1, t_1 + \tau_2, t_1 + \tau_3) \; = \; & m_x^4(t_1, t_1 + \tau_1, t_1 + \tau_2, t_1 + \tau_3) \quad (8.3) \\
- \; & m_x^2(t_1, t_1 + \tau_1) \, m_x^2(t_1 + \tau_2, t_1 + \tau_3) \\
- \; & m_x^2(t_1, t_1 + \tau_2) \, m_x^2(t_1 + \tau_3, t_1 + \tau_1) \\
- \; & m_x^2(t_1, t_1 + \tau_3) \, m_x^2(t_1 + \tau_1, t_1 + \tau_2).
\end{aligned}
$$

The Fourier transforms of the cumulants provide the corresponding higher-order spectra or polyspectra (with as many frequency variables as the order minus one). The Fourier transforms of the second-order, third-order, and fourth-order cumulants are known as the power spectrum (PSD), bispectrum, and trispectrum, respectively. A Gaussian process possesses only first-order and second-order statistics: moments and spectra of order higher than two are zero. Higher-order moments, cumulants, and spectra may be used to characterize nonlinear, mixed-phase, and non-Gaussian signals [77, 228, 229]. Variations over time of such measures may be used to detect the related types of nonstationarity.

- **System parameters:** When a time-varying model of the system producing the signal is available in terms of its coefficients, such as $a_k(n)$ in Equation 8.1, we may follow or track changes in the coefficients over time. Significant changes in the model parameters indicate corresponding changes in the output signal.

8.4 FIXED SEGMENTATION

Given a nonstationary signal, the simplest approach to break it into quasi-stationary segments would be to consider small windows of fixed duration. Given a signal $x(i)$ for $i = 0, 1, 2, \ldots, N-1$, we could consider a fixed segment duration of M samples, with $M << N$, and break the signal into K parts as

$$
x_k(n) = x(n + (k-1)M), \;\; 0 \le n \le M - 1, \;\; 1 \le k \le K. \quad (8.4)
$$

With the assumption that the signal does not change its characteristics to any significant extent within the duration corresponding to M samples (or $\frac{M}{f_s}$ s), each segment may be considered to be quasi-stationary.

Note that the segmentation here is similar to that in the Bartlett and Welch procedures described in Sections 6.4.2 and 6.4.3. However, we will not be averaging the spectra over the segments now, but will be treating them as separate entities. The signal processing techniques we have studied so far may then be applied to analyze each segment separately.

8.4.1   The short-time Fourier transform

Once the given signal has been segmented into quasi-stationary parts $x_k(n)$ as above, we may compute the Fourier spectrum for each segment as

$$X_k(\omega) = \sum_{n=0}^{M-1} x_k(n) \exp(-j\omega n). \tag{8.5}$$

The array of spectra $X_k(\omega)$ for $k = 1, 2, \ldots, K$ will describe the time-varying spectral characteristics of the signal.

Segmentation of the given signal as above may be interpreted as the application of a moving window to the signal. The kth segment $x_k(n)$ may be expressed as the multiplication of the signal $x(n)$ with a window function $w(n)$ positioned at the beginning of the segment as

$$x_k(n) = x(n)w(n - (k-1)M), \quad 1 \leq k \leq K, \tag{8.6}$$

where

$$w(n) = \begin{cases} 1 & \text{for } 0 \leq n \leq M - 1 \\ 0 & \text{otherwise} \end{cases}. \tag{8.7}$$

Figure 8.3 (a) illustrates the PCG of a patient with systolic murmur and opening snap of the mitral valve, with a moving rectangular analysis window of duration 64 ms superimposed on the signal at three different instants of time. The duration of each window is 64 samples, equal to 64 ms with $f_s = 1\ kHz$. The three windows have been positioned approximately over the S1, systolic murmur, and S2 events in the signal. Figure 8.3 (b) shows the log PSDs of the signal segments extracted by the three analysis windows. It is seen that the PSDs differ significantly, with the second window displaying the largest amount of high-frequency power due to the murmur. The third window displays more medium-frequency content than the first. It is clear that the PCG signal is nonstationary in the PSD.

In general, the window may be positioned at any time instant m, and the resulting segment may be expressed as $x(n)w(n - m)$. We need to state how the window is moved or advanced from one segment to another; in the extreme situation, we may advance the window one sample at a time, in which case adjacent windows would have an overlap of $(M - 1)$ samples. We may then compute the Fourier transform of every segment as

$$X(m, \omega) = \sum_{n=0}^{M-1} x(n)w(n - m) \exp(-j\omega n). \tag{8.8}$$

In the case when both the time and frequency variables are continuous, we may write the expression above in a more readily understandable form as

$$X(\tau, \omega) = \int_{-\infty}^{\infty} x(t)w(t - \tau) \exp(-j\omega t)\, dt. \tag{8.9}$$

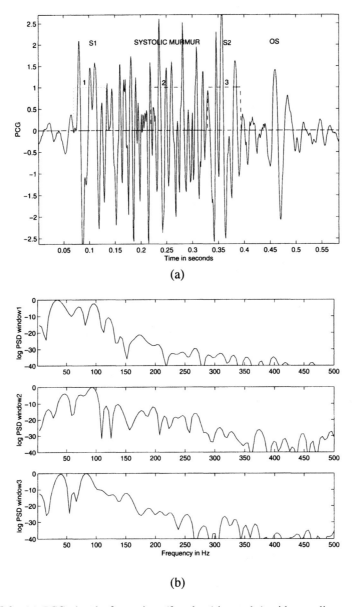

(a)

(b)

Figure 8.3 (a) PCG signal of a patient (female, 14 months) with systolic murmur and opening snap (OS) of the mitral valve. Three short-time analysis windows are superimposed, each one being a rectangular window of duration 64 *ms*. (b) Log PSDs of the three windowed signal segments. Each FFT was computed with zero-padding to a total length 256 samples. $f_s = 1\ kHz$. See also Figure 6.12.

The spectrum is now expressed not only as a function of frequency ω, but also as a function of time τ. Although the limits of the integral have been stated as $(-\infty, \infty)$, the finite duration of the window placed at time τ will perform segmentation of the signal as desired.

The spectral representation of the signal as a function of time in Equations 8.8 and 8.9 is known as a *time-frequency distribution* or TFD [230, 231, 232, 233]. Since the Fourier transform is applied, in the procedure above, to short windows of the signal in time, the result is known as the *short-time Fourier transform* or STFT of the signal. The method of analysis of a nonstationary signal in short windows is, in general, known as *short-time analysis*. The magnitude of the STFT (squared and/or with the logarithmic operation if desired) is known as the *spectrogram* of the signal.

Figure 8.4 illustrates the spectrogram of the PCG signal of a patient with systolic murmur and opening snap of the mitral valve: the signal and the window parameters are the same as in Figure 8.3, but now the spectra are plotted for every window position with a displacement of 32 *ms*. The relatively high-frequency nature of the murmur as compared to S1 and S2 is clearly evident in the spectrogram.

We have previously encountered spectrograms of speech and VAG signals: refer to Figure 3.4 and Sections 3.6.3 and 3.10. More examples of spectrograms will be provided at the end of this section and later in this chapter.

8.4.2 Considerations in short-time analysis

Short-time analysis of signals could be computationally expensive. In the case of the STFT, the Fourier transform has to be computed for each segment of the signal. In practice, there should be no need to compute the Fourier transform for every possible window position, that is, for every m in Equation 8.8. We could advance the analysis window by M samples, in which case adjacent windows will not overlap. It is common practice to advance the analysis window by $\frac{M}{2}$ samples, in which case adjacent windows will overlap for $\frac{M}{2}$ samples; some overlap is desirable in order to maintain continuity in the STFT or TFD computed.

An important question arises regarding the duration of the analysis window M to be used. The window should be short enough to ensure that the segment is stationary, but long enough to permit meaningful analysis. We have seen in Section 6.3 that a short window possesses a wide main lobe in its frequency response. Since the given signal is multiplied in the time domain with the analysis window, the spectrum of the signal gets convolved with the spectral response of the window in the frequency domain. Convolution in the frequency domain with a function having a large main lobe leads to significant loss of spectral resolution.

The limitation imposed by the use of a window is related to the uncertainty principle or time-bandwidth product, expressed as [231]

$$\Delta t \times \Delta \omega \geq \frac{1}{2}, \tag{8.10}$$

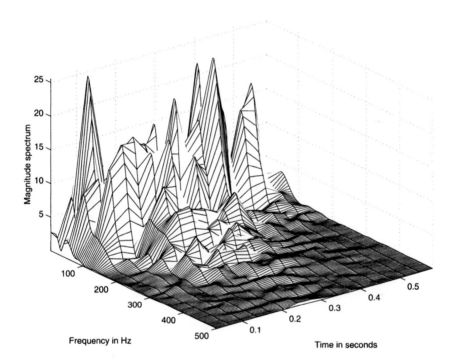

Figure 8.4 Spectrogram of the PCG signal of a patient (female, 14 months) with systolic murmur and opening snap of the mitral valve, computed with a moving short-time analysis window of duration 64 samples (64 ms with $f_s = 1\ kHz$), with the window advance interval being 32 samples. Each FFT was computed with zero-padding to a total length 256 samples. $f_s = 1\ kHz$. See also Figures 6.12 and 8.3.

where

$$(\Delta t)^2 = \int_{-\infty}^{\infty} (t - \bar{t})^2 \, |x(t)|^2 \, dt, \tag{8.11}$$

$$\bar{t} = \int_{-\infty}^{\infty} t \, |x(t)|^2 \, dt, \tag{8.12}$$

$$(\Delta \omega)^2 = \int_{-\infty}^{\infty} (\omega - \bar{\omega})^2 \, |X(\omega)|^2 \, d\omega, \tag{8.13}$$

$$\bar{\omega} = \int_{-\infty}^{\infty} \omega \, |X(\omega)|^2 \, d\omega, \tag{8.14}$$

and Δt and $\Delta \omega$ represent the time extent (duration) and frequency extent (bandwidth) of the signal $x(t)$ and its Fourier transform $X(\omega)$, respectively. The gist of the limitation stated above is that a signal and its Fourier transform cannot be made arbitrarily narrow. The effect of this limitation on the STFT and TFD-based analysis is that we cannot simultaneously obtain arbitrarily high resolution along both the time and frequency axes.

At the extremes, a continuous-time signal $x(t)$ provides infinite time resolution but no frequency resolution: the value of the signal is known at every instant of time t, but nothing is known about the frequency content of the signal. Conversely, the PSD $S_{xx}(f)$ provides infinite frequency resolution but no time resolution: the overall strength of sinusoids at every frequency f present in the signal over all time t is known, but nothing is known about where exactly in time a given frequency component begins or ends. (The phase spectrum contains this information but cannot be readily interpreted and used for the purposes of this discussion.)

In the case of sampled signals and spectra, the sampling intervals Δt in the time domain and Δf in the frequency domain will be finite, and limited by Heisenberg's inequality as stated above. Increasing the time resolution of the STFT by making the analysis window short in duration will compromise frequency resolution; on the other hand, increasing the window duration will lead to a loss in time resolution.

In general, the window function $w(n)$ included in Equation 8.8 need not be a rectangle: any of the window functions listed in Section 6.4.3 may be used. Once a window is chosen, the joint time-frequency (TF) resolution is the same over the entire TF plane.

The STFT expression in Equation 8.8 indicated the placement of a causal analysis window beginning at the time instant of reference m in the argument of the STFT. It is also common practice to use a symmetrical noncausal window defined for $-\frac{M}{2} \leq n \leq \frac{M}{2}$, in which case the reference point of the analysis window would be the center of the window.

Illustration of application: Spectrograms of the speech signal in Figure 1.29 with different window parameters are provided in Figures 8.5 and 8.6. The spectrograms are shown here as gray-scale images, with the darkness at each point being proportional to the log PSD for the corresponding temporal analysis window position and frequency coordinate. It is evident that increasing the length of the analysis window provides better frequency resolution (the definition or clarity of the frequency components) while at the same time reducing the temporal resolution (that is, causing smearing in the temporal dimension). Decreasing the window length causes the reverse effects. The spectrogram in Figure 8.5 (b) with the analysis window duration being 16 *ms* clearly illustrates the high-frequency (broad-band) nature of the fricatives; the transient and broad-band nature of the plosive /T/ is also clearly shown. The same features are not clearly depicted by the spectrogram in Figure 8.6 (b) where the analysis window is fairly long (128 *ms*); however, the formant structure of the voiced-speech components (the vowels) is clearly depicted. The formant structure of the voiced-speech components is not clearly visible in the spectrogram in Figure 8.5 (b).

8.5 ADAPTIVE SEGMENTATION

One of the limitations of short-time analysis lies with the use of a fixed window duration. A signal may remain stationary for a certain duration of time much longer than the window duration chosen, and yet the signal would be broken into many segments over such a duration. Conversely, a signal may change its characteristics within the duration of the fixed window: short-time analysis cannot guarantee stationarity of the signal over even the relatively short duration of the analysis window used. It would be desirable to adapt the analysis window to changes in the given signal, allowing the window to be as long as possible while the signal remains stationary, and to start a new window at the exact instant when the signal or the related system changes its characteristics.

Problem: *Propose methods to break a nonstationary signal into quasi-stationary segments of variable duration.*

Solution: We saw in Section 7.5 that a signal may be represented or modeled as a linear combination of a small number of past values of the signal, subject to a small error of prediction. It then follows that if a signal were to change its behavior, it would no longer be predictable from its preceding samples as they would correspond to the previous state of the time-variant system generating the nonstationary signal. Therefore, we could expect a large jump in the prediction error at instants of time when the signal changes in its characteristics. Furthermore, the AR model parameters represent the system generating the signal, and provide the poles of the system. If the system were to change in terms of the locations of its poles, the same model would no longer hold: a new model would have to be initiated at such instants of change. This suggests that we could estimate AR models on a short-time basis, and monitor the model parameters from segment to segment: a significant change in the model parameters would indicate a point of change in the signal. (We have seen in Section 7.9 how a similar approach was used by Iwata et al. [207] to detect S1 and S2 in PCGs.) Adjacent segments that have the same or similar model parameters could be concatenated to form longer segments. As the AR model provides several parameters and may be interpreted in several ways (see Section 7.5.2), tracking the behavior of the model over a moving analysis window may be accomplished in many

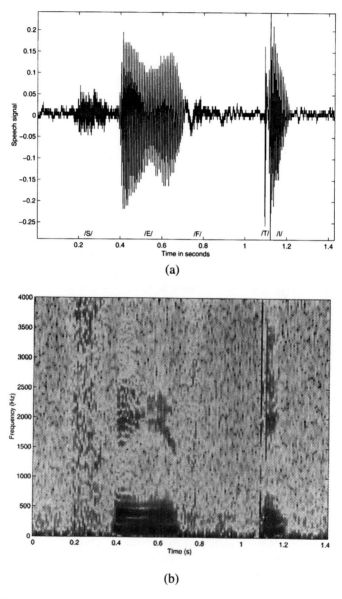

(a)

(b)

Figure 8.5 (a) Time-domain speech signal of the word "safety" uttered by a male speaker. (The signal is also illustrated in Figures 1.29, 3.1, and 3.3.) (b) Spectrogram (log PSD) of the signal computed with a moving short-time analysis window of duration 16 ms (128 samples with $f_s = 8\ kHz$), with the window advance interval being 8 ms.

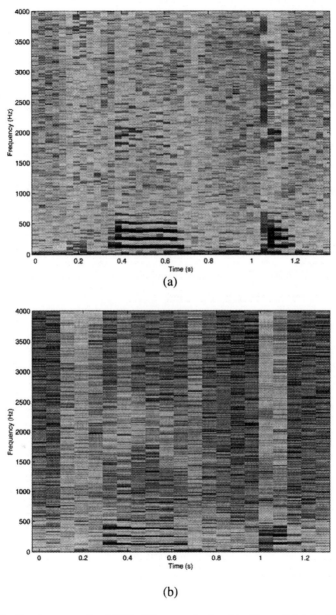

Figure 8.6 Spectrograms (log PSD) of the speech signal in Figure 8.5 (a) with a moving window of duration 64 ms (512 samples with $f_s = 8\ kHz$), with the window advance interval being 32 ms. (b) with a moving window of duration 128 ms (1024 samples), with the window advance interval being 64 ms.

ways. The following subsections provide the details of a few approaches for adaptive segmentation based upon the notions stated above.

8.5.1 Spectral error measure

Bodenstein and Praetorius [98, 234] used the all-pole LP or AR model (see Section 7.5) for adaptive segmentation of EEG signals into quasi-stationary segments and also for further feature extraction. They made the following observations about the application of AR modeling to EEG signals:

- *Time domain:* The present value of the prediction error indicates the instantaneous degree of "unexpectedness" in the signal.

- *Autocorrelation domain:* The prediction error is decorrelated.

- *Spectral domain:* The prediction error being white noise, the AR model yields an all-pole representation of the signal spectrum, which is particularly suitable for the modeling of resonance.

These properties are useful for

- detection and elimination of transients;

- segmentation of the EEG into quasi-stationary segments; and

- feature extraction and pattern recognition (diagnosis).

Ferber [235] provides a description of nonstationarities in the EEG and suggests a few approaches to treat the same.

Analysis of spectral change: Let the PSD of the given nonstationary signal be $S(0, \omega)$ at zero time, and $S(t, \omega)$ at time t. The *spectral error* of $S(t, \omega)$ with respect to $S(0, \omega)$ may be taken to be dependent upon the difference between the corresponding log PSDs, that is, to be proportional to $\log[S(t, \omega)] - \log[S(0, \omega)]$, or equivalently, to be proportional to $\frac{S(t,\omega)}{S(0,\omega)}$. Consider the state when an AR model has been adapted to the signal spectrum $S(0, \omega)$ at zero time. If we pass the signal at time t through the AR model, the prediction error will have an instantaneous spectrum given by

$$S_e(\omega) = \frac{S(t, \omega)}{S(0, \omega)}, \tag{8.15}$$

which is similar to the spectral ratio in Equation 7.50. Thus the problem of comparing two arbitrary PSDs of a nonstationary signal at two different instants of time may now be expressed as testing $S_e(\omega)$ for deviation from a uniform PSD.

Let $a_R(k)$, $k = 1, 2, \ldots, P$, represent the reference AR model. When the current signal $y(n)$ is passed through the filter represented by the AR model, we obtain the prediction error

$$e(n) = \sum_{k=0}^{P} a_R(k) \, y(n - k). \tag{8.16}$$

The error indicates the deviation of the current signal from the previously computed model. Consider the integral

$$\varepsilon = \int_{-\infty}^{\infty} [1 - S_e(\omega)]^2 \, d\omega, \tag{8.17}$$

where $S_e(\omega)$ is the PSD of the prediction error. Ideally, when the AR model has been optimized for the signal on hand, the prediction error is expected to have a uniform PSD. However, if the signal is nonstationary, some changes would have occurred in the spectral characteristics of the signal, which would be reflected in the PSD of the error. If $\phi_e(k)$ is the ACF corresponding to $S_e(\omega)$, the latter is given by the Fourier transform of the former. However, since both functions are real and even, we have

$$S_e(\omega) = \phi_e(0) + 2 \sum_{k=1}^{\infty} \phi_e(k) \, \cos(2\pi\omega k). \tag{8.18}$$

Then,

$$\varepsilon = \int_{-\infty}^{\infty} \left[1 - \phi_e(0) - 2 \sum_{k=1}^{\infty} \phi_e(k) \, \cos(2\pi\omega k) \right]^2 d\omega. \tag{8.19}$$

Due to the orthonormality of the trigonometric functions, we get

$$\varepsilon = [1 - \phi_e(0)]^2 + 2 \sum_{k=1}^{\infty} \phi_e^2(k). \tag{8.20}$$

In practice, the summation may be performed up to some lag, say M. Bodenstein and Praetorius [98] recommended normalization of the error measure by division by $\phi_e^2(0)$, leading to the *spectral error measure (SEM)*

$$SEM = \left[\frac{1}{\phi_e(0)} - 1 \right]^2 + 2 \sum_{k=1}^{M} \left[\frac{\phi_e(k)}{\phi_e(0)} \right]^2. \tag{8.21}$$

Here, the first term represents the change in the total power of the prediction error; the second term depends upon the change in spectral shape only. Note that the prediction error is expected to have a uniform (flat) PSD as long as the signal remains stationary with respect to the AR model designed. The SEM was shown to vary significantly in response to changes in the spectral characteristics of EEG signals, and to be useful in breaking the signals into quasi-stationary parts. Figure 8.7 shows the general scheme of EEG segmentation by using the SEM.

Algorithm for adaptive segmentation [98]:

Let $n = 0$ represent the starting point of analysis where the first reference or fixed analysis window is placed for each adaptive segment, as in Figure 8.7 (a). $(N + P)$ samples of the signal $y(n)$ should be available prior to the arbitrarily designated origin at $n = 0$, where $(2N + 1)$ is the size of the analysis window and P is the order of the AR model to be used.

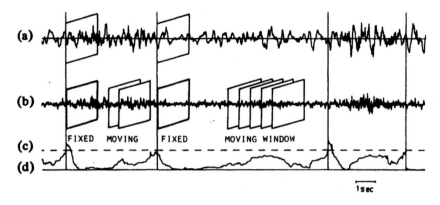

(a)

(b)

FIXED   MOVING   FIXED   MOVING WINDOW

(c)

(d)

1 sec

Figure 8.7 Adaptive segmentation of EEG signals via use of SEM. (a) Original EEG signal. The rectangular window at the beginning of each adaptive segment indicates the signal window to which the AR model has been optimized. (b) Prediction error. The initial ACF of the error is computed over the fixed window; the running ACF of the error is computed over the moving window. (c) Segmentation threshold. (d) SEM. The vertical lines represent the segmentation boundaries. Reproduced with permission from G. Bodenstein and H.M. Praetorius, Feature extraction from the electroencephalogram by adaptive segmentation, *Proceedings of the IEEE*, 65(5):642–652, 1977. ©IEEE.

1. Using the signal samples $y(-N)$ to $y(N)$, compute the signal ACF up to lag P.

2. Derive the corresponding AR model of order P.

3. Using the signal values $y(-N-P)$ to $y(n+N)$, compute the prediction error $e(-N)$ to $e(n+N)$, and compute the running short-time ACF $\phi_e(n,m)$ of the prediction error as

$$\phi_e(n,m) = \frac{1}{2N+1} \sum_{k=-N}^{N-m} e(n+k)\,e(n+k+m). \qquad (8.22)$$

Note that the ACF now has two indices: the first index n to indicate the position of the short-time analysis window, and the second index m to indicate the lag for which the ACF is computed.

4. Calculate $\phi_e(0,m)$ for $m = 0, 1, \ldots, M$. This represents the fixed window at the beginning of each adaptive segment in Figure 8.7 (a).

Perform the following three steps for each data point:

5. Compute $\phi_e(n,m)$ for the moving window [see Figure 8.7 (b)] by the recursive relationship

$$
\begin{aligned}
\phi_e(n,m) &= \phi_e(n-1,m) + e(n+N)e(n+N-m) \qquad (8.23) \\
&\quad - e(n-N-1)e(n-N-1-m).
\end{aligned}
$$

This represents the moving window in Figure 8.7 (b).

6. Compute the SEM at time n as

$$SEM(n) = \left[\frac{\phi_e(0,0)}{\phi_e(n,0)} - 1 \right]^2 + 2 \sum_{k=1}^{M} \left[\frac{\phi_e(n,k)}{\phi_e(n,0)} \right]^2, \tag{8.24}$$

where $\phi_e(0,0)$ accounts for the fact that the signal may have an arbitrary power level.

7. Test if $SEM(n) > Th_1$, where Th_1 is a threshold.

   If the condition is not satisfied, increase n by 1 and return to Step 5.

   If the condition is satisfied, a segment boundary has been detected at time n, as indicated by the vertical lines in Figure 8.7. Reset the procedure by the following step:

8. Shift the time axis by substituting $(n+k)$ with $(k-N)$ and start the procedure again with Step 1.

In the investigations of Bodenstein and Praetorius [98], SEM demonstrated sharp jumps as transients of duration less than 100 ms entered and left the moving analysis window of duration 2 s ($2N + 1 = 101$ samples with $f_s = 50$ Hz). Such jumps could lead to inappropriate segmentation, especially with burst-suppression type EEG episodes as illustrated in Figure 8.8. To overcome this problem, it was suggested that the prediction error $e(n)$ be limited (clipped) by a threshold Th_2 as

$$e(n) = \begin{cases} e(n) & \text{if } |e(n)| < Th_2 \\ \text{sgn}[e(n)]\, Th_2 & \text{if } |e(n)| \geq Th_2 \end{cases}. \tag{8.25}$$

The threshold Th_2 is shown by the dashed lines in Figure 8.8 (c). The SEM computed from the clipped $e(n)$ is shown in Figure 8.8 (d), which, when checked against the original threshold Th_1, will yield the correct segmentation boundary. The signal reconstructed from the clipped prediction error is shown in Figure 8.8 (e), which shows that the clipping procedure has suppressed the effect of the transient without affecting the rest of the signal.

In spite of the clipping procedure as in Equation 8.25, it was indicated by Bodenstein and Praetorius [98] that the procedure was too sensitive and caused false alarms. To further limit the effects of random fluctuations in the prediction error, a smoothed version $e_s(n)$ of the squared prediction error was computed as

$$e_s(n) = e^2(n-1) + 2e^2(n) + e^2(n+1) \tag{8.26}$$

for those samples of $e(n)$ that satisfied the condition $|e(n)| > Th_2$. Another threshold Th_3 was applied to $e_s(n)$, and the triplet $\{y(n-1), y(n), y(n+1)\}$ was considered to be a part of a transient only if $e_s(n) > Th_3$. The procedure of Bodenstein and Praetorius combines adaptive segmentation of EEG signals with transient detection as the two tasks are interrelated.

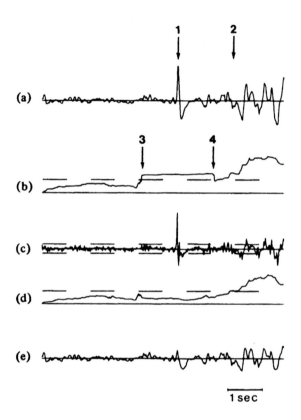

Figure 8.8   Elimination of transients by clipping the prediction error. (a) Original EEG signal of the burst-suppression type. The sharp wave marked by the arrow 1 is followed by the onset of a burst marked by the arrow 2. (b) SEM showing sudden jumps at points indicated by the arrows 3 and 4 as the sharp wave enters and leaves the analysis window. (c) Clipping of the prediction error with threshold Th_2. (d) SEM after clipping the prediction error. The dashed line represents the threshold Th_1. (e) Signal reconstructed from the clipped prediction error. Reproduced with permission from G. Bodenstein and H.M. Praetorius, Feature extraction from the electroencephalogram by adaptive segmentation, *Proceedings of the IEEE*, 65(5):642–652, 1977. ©IEEE.

Illustration of application: Figure 8.9 shows the EEG signal of a child in sleep stage I, superimposed with 14 Hz spindles. The SEM and its components are also shown in the figure. The vertical lines indicate the segment boundaries detected. Bodenstein et al. [236] and Creutzfeldt et al. [237] describe further extension of the approach to computerized pattern classification of EEG signals including clustering of similar segments and labeling of the types of activity found in an EEG record.

The SEM method was applied for adaptive segmentation of VAG signals by Tavathia et al. [55]. It was indicated that each segment could be characterized by the frequency of the most-dominant pole obtained via AR modeling and the spectral power ratio $E_{40:120}$ as per Equation 6.48; however, no classification experiments were performed. More examples of application of the SEM technique will be presented in Sections 8.5.4 and 8.7.

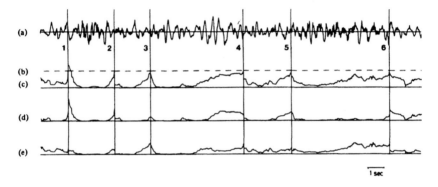

Figure 8.9   Use of the spectral error measure SEM to segment an EEG signal. (a) Original EEG signal of a child in sleep stage I with superimposed 14 Hz spindles. (b) Segmentation threshold. (c) SEM. (d) Deviation in prediction error power. (e) Deviation in prediction error spectral shape. The vertical lines represent the segmentation boundaries. Reproduced with permission from G. Bodenstein and H.M. Praetorius, Feature extraction from the electroencephalogram by adaptive segmentation, *Proceedings of the IEEE*, 65(5):642–652, 1977. ©IEEE.

8.5.2   ACF distance

Michael and Houchin [238] proposed a method comparable to that of Bodenstein and Praetorius [98], but based upon a simpler scheme using the ACF. It should be noted that the AR model coefficients are indeed derived from the ACF, and that the spectra used to compute SEM are related to the corresponding ACFs by the Fourier transform. However, direct use of the ACF removes the assumption made in AR modeling that the signal is the result of an AR process.

In the method of Michael and Houchin, the ACF is treated as a statistical measure of the given signal, and significant variations in the ACF are used to detect nonstationarity. A reference window is extracted at the beginning of each scan, and the given signal (EEG) is observed through a moving window. The duration of the window has to be chosen such that it is shorter than the shortest expected quasi-stationary segment of the given signal, but long enough to characterize the lowest frequency present. If the difference between the signal's statistics (ACF) in the moving window and the reference window is significant, a segment boundary is drawn, and the procedure is restarted.

Let $\phi_R(k)$ be the ACF of the reference window at the beginning of a new segmentation step, where k is the lag or delay. Let $\phi_T(n, k)$ be the ACF of the test window positioned at time instant n. Given that the ACF for zero lag is the power of the signal, Michael and Houchin computed a normalized power distance $d_P(n)$ between the ACFs as (see also Appel and v. Brandt [220])

$$d_P(n) = \frac{|\sqrt{\phi_T(n, 0)} - \sqrt{\phi_R(0)}|}{\min\{\sqrt{\phi_T(n, 0)}, \sqrt{\phi_R(0)}\}}. \tag{8.27}$$

A spectral distance $d_F(n)$ was computed using the ACF coefficients only up to lag q as

$$d_F(n) = \frac{\sum_{k=1}^{q} |\phi_T(n, k) - \phi_R(k)|}{0.5 + \sum_{k=1}^{q} \min\{\sqrt{\phi_T(n, 0)}, \sqrt{\phi_R(0)}\}}. \tag{8.28}$$

The lag limit q was set as the lower value of the lags at which the ACFs changed from positive to negative values for the first time. The net ACF distance $d(n)$ was then computed as

$$d(n) = \frac{d_P(n)}{Th_P} + \frac{d_F(n)}{Th_F}, \tag{8.29}$$

where Th_P and Th_F are thresholds. The condition $d(n) > 1$ was considered to represent a significant change in the ACF, and used to mark a segment boundary.

Due to the use of a moving window of finite size, the true boundary or point of change in the signal characteristics will lie within the last test window before a segment boundary is triggered. Michael and Houchin used a linear interpolation procedure based upon the steepness of the ACF distance measure to correct for such a displacement. Barlow et al. [239] provide illustrations of application of the method to clinical EEGs. Their work includes clustering of similar segments based upon mean amplitude and mean frequency measures, "dendrograms" to illustrate the clustering of segments, as well as labeling of the various states found in an EEG record. Illustration of application of the ACF method will be provided in Section 8.5.4.

8.5.3   The generalized likelihood ratio

The generalized likelihood ratio (GLR) method, proposed by Appel and v. Brandt [219], uses a reference window that is continuously grown as long as no new boundary is marked. The test window is a sliding window of constant duration as in the case of the SEM and ACF methods. Figure 8.10 illustrates the windows used. The advantage of the growing reference window is that it contains the maximum amount of information available from the beginning of the new segment to the current instant. Three different data sets are defined: the growing reference window, the sliding test window, and a pooled window formed by concatenating the two. Distance measures are then derived using AR model prediction errors computed for the three data sets.

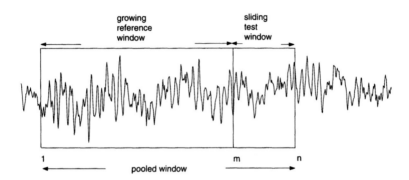

Figure 8.10 The growing reference window, the sliding test window, and the pooled window used in the GLR method for adaptive segmentation.

Let $\varepsilon(m : n)$ represent the prediction error energy (TSE ε as in Equation 7.19) within an arbitrary data set or window with boundaries m and n. The maximum log likelihood measure $H(m : n)$ for the window is defined as

$$H(m : n) = (n - m + 1) \ln \left[\frac{\varepsilon(m : n)}{(n - m + 1)} \right]. \tag{8.30}$$

Three measures are computed for the three data sets described above as $H(1 : m - 1)$ for the growing reference window, $H(m : n)$ for the test window, and $H(1 : n)$ for the composite or pooled window. Here, the reference window is denoted as commencing from the time instant or sample 1, m is the last sample of the growing reference window, and the current test window spans the duration from m to the current time instant n (see Figure 8.10). The GLR distance measure is defined as

$$d(n) = H(1 : n) - [H(1 : m - 1) + H(m : n)]. \tag{8.31}$$

Here, the first quantity represents the TSE if the test window is added to the growing reference window; the second quantity represents the TSE of the reference window grown so far; and the third quantity represents the TSE in modeling the test window itself. The measure $d(n)$ answers the question: "How much is the increase in the TSE if we add the test window to the growing reference window"?

Appel and v. Brandt [219] and Cohen [173] provide more details on the GLR. The GLR distance is a measure of the statistical similarity of the reference and test data sequences, with the assumption that their AR model coefficients have a normal

(Gaussian) distribution. The GLR distance is also a measure of the loss of information caused if no segment boundary is drawn at the position of the test window, that is, if it is assumed that the null hypothesis that the two sequences are similar is true.

Appel and v. Brandt [219] discuss issues related to the choice of the parameters involved in the GLR method, including the AR model order, the test window length, and the threshold, on the GLR distance measure. The GLR method was also used by Willsky and Jones [240] to detect abrupt changes (sporadic anomalies and failures) in the variables of stochastic linear systems, and by Basseville and Benveniste [241] for segmentation of nonstationary signals (see also Cohen [173]). Illustration of application of the GLR method will be provided in Section 8.5.4.

8.5.4   Comparative analysis of the ACF, SEM, and GLR methods

Appel and v. Brandt [220] performed a comparative analysis of the performance of the ACF, SEM, and GLR methods of adaptive segmentation using synthesized signals as well as EEG signals. A simple two-pole system was used as the basis to simulate nonstationary signals. The gain, pole radius, and pole angle were individually varied back and forth between two sets of values. Several outputs of the dynamic system were computed with random signals (Gaussian-distributed white noise) as input. The signals were processed by the ACF, SEM, and GLR methods for adaptive segmentation. The variability of the segment boundaries detected for various realizations of the nonstationary (random) output signals for the same sequences of system parameters was analyzed.

Figure 8.11 shows the results related to variations in the angles of the poles, that is, in the resonance frequency of the system. The angle of the pole in the upper-half of the z-plane was changed from $20°$ to $40°$ and back at samples 200 and 400; the conjugate pole was also varied accordingly. The same changes were repeated at samples 700 and 800. The upper panel in the figure shows the pole positions and the related PSDs. The middle panel illustrates one sample of the 200 test signals generated: the higher-frequency characteristics of the signal related to the shifted pole positioned at $40°$ is evident over the intervals $200 - 400$ and $700 - 800$ samples. The lower panel illustrates the variability in the detected segment indices (dotted curve) and the estimated segment boundary positions (solid curves) for the three methods over 200 realizations of the test signals. (The true segment indices and boundaries are 1 : 200, 2 : 400, 3 : 700, and 4 : 800; ideally, the curves should exhibit steps at the points of change.) It is evident that the GLR method has provided the most consistent and accurate segmentation results, although at the price of increased computational load. The SEM method has performed better than the ACF method, the latter showing the poorest results.

Figure 8.12 shows the results related to variations in the distance of the poles from the origin, that is, in the bandwidth of the resonance frequency of the system. The distance of the poles from the origin was changed from 0.7 to 0.9 and back at samples 200 and 400. The same changes were repeated at samples 700 and 800. The PSDs display the increased prominence of the spectral peak when the poles are pushed toward the unit circle. The ACF method has not performed well in recognizing the

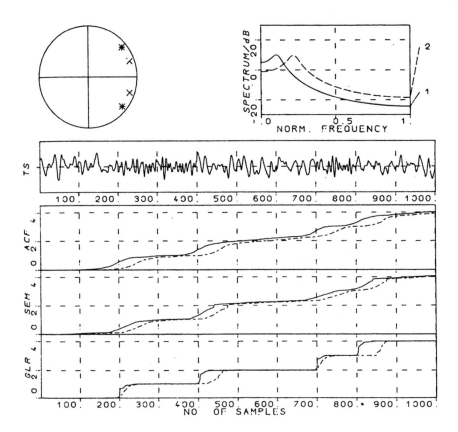

Figure 8.11 Comparative analysis of the ACF, SEM, and GLR methods for adaptive segmentation with the pole angle varied. Upper panel: pole positions and the related PSDs. *Note:* Norm. Frequency is normalized frequency such that the maximum frequency present in the sampled signal is unity. Middle panel: sample test signal; TS = time series. Lower panel: variability in the detected segment indices (dotted curve) and the estimated segment boundary positions (solid curves) for the three methods over 200 realizations of the test signals. See the text for more details. Reproduced with permission from U. Appel and A. v. Brandt, A comparative analysis of three sequential time series segmentation algorithms, *Signal Processing,* 6:45–60, 1984. ©Elsevier Science Publishers B.V. (North Holland).

nonstationarities of this type in the test signals. The GLR method has performed better than the ACF method in segmentation.

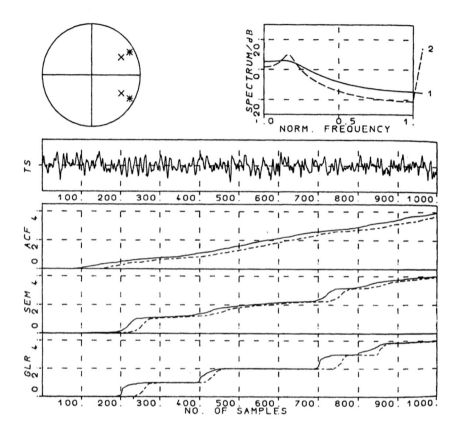

Figure 8.12 Comparative analysis of the ACF, SEM, and GLR methods for adaptive segmentation with the pole radius varied. Upper panel: pole positions and the related PSDs. *Note:* Norm. Frequency is normalized frequency such that the maximum frequency present in the sampled signal is unity. Middle panel: sample test signal; TS = time series. Lower panel: variability in the detected segment indices (dotted curve) and the estimated segment boundary positions (solid curves) for the three methods over 200 realizations of the test signals. See the text for more details. Reproduced with permission from U. Appel and A. v. Brandt, A comparative analysis of three sequential time series segmentation algorithms, *Signal Processing,* 6:45–60, 1984. ©Elsevier Science Publishers B.V. (North Holland).

Figure 8.13 shows the results of application of the three methods to an EEG signal. Although the exact locations where the signal changes its characteristics are not known for the EEG signal, the boundaries indicated by the GLR method appear to be the most accurate. It may be desirable in real-life applications to err on the side of superfluous segmentation; a subsequent clustering step could merge adjacent segments with similar model parameters.

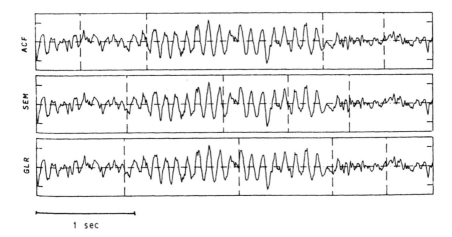

1 sec

Figure 8.13 Comparative analysis of the ACF, SEM, and GLR methods for adaptive segmentation of an EEG signal. Reproduced with permission from U. Appel and A. v. Brandt, A comparative analysis of three sequential time series segmentation algorithms, *Signal Processing*, 6:45–60, 1984. ©Elsevier Science Publishers B.V. (North Holland).

8.6   USE OF ADAPTIVE FILTERS FOR SEGMENTATION

We saw in Sections 3.6.2 and 3.6.3 that the coefficient (tap-weight) vectors of the adaptive LMS and RLS filters are expressed as functions of time. The filters adapt to changes in the statistics of the primary and reference signals. Could we, therefore, use the tap-weight vector $\mathbf{w}(n)$ to detect nonstationarities in a signal?

Problem: *Investigate the potential use of the RLS adaptive filter for adaptive segmentation of nonstationary signals.*

Solution: When we have only one signal to work with — the signal that is to be segmented — the question arises as to how we may provide two inputs, namely, the primary and reference signals, to the adaptive filter. If we assume that the signal to be segmented (applied at the primary input) was generated by an AR system, then we may provide the same signal with a delay as the reference input to the adaptive filter. The delay is to be set such that the reference input at a given instant of time is uncorrelated with the primary input; the delay may also be set on the basis of the order of the filter. (It is also possible to apply white noise at the reference input.) In essence, the adaptive filter then acts the role of an adaptive AR model. The filter tap-weight vector is continually adapted to changes in the statistics (ACF) of the input signal. The output represents the prediction error. Significant changes in the tap-weight vector or the prediction error may be used to mark points of prominent nonstationarities in the signal. Figure 8.14 shows a signal-flow diagram of the adaptive filter as described above; the filter structure is only slightly different from that in Figure 3.51.

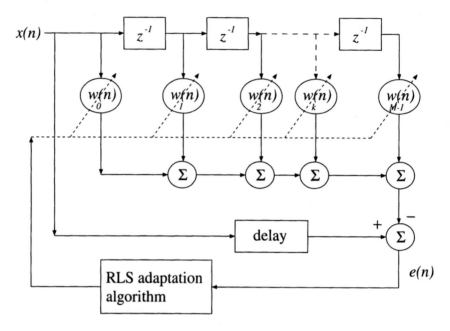

Figure 8.14 Adaptive RLS filter for segmentation of nonstationary signals.

8.6.1 Monitoring the RLS filter

The RLS filter as in Figure 8.14 attempts to predict the current signal sample from the available knowledge of the previous samples stored in the filter's memory units. If a large change occurs in the signal, the prediction error exhibits a correspondingly large value. In response, the adaptive filter's tap-weight vector is modified by the RLS algorithm.

Moussavi et al. [56] applied the RLS filter for segmentation of VAG signals. The order of the filter was set to be 5 in order to be low enough to detect transient changes and also to provide fast convergence. The forgetting factor was defined as $\lambda = 0.98$ so that the filter may be assumed to operate in an almost-stationary situation. The delay between the input and the reference input was set to be 7 samples (which corresponds to $3.5 \, ms$ with $f_s = 2 \, kHz$).

The adaptive segmentation algorithm of Moussavi et al. is as follows:

1. Initialize the RLS algorithm.

2. Find the squared Euclidean distance between the current tap-weight vector $\mathbf{w}(n)$ and the preceding vector $\mathbf{w}(n-1)$ as

$$\Delta(n) = |\mathbf{w}(n) - \mathbf{w}(n-1)|^2. \tag{8.32}$$

3. After computing $\Delta(n)$ for all samples of the signal available (in off-line processing), compute the standard deviation of the $\Delta(n)$ values. Define a threshold as three times the standard deviation.

4. Label all samples n for which $\Delta(n)$ exceeds the threshold as primary segment boundaries.

5. Compute the primary segment lengths (durations) as the differences between successive primary segment boundaries. Reject all primary segment boundaries that result in segment duration less than a preset minimum (defined in the work of Moussavi et al. [56] as 120 samples or 60 ms, corresponding to a knee-joint angle range of approximately 4°).

6. The remaining boundary points are the final segment boundaries.

The main advantage of the RLS method is that there are no explicit reference and test windows as in the case of the ACF, SEM, and GLR methods. The RLS method computes a new filter tap-weight vector at each sample of the incoming signal. The method was found to perform well in the detection of trend-wise or gradual changes as well as sudden variations in VAG signals.

Illustration of application: Figures 8.15 and 8.16 illustrate the segmentation of the VAG signals of a normal subject and a patient with arthroscopically confirmed cartilage pathology, respectively. The figures also illustrate the spectrograms of the two signals. While the segmentation of the abnormal signal in Figure 8.16 may appear to be superfluous at first sight, close inspection of the corresponding spectrogram indicates that the spectral characteristics of the signal do indeed change within short intervals. It is evident that the RLS method has detected the different types of nonstationarity present in the signals. Moussavi et al. [56] tested the method with 46 VAG signals and observed that the segmentation boundaries agreed well with the nature of the joint sounds heard via auscultation with a stethoscope as well as with the spectral changes observed in the spectrograms of the signals.

8.6.2   The RLS lattice filter

In order to apply the RLS method for adaptive segmentation in a nonstationary environment, it is necessary to solve the least-squares problem recursively and rapidly. The *recursive least-squares lattice* (RLSL) algorithm is well suited for such purposes. Since the RLSL method uses a lattice filter, and is based upon forward and backward prediction and time-varying reflection coefficients, it is necessary to define some of the related procedures.

Forward and backward prediction: Let us rewrite Equation 7.17 related to LP or AR modeling as

$$\tilde{y}(n) = -\sum_{k=1}^{M} a_{M,k}\, y(n-k), \tag{8.33}$$

with the inclusion of the order of the model M as a subscript for the model coefficients a_k. In this procedure, M past samples of the signal $y(n-1), y(n-2), \ldots, y(n-M)$ are used in a linear combination to predict the current sample $y(n)$ in the *forward*

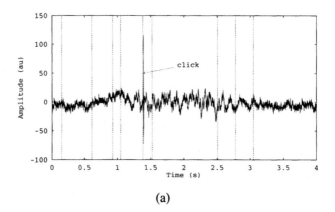

(a)

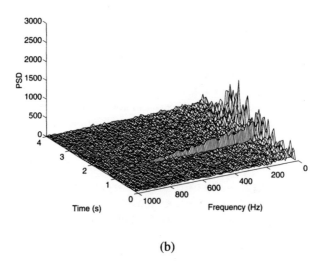

(b)

Figure 8.15   (a) Segmentation of the VAG signal of a normal subject using the RLS method. A click heard in auscultation of the knee joint is labeled. (b) Spectrogram (STFT) of the signal. Reproduced with permission from Z.M.K. Moussavi, R.M. Rangayyan, G.D. Bell, C.B. Frank, K.O. Ladly, and Y.T. Zhang, Screening of vibroarthrographic signals via adaptive segmentation and linear prediction modeling, *IEEE Transactions on Biomedical Engineering,* 43(1):15–23, 1996. ©IEEE.

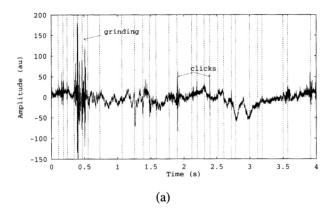

(a)

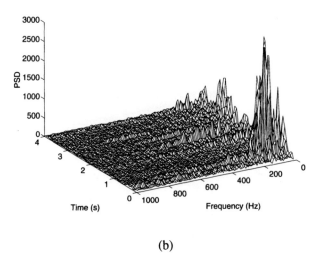

(b)

Figure 8.16 (a) Segmentation of the VAG signal of a subject with cartilage pathology using the RLS method. Clicking and grinding sounds heard during auscultation of the knee joint are labeled. (b) Spectrogram (STFT) of the signal. Reproduced with permission from Z.M.K. Moussavi, R.M. Rangayyan, G.D. Bell, C.B. Frank, K.O. Ladly, and Y.T. Zhang, Screening of vibroarthrographic signals via adaptive segmentation and linear prediction modeling, *IEEE Transactions on Biomedical Engineering*, 43(1):15–23, 1996. ©IEEE.

direction. The *forward prediction error* is

$$e_{M,f}(n) = y(n) - \tilde{y}(n) = \sum_{k=0}^{M} a_{M,k}\, y(n-k), \tag{8.34}$$

with $a_{M,0} = 1$. This equation is a restatement of Equation 7.18 with the inclusion of the order of the model M as a subscript for the error e as well as the subscript f to indicate that the prediction is being performed in the forward direction.

The term *backward prediction* refers to the estimation of $y(n - M)$ from the samples $y(n), y(n-1), \ldots, y(n - M + 1)$ as

$$\tilde{y}(n - M) = -\sum_{k=0}^{M-1} a_{M,k}^{\#}\, y(n-k), \tag{8.35}$$

where $a_{M,k}^{\#}$ are the backward prediction coefficients. Application of the least-squares method described in Section 7.5 for a stationary signal leads to the result

$$a_{M,k}^{\#} = a_{M,M-k}, \quad k = 0, 1, 2, \ldots, M, \tag{8.36}$$

that is, the backward prediction coefficients are the same as the forward prediction coefficients, but in reverse order [77]. The *backward prediction error* is, therefore, given by

$$\begin{aligned} e_{M,b}(n) &= y(n - M) - \tilde{y}(n - M) \tag{8.37} \\ &= \sum_{k=0}^{M} a_{M,k}^{\#}\, y(n-k) = \sum_{k=0}^{M} a_{M,M-k}\, y(n-k). \end{aligned}$$

The Burg-lattice method: The Burg-lattice method [77] is based on minimizing the sum of the squared forward and backward prediction errors. Assuming that the input $y(n)$ is ergodic, the *performance index* ξ_m is given by

$$\xi_m = \sum_{n=m+1}^{N} [e_{m,f}^2(n) + e_{m,b}^2(n)], \tag{8.38}$$

where $e_{m,f}(n)$ is the forward prediction error and $e_{m,b}(n)$ is the backward prediction error, with the model order m being recursively updated as $m = 1, 2, \ldots, M$. The length of the available block of data is N samples.

If we use the Levinson-Durbin method to estimate the forward prediction coefficients, we get (see Section 7.5 and Equation 7.38)

$$a_{m,k} = a_{m-1,k} + \gamma_m\, a_{m-1,m-k}, \tag{8.39}$$

where γ_m is the reflection coefficient for order m. Similarly, for the case of backward prediction, we get

$$a_{m,m-k} = a_{m-1,m-k} + \gamma_m\, a_{m-1,m}, \tag{8.40}$$

including the substitution $a_{m,k}^{\#} = a_{m,m-k}$.

Combining the relationships in Equations 8.34, 8.38, 8.39, and 8.40 leads to the lattice structure for computation of the forward and backward prediction errors, where the two prediction error series are inter-related recursively as [77]

$$e_{m,f}(n) = e_{m-1,f}(n) + \gamma_m \, e_{m-1,b}(n-1) \tag{8.41}$$

and

$$e_{m,b}(n) = e_{m-1,b}(n-1) + \gamma_m \, e_{m-1,f}(n). \tag{8.42}$$

(All coefficients are assumed to be real-valued in this derivation; Haykin [77] allows for all coefficients to be complex-valued.) Figure 8.17 illustrates a basic unit of the lattice structure that performs the recursive operations in Equations 8.41 and 8.42. The reflection coefficient γ_m may be chosen so as to minimize the performance index given in Equation 8.38, that is, by setting

$$\frac{\partial \xi_m}{\partial \gamma_m} = 2 \sum_{n=m+1}^{N} \left[e_{m,f}(n) \, \frac{\partial e_{m,f}(n)}{\partial \gamma_m} + e_{m,b}(n) \, \frac{\partial e_{m,b}(n)}{\partial \gamma_m} \right] = 0. \tag{8.43}$$

Partial differentiation of Equations 8.41 and 8.42 with respect to γ_m yields

$$\frac{\partial e_{m,f}(n)}{\partial \gamma_m} = e_{m-1,b}(n-1) \tag{8.44}$$

and

$$\frac{\partial e_{m,b}(n)}{\partial \gamma_m} = e_{m-1,f}(n). \tag{8.45}$$

Substituting the results above in Equation 8.43, we get

$$\sum_{n=m+1}^{N} \left[e_{m,f}(n) \, e_{m-1,b}(n-1) + e_{m,b}(n) \, e_{m-1,f}(n) \right] = 0. \tag{8.46}$$

Substituting Equations 8.41 and 8.42 in Equation 8.46, we get

$$\sum_{n=m+1}^{N} \left[\{ e_{m-1,f}(n) + \gamma_m \, e_{m-1,b}(n-1) \} \, e_{m-1,b}(n-1) \right.$$
$$\left. + \, \{ e_{m-1,b}(n-1) + \gamma_m \, e_{m-1,f}(n) \} \, e_{m-1,f}(n) \right] = 0. \tag{8.47}$$

The reflection coefficients γ_m can then be calculated as

$$\gamma_m = -2 \, \frac{\sum_{n=m+1}^{N} e_{m-1,f}(n) \, e_{m-1,b}(n-1)}{\sum_{n=m+1}^{N} \left[e_{m-1,f}^2(n) + e_{m-1,b}^2(n-1) \right]}. \tag{8.48}$$

The magnitudes of the reflection coefficients are less than unity. The Burg formula always yields a minimum-phase design for the lattice predictor.

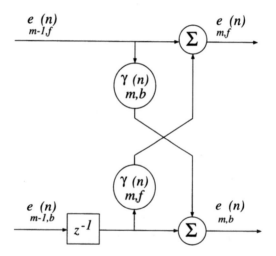

Figure 8.17   Basic unit of the lattice structure that performs the recursive operations in Equations 8.41 and 8.42 as well as the recursive operations in Equations 8.52 and 8.53. In the case of the former, due to the stationarity of the processes involved, the forward and backward reflection coefficients are the same and are independent of time. Adapted, with permission, from S. Krishnan, Adaptive Signal Processing Techniques for Analysis of Knee Joint Vibroarthrographic Signals, Ph.D. Thesis, University of Calgary, 1999.

The prediction coefficients or the AR model parameters can be computed from the reflection coefficients by using the relationship in Equation 8.39. The order m is updated recursively as $m = 1, 2, \ldots, M$, with $a_{m,0} = 1$, and $a_{m-1,k} = 0$ for $k > m - 1$. From Equation 8.39 and Figure 8.17, it can be observed that the AR coefficients can be computed for any model order by simply adding one or more lattice stages without affecting the earlier computations for lower orders. This is one of the main advantages of the Burg-lattice AR modeling algorithm, especially in situations where the order of the system being modeled is not known in advance.

RLSL algorithm for adaptive segmentation: A general schematic representation of the RLSL filter structure is given in Figure 8.18. Two levels of updating are used in the RLSL algorithm:

1. *Order-update:* This involves updating the forward prediction error $e_{m,f}(n)$, the backward prediction error $e_{m,b}(n)$, the forward prediction error power $\varepsilon_{m,f}(n)$, and the backward prediction error power $\varepsilon_{m,b}(n)$. Here, m indicates the model order, and n indicates the time instant.

2. *Time-update:* This involves time-updating of the parameters that ensure adaptation, including the forward reflection coefficients $\gamma_{m,f}(n)$ and backward reflection coefficients $\gamma_{m,b}(n)$. Note that, in the general nonstationary environment, $\gamma_{m,f}(n) \neq \gamma_{m,b}(n)$.

Order-updating and time-updating together enable the RLSL algorithm to achieve extremely fast convergence and excellent tracking capability.

The RLSL algorithm can be expressed in three stages [77, 88, 90]:

1. *Initialization of the algorithm and lattice for filter order M:* The parameters of the algorithm are initialized at $n = 0$ and for each order $m = 1, 2, \ldots, M$ by setting the forward prediction error power $\varepsilon_{m-1,f}(0)$ and the backward prediction error power $\varepsilon_{m-1,b}(0)$ equal to a small positive constant; the forward reflection coefficients $\gamma_{m,f}(0) = 0$; the backward reflection coefficients $\gamma_{m,b}(0) = 0$; the conversion factor $\gamma_{0,c}(0) = 1$; and an auxiliary variable $\Delta_{m-1}(0) = 0$.

   For each time instant $n \geq 1$, the following zeroth-order variables are generated: the forward prediction error $e_{0,f}(n)$ equal to the data input $y(n)$; the backward prediction error $e_{0,b}(n) = y(n)$; $\varepsilon_{0,f}(n) = \varepsilon_{0,b}(n) = \lambda \varepsilon_{0,f}(n-1) + |y(n)|^2$, where λ is the forgetting factor, and $\gamma_{0,c}(n) = 1$.

   The variables involved in joint process estimation, for each order $m = 0, 1, \ldots, M$ at time $n = 0$, are initialized by setting the scalar $\rho_m(0) = 0$, and for each instant $n \geq 1$ the zeroth-order variable of *a priori* estimation error $e_0 = d(n)$, where $d(n)$ is the desired response of the system.

2. *Prediction part of the RLSL algorithm:* For $n = 1, 2, \ldots, N_s$, where N_s is the number of signal samples available, the various order-updates are computed in the sequence $m = 1, 2, \ldots, M$, where M is the final order of the least squares

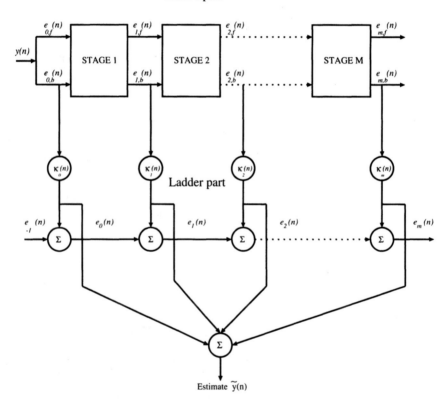

Figure 8.18 General schematic representation of the RLSL filter structure for adaptive segmentation of nonstationary signals. Adapted, with permission, from S. Krishnan, Adaptive Signal Processing Techniques for Analysis of Knee Joint Vibroarthrographic Signals, Ph.D. Thesis, University of Calgary, 1999.

predictor, as follows:

$$\Delta_{m-1}(n) = \lambda \Delta_{m-1}(n-1) + \frac{e_{m-1,b}(n-1)\, e_{m-1,f}(n)}{\gamma_{m-1,c}(n-1)}, \qquad (8.49)$$

where $\Delta_{m-1}(n)$ is the cross-correlation between the delayed backward prediction error $e_{m-1,b}(n-1)$ and the forward prediction error $e_{m-1,f}(n)$ of the lattice filter.

The forward reflection coefficient $\gamma_{m,f}(n)$ is then updated as

$$\gamma_{m,f}(n) = -\frac{\Delta_{m-1}(n)}{\varepsilon_{m-1,b}(n)}. \qquad (8.50)$$

Similarly, the backward reflection coefficient is updated as

$$\gamma_{m,b}(n) = -\frac{\Delta_{m-1}(n)}{\varepsilon_{m-1,f}(n-1)}. \qquad (8.51)$$

In general, $\varepsilon_{m-1,f}(n)$ and $\varepsilon_{m-1,b}(n-1)$ are unequal, so that in the RLSL algorithm, unlike in the Burg algorithm described earlier in this section, we have $\gamma_{m,f}(n) \neq \gamma_{m,b}(n)$.

From the lattice structure as described earlier in the context of Equations 8.41 and 8.42 and depicted in Figure 8.17, and noting that the reflection coefficients $\gamma_{m,f}(n)$ and $\gamma_{m,b}(n)$ are now different and time-variant parameters, we can write the order-update recursion of the forward prediction error as (see Figure 8.17)

$$e_{m,f}(n) = e_{m-1,f}(n) + \gamma_{m,f}(n)\, e_{m-1,b}(n-1), \qquad (8.52)$$

and the order-update recursion of the backward prediction error as

$$e_{m,b}(n) = e_{m-1,b}(n-1) + \gamma_{m,b}(n)\, e_{m-1,f}(n). \qquad (8.53)$$

The prediction error powers are updated as

$$\varepsilon_{m,f}(n) = \varepsilon_{m-1,f}(n) + \gamma_{m,f}(n)\, \Delta_{m-1}(n), \qquad (8.54)$$

and

$$\varepsilon_{m,b}(n) = \varepsilon_{m-1,b}(n-1) + \gamma_{m,b}(n)\, \Delta_{m-1}(n). \qquad (8.55)$$

The conversion factor $\gamma_{m,c}(n-1)$ is updated as

$$\gamma_{m,c}(n) = \gamma_{m-1,c}(n) - \frac{e_{m-1,b}^2(n)}{\varepsilon_{m-1,b}(n)}. \qquad (8.56)$$

The equations in this step constitute the basic order-update recursions for the RLSL predictor. The recursions generate two sequences of prediction errors:

the forward prediction error and the backward prediction error. The two error sequences play key roles in the recursive solution of the linear least-squares problem.

3. *Filtering part of the RLSL algorithm:* For $n = 1, 2, \ldots, N_s$, the various order-updates are computed in the sequence $m = 0, 1, \ldots, M$ as follows:

$$\rho_m(n) = \lambda \rho_m(n-1) + \frac{e_{m,b}(n)}{\gamma_{m,c}(n)} e_{m-1}(n). \qquad (8.57)$$

The regression coefficients $\kappa_m(n)$ of the joint process estimator are defined in terms of the scalar $\rho_m(n)$ as

$$\kappa_m(n) = \frac{\rho_m(n)}{\varepsilon_{m,b}(n)}. \qquad (8.58)$$

The order-update recursion of the *a posteriori* estimation error $e_m(n)$ is then given as

$$e_m(n) = e_{m-1}(n) - \kappa_m(n)\, e_{m,b}(n). \qquad (8.59)$$

The dynamics of the input signal, that is, the statistical changes occurring in the signal, are reflected in the lattice filter parameters. Parameters such as the reflection coefficients (γ_f and γ_b) and the MS value of the estimation error (that is, $E[e_m^2(n)]$) may therefore be used to monitor the statistical changes.

The conversion factor γ_c that appears in the algorithm can be used as a good statistical detection measure of the "unexpectedness" of the recent data samples. As long as the data belong to the same distribution, the variable γ_c will be near unity. If the recent data samples belong to a different distribution, γ_c will tend to fall from unity. This will cause the factor $\frac{1}{\gamma_c}$ appearing in the time-update formula (Equation 8.49) to be large, which leads to abrupt changes in the lattice parameters. The quantities γ_c, $\frac{1}{\gamma_c}$, or $\frac{1}{1-\gamma_c}$ may be used for fast tracking of changes in the input data, and to test for segment boundaries in a nonstationary environment.

Illustration of application: The advantage in using the RLSL filter for segmentation of VAG signals is that the statistical changes in the signals are well reflected in the filter parameters, and hence segment boundaries can be detected by monitoring any one of the filter parameters such as the MSE, conversion factor, or the reflection coefficients. Krishnan et al. [57, 88] used the conversion factor (γ_c) to monitor statistical changes in VAG signals. In a stationary environment, γ_c starts with a low initial value, and remains small during the early part of the initialization period. After a few iterations, γ_c begins to increase rapidly toward the final value of unity. In the case of nonstationary signals such as VAG, γ_c will fall from its steady-state value of unity whenever a change occurs in the statistics of the signal. This can be used in segmenting VAG signals into quasi-stationary components. The segmentation procedure proposed by Krishnan et al. [57, 88] is summarized as follows:

1. The VAG signal is passed twice through the segmentation filter: the first pass is used to allow the filter to converge, and the second pass is used to test the γ_c value at each sample against a threshold value for the detection of segment boundaries.

2. Whenever γ_c at a particular sample during the second pass is less than the threshold, a primary segment boundary is marked.

3. If the difference between two successive primary segment boundaries is less than the minimum desired segment length (120 samples in the work of Krishnan et al.), the later of the two boundaries is deleted.

Figures 8.19 and 8.20 show the results of application of the RLSL segmentation method to two VAG signals. Plots of $\gamma_c(n)$ are also included in the figures. It may be observed that the value of $\gamma_c(n)$ drops whenever there is a significant change in the characteristics of the signal. Whereas the direct application of a threshold on $\gamma_c(n)$ would result in superfluous segmentation, inclusion of the condition on the minimum segment length that is meaningful in the application is seen to provide practically useful segmentation. The number of segments was observed to be, on the average, eight segments per VAG signal. Signals of patients with cartilage pathology were observed to result in more segments than normal signals.

An advantage of the RLSL method of adaptive segmentation is that a fixed threshold may be used; Krishnan et al. found a fixed threshold value of 0.9985 to give good segmentation results with VAG signals. The adaptive segmentation procedure was found to provide segments that agreed well with manual segmentation based upon auscultation and/or arthroscopy. Adaptive analysis of VAG signals will be further described in Section 9.13.

8.7 APPLICATION: ADAPTIVE SEGMENTATION OF EEG SIGNALS

Problem: *Propose a method for parametric representation of nonstationary EEG signals.*

Solution: Bodenstein and Praetorius [98] applied their adaptive segmentation procedure based upon the SEM (see Section 8.5.1) for representation and analysis of EEG signals with the following propositions.

1. An EEG signal consists of quasi-stationary segments upon which transients may be superimposed.

2. A segment is specified by its time of occurrence, duration, and PSD (represented by its AR model coefficients). A transient is specified by its time of occurrence and a set of grapho-elements (or directly by its samples).

3. An EEG signal consists of a finite number of recurrent states.

It should be noted that whereas the adaptive segments have variable length, each adaptive segment is represented by the same number of AR model coefficients.

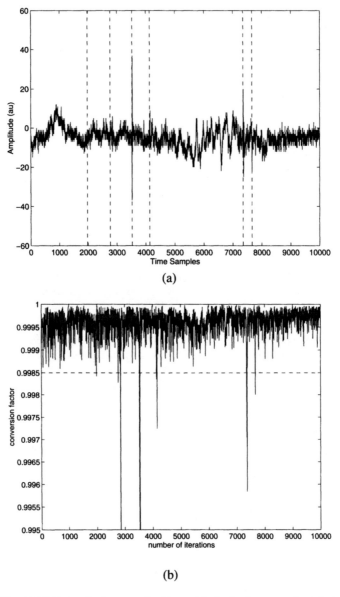

Figure 8.19 (a) VAG signal of a normal subject with the final segment boundaries given by the RLSL method shown by vertical dashed lines. (b) Plot of the conversion factor $\gamma_c(n)$; the horizontal dashed line represents the fixed threshold used to detect segment boundaries. The duration of the signal is 5 s, with $f_s = 2\ kHz$. Reproduced with permission from S. Krishnan, R.M. Rangayyan, G.D. Bell, C.B. Frank, and K.O. Ladly, Adaptive filtering, modelling, and classification of knee joint vibroarthrographic signals for non-invasive diagnosis of articular cartilage pathology, *Medical and Biological Engineering and Computing*, 35(6):677–684, 1997. ©IFMBE.

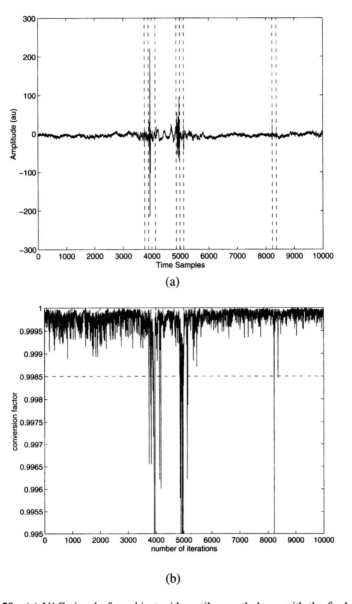

(a)

(b)

Figure 8.20 (a) VAG signal of a subject with cartilage pathology, with the final segment boundaries given by the RLSL method shown by vertical dashed lines. (b) Plot of the conversion factor $\gamma_c(n)$; the horizontal dashed line represents the fixed threshold used to detect segment boundaries. The duration of the signal is 5 s, with $f_s = 2\ kHz$. Reproduced with permission from S. Krishnan, R.M. Rangayyan, G.D. Bell, C.B. Frank, and K.O. Ladly, Adaptive filtering, modelling, and classification of knee joint vibroarthrographic signals for non-invasive diagnosis of articular cartilage pathology, *Medical and Biological Engineering and Computing,* 35(6):677–684, 1997. ©IFMBE.

The number of parameters is therefore independent of segment duration, which is convenient when pattern classification techniques are applied to the segments. Since the AR model is computed once at the beginning of each segment and some prediction error is permitted in the moving analysis window, the initial AR model may not adequately represent the entire adaptive segment. A new model may be computed using the signal samples over the entire duration of each adaptive segment. Instead, Bodenstein and Praetorius maintained the initial AR model of order P of each adaptive segment, and an additional *corrective predictor* of order M was derived for each adaptive segment using the ACF of the prediction error which is computed and readily available in the segmentation procedure. Each adaptive segment was then represented by the $(P + M)$ AR model coefficients, the associated prediction error RMS values, and the segment length. The PSD of the segment may be derived from the two sets of AR model coefficients.

With the EEG signals bandpass filtered to the range $1 - 25$ Hz and sampled at 50 Hz in the work of Bodenstein and Praetorius [98], the ACF window length was set to be 2 s with $2N + 1 = 101$ samples. Bodenstein and Praetorius used the rule of thumb that the AR model order should be at least twice the number of expected resonances in the PSD of the signal. Short segments of EEG signals rarely demonstrate more than two spectral peaks, which suggests that an AR model order of $P = 5$ should be adequate. Regardless, Bodenstein and Praetorius used $P = 8$, which met the Akaike criterion as well (see Section 7.5.2). The order of the ACF of the prediction error and the associated corrective predictor was set to a low value of $M = 3$, allowing for one spectral peak (the error should ideally have a flat PSD). The thresholds were defined as $Th_1 = 0.5$ (empirical), and $Th_2 = 2.5\sigma$, where σ is the RMS value of the prediction error (see Section 8.5.1). The range of $20\sigma^2$ to $40\sigma^2$ was recommended for Th_3. A transitional delay of 25 samples was allowed between each segmentation boundary and the starting point of the following fixed window to prevent the inclusion of the spectral components of one segment into the following segment.

Figure 8.21 shows a few examples of adaptive segmentation of EEG signals. A clustering procedure was included to remove spurious boundaries, some examples of which may be seen in Figure 8.21 (d): neighboring segments with similar parameters were merged in a subsequent step. Visual inspection of the results indicates that most of the adaptive segments are stationary (that is, they have the same appearance) over their durations. It is worth noting that the longest segment in Figure 8.21 (d) of duration 16 s or 800 samples is represented by just 12 parameters.

Figure 8.22 shows examples of detection of transients in two contralateral channels of the EEG of a patient with epilepsy. The EEG signal between seizures (inter-ictal periods) is expected to exhibit a large number of sharp waves. The length of the arrows shown in the figure was made proportional to the cumulated supra-threshold part of the squared prediction error in order to indicate how pronounced the event was regarded to be by the algorithm.

The method was further extended to parallel analysis of multichannel EEG signals by Bodenstein et al. [236] and Creutzfeldt et al. [237]. Procedures were proposed for computerized pattern classification and labeling of EEG signals, including clustering

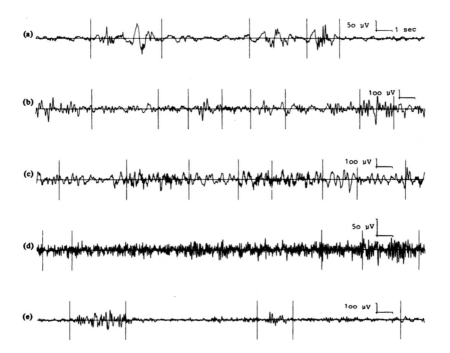

Figure 8.21 Examples of segmentation of EEG signals. (a) Newborn in non-REM sleep. REM = rapid eye movement. (b) Child of age 7 years in sleep stage I. (c) Child of age 8 years in sleep stage III. (d) Alpha rhythm of an adult. (e) EEG of an adult with paroxysms. The vertical lines represent the segmentation boundaries. Reproduced with permission from G. Bodenstein and H.M. Praetorius, Feature extraction from the electroencephalogram by adaptive segmentation, *Proceedings of the IEEE,* 65(5):642–652, 1977. ©IEEE.

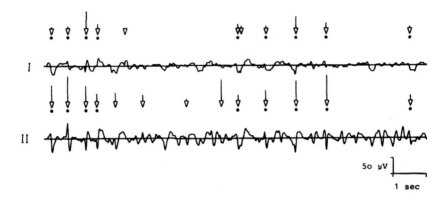

Figure 8.22 Example of detection of transients in the EEG signal of a patient with epilepsy. The signals shown are from contralateral channels between seizures (inter-ictal period). The longer the arrow the more pronounced is the transient detected at the corresponding time instant. Transients detected simultaneously in the two contralateral channels are marked with dots. Reproduced with permission from G. Bodenstein and H.M. Praetorius, Feature extraction from the electroencephalogram by adaptive segmentation, *Proceedings of the IEEE*, 65(5):642–652, 1977. ©IEEE.

of similar segments and state diagrams indicating the sequence of the types of activity found in an EEG record. Figure 8.23 illustrates the record produced by the application of the procedure to two channels of an EEG signal. Typical EEG segments belonging to the four clusters detected in the signal are shown on the left-hand and right-hand sides of the upper portion of the figure. Each signal segment is labeled with the frequencies (FRQ, in Hz) and amplitudes (AMP, in μV) of the resonances detected using an eighth-order AR model. The central column of the upper portion of the figure illustrates the PSDs of the corresponding segments on the left-hand side (solid line) and right-hand side (dashed line). The middle portion of the figure provides the state diagram, indicating the transitions between the four states (represented by the four clusters of the EEG segments) detected in the two channels of the signal. The states represent 1: background, 2: eyes open, 3: paroxysm, and 4: epileptiform spike-and-wave complexes. The values on the right-hand side of the state diagram give the percentage of the total duration of the signal for which the EEG was in the corresponding states. The bottom portion of the figure illustrates singular events, that is, segments that could not be grouped with any of the four clusters. It was indicated that the segments of most EEG signals could be clustered into at most five states, and that the summarized record as illustrated in Figure 8.23 could assist clinicians in analyzing lengthy EEG records in an efficient manner.

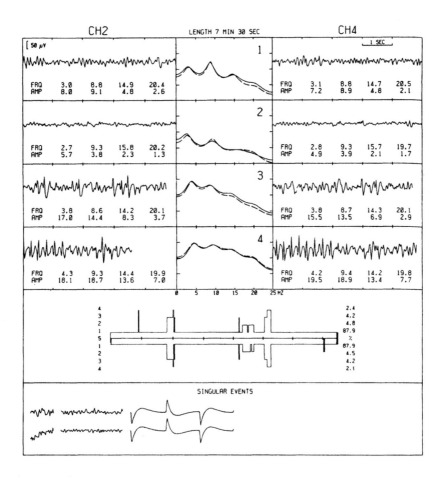

Figure 8.23 Example of application of segmentation and pattern analysis to the EEG signal of a patient with epileptiform activity. Refer to the text for details. Reproduced with permission from G. Bodenstein, W. Schneider, and C.V.D. Malsburg, Computerized EEG pattern classification by adaptive segmentation and probability-density-function classification. Description of the method, *Computers in Biology and Medicine,* 15(5):297–313, 1985. ©Elsevier Science.

8.8 APPLICATION: ADAPTIVE SEGMENTATION OF PCG SIGNALS

We have noted several times that the PCG signal is nonstationary. Let us now assess the feasibility of adaptive segmentation of PCG signals using the RLSL method, with no other signal being used as a reference.

Figure 8.24 illustrates the results of segmentation of the PCG signal of a normal subject. The top trace shows the PCG signal over three cardiac cycles; the segment boundaries detected are indicated by the vertical dotted lines as well as by the triangular markers on the time axis. The second trace illustrates a plot of the conversion factor γ_c: the conversion factor drops from unity whenever there is a change in the signal characteristics, in particular at the boundaries of S1 and S2. A threshold of 0.995 (indicated by the horizontal line overlaid on the second trace) applied to γ_c and a condition imposing a minimum segment length of 50 samples (50 ms) were used to obtain the segment boundaries. The third and fourth traces illustrate the ECG and carotid pulse signals of the subject acquired simultaneously with the PCG. The segment boundaries obtained by the RLSL method agree very well with the readily noticeable S1 and S2 boundaries as well as the QRS and dicrotic notch positions. (See also Sections 1.2.8, 2.3, and 4.10.)

Figure 8.25 illustrates the results of adaptive segmentation of the PCG signal of a subject with systolic murmur due to aortic stenosis. The results in this case, however, are not as clear or as easy to interpret as in the preceding case. The method has indeed identified the beginning of S1 and S2; furthermore, the split nature of S2 has been identified by an additional segment boundary within each S2. However, the method has not reliably identified the boundaries between the episodes of S1 and systolic murmur illustrated: the condition on the minimum segment length has affected the placement of the segment boundary after the beginning of S1. Use of other conditions on γ_c may provide better segmentation results.

8.9 APPLICATION: TIME-VARYING ANALYSIS OF HEART-RATE VARIABILITY

The heart rate is controlled by the autonomous and central nervous systems: the vagal and sympathetic activities lead to a decrease or increase, respectively, in the heart rate (see Section 1.2.4). We saw in Section 7.8 how respiration affects heart rate, and how Fourier analysis may be extended to analyze HRV. When heart rate data such as beat-to-beat RR intervals are collected over long periods of time (several hours), the signal could be expected to be nonstationary.

Bianchi et al. [225] extended AR modeling techniques for time-variant PSD analysis of HRV data in order to study transient episodes related to ischemic attacks. The prediction error was weighted with a forgetting factor, and a time-varying AR model was derived. The RLS algorithm was used to update the AR model coefficients at every RR interval sample (every cardiac cycle). The AR coefficients were then used to compute a time-varying PSD. The following frequency bands were indicated to be

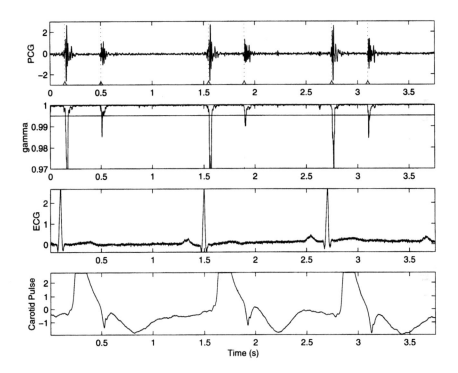

Figure 8.24   Adaptive segmentation of the PCG signal of a normal subject using the RLSL method. Top to bottom: PCG signal (the vertical dotted lines and triangular markers represent the segmentation boundaries); conversion factor γ_c (the horizontal line is the threshold used); ECG; carotid pulse (clipped due to saturation).

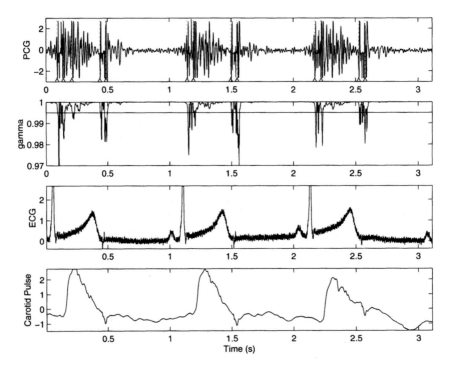

Figure 8.25 Adaptive segmentation of the PCG signal of a subject (female, 11 years) with systolic murmur due to aortic stenosis. Top to bottom: PCG signal (the vertical lines and triangular markers represent the segmentation boundaries); conversion factor γ_c (the horizontal line is the threshold used); ECG; carotid pulse.

of interest in the analysis of RR interval PSDs: very-low-frequency (VLF) band in the range $0 - 0.03$ Hz related to humoral and thermoregulatory factors; low-frequency (LF) band in the range $0.03 - 0.15$ Hz related to sympathetic activity; high-frequency (HF) band in the range $0.18 - 0.4$ Hz related to respiration and vagal activity.

Figure 8.26 shows an RR interval series including an ischemic episode (delineated by B for beginning and E for ending points, respectively). Figure 8.27 shows the time-varying PSD in the form of a spectrogram. Figure 8.28 shows a segment of RR interval data and a few measures derived from the data.

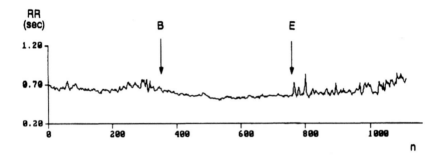

Figure 8.26   RR interval series including an ischemic episode. B: beginning and E: end of the episode. Reproduced with permission from A.M. Bianchi, L. Mainardi, E. Petrucci, M.G. Signorini, M. Mainardi, and S. Cerutti, Time-variant power spectrum analysis for the detection of transient episodes in HRV signal, *IEEE Transactions on Biomedical Engineering*, 40(2):136–144, 1993. ©IEEE.

Some of the important observations made by Bianchi et al. (and illustrated by the spectrogram in Figure 8.27 and the parameters in Figure 8.28) are:

- There is an increase in LF power about $1.5 - 2$ *minutes* before an ischemic event.

- The RR variance decreases as an episode begins.

- There is a predominant rise in LF power at the end of an ischemic episode.

- A small HF component appears toward the end of an episode.

- Early activation of an LF component precedes tachycardia and ST displacement in the ECG that are generally indicative of the onset of an ischemic episode.

- The results suggest an arousal of the sympathetic system before an acute ischemic attack.

Time-varying AR modeling techniques have also been applied for the analysis of EEG signals [224]. Time-varying ARMA modeling techniques have been applied to analyze EGG signals [38].

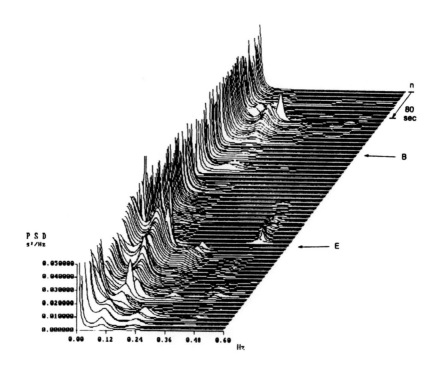

Figure 8.27 Spectrogram of the RR interval series in Figure 8.26. Time progresses from the top to the bottom. B: beginning and E: end of an ischemic episode. Reproduced with permission from A.M. Bianchi, L. Mainardi, E. Petrucci, M.G. Signorini, M. Mainardi, and S. Cerutti, Time-variant power spectrum analysis for the detection of transient episodes in HRV signal, *IEEE Transactions on Biomedical Engineering*, 40(2):136–144, 1993. ©IEEE.

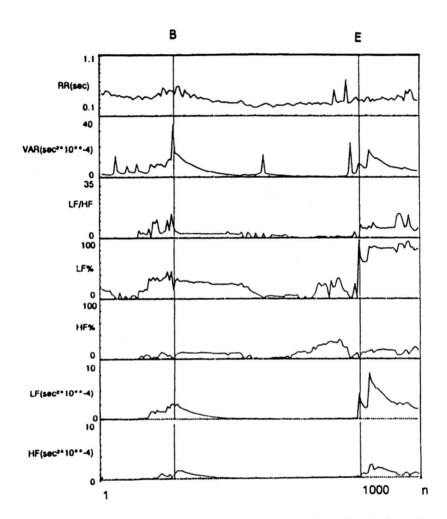

Figure 8.28 Top to bottom: RR interval series including an ischemic episode; variance; low-frequency (LF) to high-frequency (HF) power ratio; percentage of LF power; percentage of HF power; LF power; and HF power. B: beginning and E: end of the episode. Reproduced with permission from A.M. Bianchi, L. Mainardi, E. Petrucci, M.G. Signorini, M. Mainardi, and S. Cerutti, Time-variant power spectrum analysis for the detection of transient episodes in HRV signal, *IEEE Transactions on Biomedical Engineering*, 40(2):136–144, 1993. ©IEEE.

8.10   REMARKS

We have now reached the stage where we have extended the application of a number of signal processing, modeling, and analysis techniques to nonstationary biomedical signals. Fixed or adaptive segmentation of the signals into quasi-stationary segments was seen to be a pre-requisite step, and we studied several approaches for segmentation. Adaptive segmentation facilitates not only the identification of distinct and separate events at unknown time instants in the given signal, but also the characterization of events of variable duration using the same number of parameters. This is advantageous in pattern classification tasks (to be studied in Chapter 9) as well as for efficient data compression.

8.11   STUDY QUESTIONS AND PROBLEMS

1. Describe the characteristics of PCG signals that would make them nonstationary. Propose signal processing strategies to break a PCG signal into quasi-stationary segments.

2. Discuss features of the EEG that make the signal nonstationary. Propose signal processing strategies to detect each type of nonstationarity and to break an EEG signal into quasi-stationary segments.

3. Investigate features of the EMG that make the signal nonstationary. Propose signal processing strategies to track the time-varying characteristics of the signal. Under what conditions can the signal be partitioned into quasi-stationary segments? What are the physiological features that you would be able to derive from each segment?

8.12   LABORATORY EXERCISES AND PROJECTS

Note: Data files related to the exercises are available at the site
ftp://ftp.ieee.org/uploads/press/rangayyan/

1. The speech signal of the word "safety" is given in the file safety.wav. You may use the program safety.m to read the data. Explore the use of short-time statistics such as ZCR and RMS values for segmentation of the signal. Study the effect of the duration of the short-time analysis window on the trends in the parameters computed and on segmentation.

2. The files pec1.dat, pec22.dat, pec33.dat, and pec52.dat give the PCG, ECG, and carotid pulse signals of two normal subjects and two patients with systolic murmur. You may use the program plotpec.m to read the data. Explore the use of short-time ZCR, RMS, and AR model coefficients for segmentation of the signals. Evaluate the segment boundaries obtained in relation to the events in the PCG signals as well as the corresponding events in the ECG and carotid pulse channels.

9

Pattern Classification and Diagnostic Decision

The final purpose of biomedical signal analysis is to classify a given signal into one of a few known categories, and to arrive at a diagnostic decision regarding the condition of the patient. A physician or medical specialist may achieve this goal via visual or auditory analysis of the signal presented: comparative analysis of the given signal with others of known diagnoses or established protocols and sets of rules assist in such a decision-making process. The basic knowledge, clinical experience, expertise, and intuition of the physician play significant roles in this process. Some measurements may also be made from the given signal to assist in its analysis, such as the QRS width from an ECG signal plot.

When signal analysis is performed via the application of computer algorithms, the typical result is the extraction of a number of numerical features. When the numerical features relate directly to measures of the signal such as the QRS width and RR interval of an ECG signal, the clinical specialist may be able to use the features in his or her diagnostic logic. Even indirect measures such as the frequency content of PCG signals and murmurs may find such direct use. However, when parameters such as AR model coefficients and spectral statistics are derived, a human analyst is not likely to be able to comprehend and analyze the features. Furthermore, as the number of the computed features increases, the associated diagnostic logic may become too complicated and unwieldy for human analysis. Computer methods would then be desirable for performing the classification and decision process.

At the outset, it should be borne in mind that a biomedical signal forms but one piece of information in arriving at a diagnosis: the classification of a given signal into one of many categories may assist in the diagnostic procedure, but will almost never be the only factor. Regardless, pattern classification based upon signal analysis is indeed an important aspect of biomedical signal analysis, and forms the theme of this chapter. Remaining within the realm of computer-aided diagnosis as introduced in Figure 1.32 and Section 1.5, it would be preferable to design methods so as to assist a medical specialist in arriving at a diagnosis rather than to provide a decision.

9.1 PROBLEM STATEMENT

A number of measures and features have been derived from a biomedical signal. Explore methods to classify the signal into one of a few specified categories. Investigate the relevance of the features and the classification methods in arriving at a diagnostic decision about the patient.

Observe that the features may have been derived manually or by computer methods. Note the distinction between classifying the given signal and arriving at a diagnosis regarding the patient: the connection between the two tasks or steps may not always be direct. In other words, a pattern classification method may facilitate the labeling of a given signal as being a member of a particular class; arriving at a diagnosis of the condition of the patient will most likely require the analysis of several other items of clinical information. Although it is common to work with a pre-specified number of pattern classes, many problems do exist where the number of classes is not known *a priori*.

The following sections present a few illustrative case-studies. A number of methods for pattern classification, decision making, and evaluation of the results of classification will be reviewed and illustrated.

9.2 ILLUSTRATION OF THE PROBLEM WITH CASE-STUDIES

9.2.1 Diagnosis of bundle-branch block

Bundle-branch block affects the propagation of the excitation pulse through the conduction system of the heart to the ventricles. A block in the left bundle branch results in delayed activation of the left ventricle as compared to the right; a block in the right bundle branch has the opposite effect. Essentially, contraction of the two ventricles becomes asynchronous. The resulting ECG typically displays a wider-than-normal QRS complex ($100 - 120$ ms or more), which could have a jagged or slurred shape as well [23]; see Figure 1.15.

The orientation of the cardiac electromotive forces will be affected by bundle-branch block. The initial forces in left bundle-branch block are directed more markedly to the left-posterior, whereas the terminal forces are directed to the superior-left and posterior [23]. Left bundle-branch block results in the loss of Q waves in leads I, V5, and V6.

The following logic assists in the diagnosis of incomplete left bundle-branch block [242]:

IF (QRS duration $\geq$ 105 ms and $\leq$ 120 ms) AND
(QRS amplitude is negative in leads V1 and V2) AND
(Q or S duration $\geq$ 80 ms in leads V1 and V2) AND
(no Q wave is present in any two of leads I, V5, and V6) AND
(R duration $>$ 60 ms in any two of leads I, aVL, V5, and V6) THEN
the patient has incomplete left bundle-branch block.

Incomplete right bundle-branch block is indicated by the following conditions [242]:

IF (QRS duration $\geq$ 91 ms and $\leq$ 120 ms) AND
(S duration $\geq$ 40 ms in any two of leads I, aVL, V4, V5, and V6) AND
in lead V1 or V2 EITHER
[(R duration $>$ 30 ms) AND (R amplitude $>$ 100 μV) AND
(no S wave is present)] OR
[(R' duration $>$ 30 ms) AND (R' amplitude $>$ 100 μV) AND
(no S' wave is present)] THEN
the patient has incomplete right bundle-branch block.

(*Note:* The first positive deflection of a QRS complex is referred to as the R wave and the second positive deflection is referred to as the R' wave. Similarly, S and S' indicate the first and second negative deflections, respectively, of a QRS wave.)

Note that the logic or decision rules above may be used either by a human analyst or in a computer algorithm after the durations and amplitudes of the various waves mentioned have been measured or computed. Cardiologists with extensive training and experience may perform such decisions via visual analysis of an ECG record without resorting to actual measurements.

9.2.2   Normal or ectopic ECG beat?

Premature ventricular contractions caused by ectopic foci could be precursors of more serious arrhythmia, and hence detection of such beats is important in cardiac monitoring. As illustrated in Sections 5.4.2 and 5.7 as well as in Figures 5.1 and 5.10, PVCs possess shorter preceding RR intervals than normal beats and display bizarre waveshapes that are markedly different from those of the normal QRS complexes of the same subject. Therefore, a simple rule to detect PVCs or ectopic beats could be as follows:

IF (the RR interval of the beat is less than the normal at the current heart rate) AND (the QRS waveshape is markedly different from the normal QRS of the patient) THEN *the beat is a PVC.*

As in the preceding case-study of bundle-branch block, the logic above may be easily applied for visual analysis of an ECG signal by a physician or a trained observer. Computer implementation of the first part of the rule relating in an objective or quantitative manner to the RR interval is simple. However, implementation of the second condition on waveshape, being qualitative and subjective, is neither direct nor easy. Regardless, we have seen in Chapter 5 how we may characterize waveshape. Figures 5.1 and 5.10 illustrate the application of waveshape analysis to quantify the differences between the shapes of normal QRS complexes and ectopic beats. Figure 5.2 suggests how a 2D feature space may be divided by a simple linear decision boundary to categorize beats as normal or ectopic. We shall study the details of such methods later in this chapter.

9.2.3 Is there an alpha rhythm?

The alpha rhythm appears in an EEG record as an almost-sinusoidal wave (see Figure 1.22); a trained EEG technologist or physician can readily recognize the pattern at a glance from an EEG record plotted at the standard scale. The number of cycles of the wave may be counted over one or two seconds of the plot if an estimate of the dominant frequency of the rhythm is required.

In computer analysis of EEG signals, the ACF and PSD may be used to detect the presence of the alpha rhythm. We saw in Chapter 4 how these two functions demonstrate peaks at the basic period or dominant frequency of the rhythm, respectively (see Figure 4.8). A peak-detection algorithm may be applied to the ACF, and the presence of a significant peak in the range $75 - 125\ ms$ may be used as an indication of the existence of the alpha rhythm. If the PSD is available, the fractional power of the signal in the band $8 - 12\ Hz$ (see Equation 6.48) may be computed: a high value of the fraction indicates the presence of the alpha rhythm. Note that the logic described above includes the qualifier "significant"; experimentation with a number of signals that have been categorized by experts should assist in assigning a numerical value to represent the significance of the features described.

9.2.4 Is a murmur present?

Detection of the presence of a heart murmur is a fairly simple task for a trained physician or cardiologist: in performing auscultation of a patient with a stethoscope, the cardiologist needs to determine the existence of noise-like, high-frequency sounds between the low-frequency S1 and S2. It is necessary to exercise adequate care to reject high-frequency noise from other sources such as breathing, wheezing, and scraping of the stethoscope against the skin or hair. The cardiologist also has to distinguish between innocent physiological murmurs and those due to cardiovascular defects and diseases. Further discrimination between different types of murmurs requires more careful analysis: Figure 5.5 illustrates a decision tree to classify systolic murmurs based upon envelope analysis.

We have seen in Chapters 6 and 7 how we may derive frequency-domain parameters that relate to the presence of murmurs in the PCG signal. Once we have derived such numerical features for a number of signals of known categories of diseases (diagnoses), it becomes possible to design and train classifiers to categorize new signals into one of a few pre-specified classes.

The preceding case-studies suggest that the classification of patterns in a signal may, in some cases, be based upon thresholds applied to quantitative measurements obtained from the signal; in some other cases, it may be based upon objective measures derived from the signal that attempt to quantify certain notions regarding the characteristics of signals belonging to various categories. Classification may also be based upon the differences between certain measures derived from the signal on hand and those of established examples with known categorization. The succeeding sections of this chapter describe procedures for classification of signals based upon the approaches suggested above.

9.3 PATTERN CLASSIFICATION

Pattern recognition or classification may be defined as categorization of input data into identifiable classes via the extraction of significant features or attributes of the data from a background of irrelevant detail [243, 244, 245, 246, 247, 248, 249]. In biomedical signal analysis, after quantitative features have been extracted from the given signals, each signal may be represented by a feature vector $\mathbf{x} = (x_1, x_2, \ldots, x_n)^T$, which is also known as a measurement vector or a pattern vector. When the values x_i are real numbers, $\mathbf{x}$ is a point in an n-dimensional Euclidean space: vectors of similar objects may be expected to form clusters as illustrated in Figure 9.1.

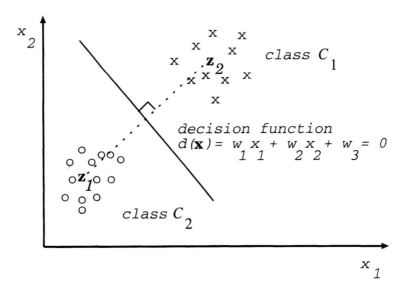

Figure 9.1   Two-dimensional feature vectors of two classes C_1 and C_2. The prototypes of the two classes are indicated by the vectors $\mathbf{z}_1$ and $\mathbf{z}_2$. The optimal linear decision function shown $d(\mathbf{x})$ (solid line) is the perpendicular bisector of the straight line joining the two class prototypes (dashed line).

For efficient pattern classification, measurements that could lead to disjoint sets or clusters of feature vectors are desired. This point underlines the importance of appropriate design of the preprocessing and feature extraction procedures. Features or characterizing attributes that are common to all patterns belonging to a particular class are known as *intraset* or *intraclass features*. Discriminant features that represent differences between pattern classes are called *interset* or *interclass features*.

The pattern classification problem is that of generating optimal decision boundaries or decision procedures to separate the data into pattern classes based on the feature vectors. Figure 9.1 illustrates a simple linear decision function or boundary to separate 2D feature vectors into two classes.

9.4   SUPERVISED PATTERN CLASSIFICATION

Problem: *You are provided with a number of feature vectors with classes assigned to them. Propose techniques to characterize the boundaries that separate the classes.*

Solution: A given set of feature vectors of known categorization is often referred to as a *training set*. The availability of a training set facilitates the development of mathematical functions that can characterize the separation between the classes. The functions may then be applied to new feature vectors of unknown classes to classify or recognize them. This approach is known as *supervised pattern classification*. A set of feature vectors of known categorization that is used to evaluate a classifier designed in this manner is referred to as a *test set*. The following subsections describe a few methods that can assist in the development of discriminant and decision functions.

9.4.1   Discriminant and decision functions

A general linear discriminant or decision function is of the form

$$d(\mathbf{x}) = w_1 x_1 + w_2 x_2 + \cdots + w_n x_n + w_{n+1} = \mathbf{w}^T \mathbf{x}, \tag{9.1}$$

where $\mathbf{x} = (x_1, x_2, \ldots, x_n, 1)^T$ is the feature vector augmented by an additional entry equal to unity, and $\mathbf{w} = (w_1, w_2, \ldots, w_n, w_{n+1})^T$ is a correspondingly augmented weight vector. A two-class pattern classification problem may be stated as

$$d(\mathbf{x}) = \mathbf{w}^T \mathbf{x} \begin{cases} > 0 & \text{if } \mathbf{x} \in C_1 \\ \leq 0 & \text{if } \mathbf{x} \in C_2 \end{cases}, \tag{9.2}$$

where C_1 and C_2 represent the two classes. The discriminant function may be interpreted as the boundary separating the classes C_1 and C_2, as illustrated in Figure 9.1.

In the general case of an M-class pattern classification problem, we will need M weight vectors and M decision functions to perform the following decisions:

$$d_i(\mathbf{x}) = \mathbf{w}_i^T \mathbf{x} \begin{cases} > 0 & \text{if } \mathbf{x} \in C_i \\ \leq 0 & \text{otherwise} \end{cases}, \quad i = 1, 2, \ldots, M, \tag{9.3}$$

where $\mathbf{w}_i = (w_{i1}, w_{i2}, \ldots, w_{in}, w_{i,n+1})^T$ is the weight vector for the class C_i.

Three cases arise in solving this problem [243]:

Case 1: Each class is separable from the rest by a single decision surface:

$$\text{if } d_i(\mathbf{x}) > 0 \text{ then } \mathbf{x} \in C_i. \tag{9.4}$$

Case 2: Each class is separable from every other individual class by a distinct decision surface, that is, the classes are pair-wise separable. There are $M(M-1)/2$ decision surfaces given by $d_{ij}(\mathbf{x}) = \mathbf{w}_{ij}^T \mathbf{x}$.

$$\text{if } d_{ij}(\mathbf{x}) > 0 \ \forall \ j \neq i \text{ then } \mathbf{x} \in C_i. \tag{9.5}$$

[*Note:* $d_{ij}(\mathbf{x}) = -d_{ji}(\mathbf{x})$.]

Case 3: There exist M decision functions $d_k(\mathbf{x}) = \mathbf{w}_k^T \mathbf{x}$, $k = 1, 2, \ldots, M$, with the property that

$$\text{if } d_i(\mathbf{x}) > d_j(\mathbf{x}) \ \forall \ j \neq i, \text{ then } \mathbf{x} \in C_i. \tag{9.6}$$

This is a special instance of Case 2. We may define

$$d_{ij}(\mathbf{x}) = d_i(\mathbf{x}) - d_j(\mathbf{x}) = (\mathbf{w}_i - \mathbf{w}_j)^T \mathbf{x} = \mathbf{w}_{ij}^T \mathbf{x}. \tag{9.7}$$

If the classes are separable under Case 3, they are separable under Case 2; the converse is, in general, not true.

Patterns that may be separated by linear decision functions as above are said to be *linearly separable*. In other situations, an infinite variety of complex decision boundaries may be formulated by using generalized decision functions based upon nonlinear functions of the feature vectors as

$$d(\mathbf{x}) = w_1 f_1(\mathbf{x}) + w_2 f_2(\mathbf{x}) + \cdots + w_K f_K(\mathbf{x}) + w_{K+1} \tag{9.8}$$

$$= \sum_{i=1}^{K+1} w_i \, f_i(\mathbf{x}). \tag{9.9}$$

Here, $\{f_i(\mathbf{x})\}$, $i = 1, 2, \ldots, K$, are real, single-valued functions of $\mathbf{x}$; $f_{K+1}(\mathbf{x}) = 1$.
Whereas the functions $f_i(\mathbf{x})$ may be nonlinear in the n-dimensional space of $\mathbf{x}$, the decision function may be formulated as a linear function by defining a transformed feature vector $\mathbf{x}^\dagger = (f_1(\mathbf{x}), f_2(\mathbf{x}), \ldots, f_K(\mathbf{x}), 1)^T$. Then, $d(\mathbf{x}) = \mathbf{w}^T \mathbf{x}^\dagger$, with $\mathbf{w} = (w_1, w_2, \ldots, w_K, w_{K+1})^T$. Once evaluated, $\{f_i(\mathbf{x})\}$ is just a set of numerical values, and $\mathbf{x}^\dagger$ is simply a K-dimensional vector augmented by an entry equal to unity.

9.4.2 Distance functions

Consider M pattern classes represented by their prototype patterns $\mathbf{z}_1, \mathbf{z}_2, \ldots, \mathbf{z}_M$. The prototype of a class is typically computed as the average of all the feature vectors belonging to the class. Figure 9.1 illustrates schematically the prototypes z_1 and z_2 of the two classes shown.

The Euclidean distance between an arbitrary pattern vector x and the ith prototype is given as

$$D_i = \|x - z_i\| = \sqrt{(x - z_i)^T (x - z_i)}. \tag{9.10}$$

A simple rule to classify the pattern vector x would be to choose that class for which the vector has the smallest distance:

$$\text{if } D_i < D_j \; \forall \; j \neq i, \text{ then } x \in C_i. \tag{9.11}$$

A simple relationship may be established between discriminant functions and distance functions as follows [243]:

$$D_i^2 = \|x - z_i\|^2 = (x - z_i)^T(x - z_i) \tag{9.12}$$
$$= x^Tx - 2x^Tz_i + z_i^Tz_i = x^Tx - 2(x^Tz_i - \frac{1}{2}z_i^Tz_i).$$

Choosing the minimum of D_i^2 is equivalent to choosing the minimum of D_i (as all $D_i > 0$). Furthermore, from the equation above, it follows that choosing the minimum of D_i^2 is equivalent to choosing the maximum of $(x^Tz_i - \frac{1}{2}z_i^Tz_i)$. Therefore, we may define the decision function

$$d_i(x) = (x^Tz_i - \frac{1}{2}z_i^Tz_i), \; i = 1, 2, \ldots, M. \tag{9.13}$$

A decision rule may then be stated as

$$\text{if } d_i(x) > d_j(x) \; \forall \; j \neq i, \text{ then } x \in C_i. \tag{9.14}$$

This is a linear discriminant function, which becomes obvious from the following representation: If z_{ij}, $j = 1, 2, \ldots, n$, are the components of z_i, let $w_{ij} = z_{ij}$, $j = 1, 2, \ldots, n$; $w_{i,n+1} = -\frac{1}{2}z_i^Tz_i$; and $x = (x_1, x_2, \ldots, x_n, 1)^T$. Then, $d_i(x) = w_i^Tx$, $i = 1, 2, \ldots, M$, where $w_i = (w_{i1}, w_{i2}, \ldots, w_{i,n+1})^T$. Therefore, distance functions may be formulated as linear discriminant or decision functions.

9.4.3   The nearest-neighbor rule

Suppose that we are provided with a set of N sample patterns $\{s_1, s_2, \ldots, s_N\}$ of known classification: each pattern belongs to one of M classes $\{C_1, C_2, \ldots, C_M\}$. We are then given a new feature vector x whose class needs to be determined. Let us compute a distance measure $D(s_i, x)$ between the vector x and each sample pattern. Then, the nearest-neighbor rule states that the vector x is to be assigned to the class of the sample that is the closest to x:

$$x \in C_i \text{ if } D(s_i, x) = \min\{D(s_l, x)\}, \; l = 1, 2, \ldots, N. \tag{9.15}$$

A major disadvantage of the above method is that the classification decision is made based upon a single sample vector of known classification. The nearest neighbor may happen to be an outlier that is not representative of its class. It would be more reliable to base the classification upon several samples: we may consider a certain number k of the nearest neighbors of the sample to be classified, and then seek a majority opinion. This leads to the so-called *k-nearest-neighbor* or *k-NN rule*: Determine the k nearest neighbors of $\mathbf{x}$, and use the majority of equal classifications in this group as the classification of $\mathbf{x}$.

9.5   UNSUPERVISED PATTERN CLASSIFICATION

Problem: *We are given a set of feature vectors with no categorization or classes attached to them. No prior training information is available. How may we group the vectors into multiple categories?*

Solution: The design of distance functions and decision boundaries requires a training set of feature vectors of known classes. The functions so designed may then be applied to a new set of feature vectors or samples to perform pattern classification. Such a procedure is known as *supervised* pattern classification due to the initial training step. In some situations a training step may not be possible, and we may be required to classify a given set of feature vectors into either a pre-specified or unknown number of categories. Such a problem is labeled as *unsupervised* pattern classification, and may be solved by cluster-seeking methods.

9.5.1   Cluster-seeking methods

Given a set of feature vectors, we may examine them for the formation of inherent groups or clusters. This is a simple task in the case of 2D vectors, where we may plot them, visually identify groups, and label each group with a pattern class. Allowance may have to be made to assign the same class to multiple disjoint groups. Such an approach may be used even when the number of classes is not known at the outset. When the vectors have a dimension higher than three, visual analysis will not be feasible. It then becomes necessary to define criteria to group the given vectors on the basis of similarity, dissimilarity, or distance measures. A few examples of such measures are [243]:

- Euclidean distance

$$D_E^2 = \|\mathbf{x} - \mathbf{z}\|^2 = (\mathbf{x} - \mathbf{z})^T(\mathbf{x} - \mathbf{z}) = \sum_{i=1}^{n} (x_i - z_i)^2. \qquad (9.16)$$

  Here, $\mathbf{x}$ and $\mathbf{z}$ are two feature vectors; the latter could be a class prototype if available. A small value of D_E indicates greater similarity between the two vectors than a large value of D_E.

- Mahalanobis distance

$$D_M^2 = (\mathbf{x} - \mathbf{m})^T\mathbf{C}^{-1}(\mathbf{x} - \mathbf{m}), \qquad (9.17)$$

where $\mathbf{x}$ is a feature vector being compared to a pattern class for which $\mathbf{m}$ is the class mean vector and $\mathbf{C}$ is the covariance matrix. A small value of D_M indicates a higher potential membership of the vector $\mathbf{x}$ in the class than a large value of D_M.

- Normalized dot product (cosine of the angle between the vectors $\mathbf{x}$ and $\mathbf{z}$)

$$D_d = \frac{\mathbf{x}^T \mathbf{z}}{\|\mathbf{x}\|\|\mathbf{z}\|}. \tag{9.18}$$

A large dot product value indicates a greater degree of similarity between the two vectors than a small value.

The covariance matrix is defined as

$$\mathbf{C} = E[(\mathbf{y} - \mathbf{m})(\mathbf{y} - \mathbf{m})^T], \tag{9.19}$$

where the expectation operation is performed over all feature vectors $\mathbf{y}$ that belong to the class. The covariance matrix provides the covariance of all possible pairs of the features in the feature vector over all samples belonging to the given class. The elements along the main diagonal of the covariance matrix provide the variance of the individual features that make up the feature vector. The covariance matrix represents the scatter of the features that belong to the given class. The mean and covariance need to be updated as more samples are added to a given class in a clustering procedure.

When the Mahanalobis distance needs to be calculated between a sample vector and a number of classes represented by their mean and covariance matrices, a pooled covariance matrix may be used if the numbers of members in the various classes are unequal and low [246]. For example, if the covariance matrices of two classes are $\mathbf{C}_1$ and $\mathbf{C}_2$, and the numbers of members in the two classes are N_1 and N_2, the pooled covariance matrix is given by

$$\mathbf{C} = \frac{(N_1 - 1)\mathbf{C}_1 + (N_2 - 1)\mathbf{C}_2}{N_1 + N_2 - 2}. \tag{9.20}$$

Various performance indices may be designed to measure the success of a clustering procedure [243]. A measure of the tightness of a cluster is the sum of the squared errors performance index:

$$J = \sum_{j=1}^{N_c} \sum_{\mathbf{x} \in S_j} \|\mathbf{x} - \mathbf{m}_j\|^2, \tag{9.21}$$

where N_c is the number of cluster domains, S_j is the set of samples in the jth cluster,

$$\mathbf{m}_j = \frac{1}{N_j} \sum_{\mathbf{x} \in S_j} \mathbf{x} \tag{9.22}$$

is the sample mean vector of S_j, and N_j is the number of samples in S_j.

A few other examples of performance indices are:

- Average of the squared distances between the samples in a cluster domain.

- Intra-cluster variance.

- Average of the squared distances between the samples in different cluster domains.

- Inter-cluster distances.

- Scatter matrices.

- Covariance matrices.

A simple cluster-seeking algorithm [243]: Suppose we have N sample patterns $\{x_1, x_2, \ldots, x_N\}$.

1. Let the first cluster center z_1 be equal to any one of the samples, say $z_1 = x_1$.

2. Choose a non-negative threshold θ.

3. Compute the distance D_{21} between x_2 and z_1. If $D_{21} < \theta$, assign x_2 to the domain (class) of cluster center z_1; otherwise, start a new cluster with its center as $z_2 = x_2$. For the subsequent steps, let us assume that a new cluster with center z_2 has been established.

4. Compute the distances D_{31} and D_{32} from the next sample x_3 to z_1 and z_2, respectively. If D_{31} and D_{32} are both greater than θ, start a new cluster with its center as $z_3 = x_3$; otherwise, assign x_3 to the domain of the closer cluster.

5. Continue to apply steps 3 and 4 by computing and checking the distance from *every* new (unclassified) pattern vector to *every* established cluster center and applying the assignment or cluster-creation rule.

6. Stop when every given pattern vector has been assigned to a cluster.

Note that the procedure does not require knowledge of the number of classes *a priori*. Note also that the procedure does not assign a real-world class to each cluster: it merely groups the given vectors into disjoint clusters. A subsequent step is required to label each cluster with a class related to the actual problem. Multiple clusters may relate to the same real-world class, and may have to be merged.

A major disadvantage of the simple cluster-seeking algorithm is that the results depend upon

- the first cluster center chosen for each domain or class,

- the order in which the sample patterns are considered,

- the value of the threshold θ, and

- the geometrical properties (distributions) of the data (or the feature-vector space).

The maximin-distance clustering algorithm [243]: This method is similar to the previous "simple" algorithm, but first identifies the cluster regions that are the farthest apart. The term "maximin" refers to the combined use of maximum and minimum distances between the given vectors and the centers of the clusters already formed.

1. Let $\mathbf{x}_1$ be the first cluster center $\mathbf{z}_1$.

2. Determine the farthest sample from $\mathbf{x}_1$, and call it cluster center $\mathbf{z}_2$.

3. Compute the distance from each remaining sample to $\mathbf{z}_1$ and to $\mathbf{z}_2$. For every pair of these computations, save the minimum distance, and select the maximum of the minimum distances. If this "maximin" distance is an appreciable fraction of the distance between the cluster centers $\mathbf{z}_1$ and $\mathbf{z}_2$, label the corresponding sample as a new cluster center $\mathbf{z}_3$; otherwise stop forming new clusters and go to Step 5.

4. If a new cluster center was formed in Step 3, repeat Step 3 using a "typical" or the average distance between the established cluster centers for comparison.

5. Assign each remaining sample to the domain of its nearest cluster center.

The K-means algorithm [243]: The preceding "simple" and "maximin" algorithms are intuitive procedures. The K-means algorithm is based on iterative minimization of a performance index that is defined as the sum of the squared distances from all points in a cluster domain to the cluster center.

1. Choose K initial cluster centers $\mathbf{z}_1(1), \mathbf{z}_2(1), \ldots, \mathbf{z}_K(1)$. The index in parentheses represents the iteration number.

2. At the kth iterative step, distribute the samples $\{\mathbf{x}\}$ among the K cluster domains, using the relation

$$\mathbf{x} \in S_j(k) \text{ if } \|\mathbf{x} - \mathbf{z}_j(k)\| < \|\mathbf{x} - \mathbf{z}_i(k)\| \ \forall \ i = 1, 2, \ldots, K, \ i \neq j, \quad (9.23)$$

where $S_j(k)$ denotes the set of samples whose cluster center is $\mathbf{z}_j(k)$.

3. From the results of Step 2, compute the new cluster centers $\mathbf{z}_j(k+1)$, $j = 1, 2, \ldots, K$, such that the sum of the squared distances from all points in $S_j(k)$ to the new cluster center is minimized. In other words, the new cluster center $\mathbf{z}_j(k+1)$ is computed so that the performance index

$$J_j = \sum_{\mathbf{x} \in S_j(k)} \|\mathbf{x} - \mathbf{z}_j(k+1)\|^2, \ j = 1, 2, \ldots, K, \quad (9.24)$$

is minimized. The $z_j(k+1)$ that minimizes this performance index is simply the sample mean of $S_j(k)$. Therefore, the new cluster center is given by

$$z_j(k+1) = \frac{1}{N_j} \sum_{x \in S_j(k)} x, \quad j = 1, 2, \ldots, K, \tag{9.25}$$

where N_j is the number of samples in $S_j(k)$. The name "K-means" is derived from the manner in which cluster centers are sequentially updated.

4. If $z_j(k+1) = z_j(k)$ for $j = 1, 2, \ldots, K$, the algorithm has converged: terminate the procedure; otherwise go to Step 2.

The behavior of the K-means algorithm is influenced by:

- the number of cluster centers specified,

- the choice of the initial cluster centers,

- the order in which the sample patterns are considered, and

- the geometrical properties (distributions) of the data (or the feature-vector space).

9.6 PROBABILISTIC MODELS AND STATISTICAL DECISION

Problem: *Pattern classification methods such as discriminant functions are dependent upon the set of training samples provided. Their success, when applied to new cases, will depend upon the accuracy of representation of the various pattern classes by the training samples. How can we design pattern classification techniques that are independent of specific training samples and optimal in a broad sense?*

Solution: Probability functions and probabilistic models may be developed to represent the occurrence and statistical attributes of classes of patterns. Such functions may be based upon large collections of data, historical records, or mathematical models of pattern generation. In the absence of information as above, a training step with samples of known categorization will be required to estimate the required model parameters. It is common practice to assume a Gaussian PDF to represent the distribution of the features for each class, and estimate the required mean and variance parameters from the training sets. When PDFs are available to characterize pattern classes and their features, optimal decision functions may be designed based upon statistical functions and decision theory. The following subsections describe a few methods that fall into this category.

9.6.1 Likelihood functions and statistical decision

Let $P(C_i)$ be the probability of occurrence of class C_i, $i = 1, 2, \ldots, M$; this is known as the *a priori*, *prior*, or unconditional probability. The *a posteriori* or *posterior* probability that an observed sample pattern $\mathbf{x}$ comes from C_i is expressed as $P(C_i|\mathbf{x})$. If a classifier decides that $\mathbf{x}$ comes from C_j when it actually came from C_i, then the classifier is said to incur a *loss* L_{ij}, with $L_{ii} = 0$ or a fixed operational cost and $L_{ij} > L_{ii} \; \forall \; j \neq i$.

Since $\mathbf{x}$ may belong to any of M classes under consideration, the expected loss, known as the *conditional average risk* or *loss*, in assigning $\mathbf{x}$ to C_j is [243]

$$R_j(\mathbf{x}) = \sum_{i=1}^{M} L_{ij} \, P(C_i|\mathbf{x}). \tag{9.26}$$

A classifier could compute $R_j(\mathbf{x})$, $j = 1, 2, \ldots, M$, for each sample $\mathbf{x}$, and then assign $\mathbf{x}$ to the class with the smallest conditional loss. Such a classifier will minimize the total expected loss over all decisions, and is called the *Bayes classifier*. From a statistical point of view, the Bayes classifier represents the optimal classifier.

According to Bayes formula, we have [243, 244]

$$P(C_i|\mathbf{x}) = \frac{P(C_i) \, p(\mathbf{x}|C_i)}{p(\mathbf{x})}, \tag{9.27}$$

where $p(\mathbf{x}|C_i)$ is called the *likelihood function* of class C_i or the *state-conditional PDF* of $\mathbf{x}$, and $p(\mathbf{x})$ is the PDF of $\mathbf{x}$ regardless of class membership (unconditional). [*Note:* $P(y)$ is used to represent the probability of occurrence of an event y; $p(y)$ is used to represent the PDF of a random variable y. Probabilities and PDFs involving a multi-dimensional feature vector are multivariate functions with dimension equal to that of the feature vector.] Bayes formula shows how observing the sample $\mathbf{x}$ changes the *a priori* probability $P(C_i)$ to the *a posteriori* probability $P(C_i|\mathbf{x})$. In other words, Bayes formula provides a mechanism to update the *a priori* probability $P(C_i)$ to the *a posteriori* probability $P(C_i|\mathbf{x})$ due to the observation of the sample $\mathbf{x}$. Then, we can express the expected loss as [243]

$$R_j(\mathbf{x}) = \frac{1}{p(\mathbf{x})} \sum_{i=1}^{M} L_{ij} \, p(\mathbf{x}|C_i) \, P(C_i). \tag{9.28}$$

As $\frac{1}{p(\mathbf{x})}$ is common for all j, we could modify $R_j(\mathbf{x})$ to

$$r_j(\mathbf{x}) = \sum_{i=1}^{M} L_{ij} \, p(\mathbf{x}|C_i) \, P(C_i). \tag{9.29}$$

In a two-class case with $M = 2$, we obtain the following expressions [243]:

$$r_1(\mathbf{x}) = L_{11} \, p(\mathbf{x}|C_1) \, P(C_1) + L_{21} \, p(\mathbf{x}|C_2) \, P(C_2). \tag{9.30}$$

$$r_2(\mathbf{x}) = L_{12} \, p(\mathbf{x}|C_1) \, P(C_1) + L_{22} \, p(\mathbf{x}|C_2) \, P(C_2). \tag{9.31}$$

$$\mathbf{x} \in C_1 \text{ if } r_1(\mathbf{x}) < r_2(\mathbf{x}), \tag{9.32}$$

that is,

$$\mathbf{x} \in C_1 \text{ if } \quad L_{11}\, p(\mathbf{x}|C_1)\, P(C_1) + L_{21}\, p(\mathbf{x}|C_2)\, P(C_2) \quad (9.33)$$
$$< \quad L_{12}\, p(\mathbf{x}|C_1)\, P(C_1) + L_{22}\, p(\mathbf{x}|C_2)\, P(C_2),$$

or equivalently,

$$\mathbf{x} \in C_1 \text{ if } (L_{21} - L_{22})\, p(\mathbf{x}|C_2)\, P(C_2) < (L_{12} - L_{11})\, p(\mathbf{x}|C_1)\, P(C_1). \quad (9.34)$$

This expression may be rewritten as [243]

$$\mathbf{x} \in C_1 \text{ if } \frac{p(\mathbf{x}|C_1)}{p(\mathbf{x}|C_2)} > \frac{P(C_2)}{P(C_1)} \frac{(L_{21} - L_{22})}{(L_{12} - L_{11})}. \quad (9.35)$$

The left-hand side of the inequality above, which is a ratio of two likelihood functions, is often referred to as the *likelihood ratio*:

$$l_{12}(\mathbf{x}) = \frac{p(\mathbf{x}|C_1)}{p(\mathbf{x}|C_2)}. \quad (9.36)$$

Then, Bayes decision rule for $M = 2$ is [243]:

1. Assign $\mathbf{x}$ to class C_1 if $l_{12}(\mathbf{x}) > \theta_{12}$, where θ_{12} is a threshold given by $\theta_{12} = \frac{P(C_2)}{P(C_1)} \frac{(L_{21}-L_{22})}{(L_{12}-L_{11})}$.

2. Assign $\mathbf{x}$ to class C_2 if $l_{12}(\mathbf{x}) < \theta_{12}$.

3. Make an arbitrary or heuristic decision if $l_{12}(\mathbf{x}) = \theta_{12}$.

The rule may be generalized to the M-class case as [243]:

$$\mathbf{x} \in C_i \text{ if } \sum_{k=1}^{M} L_{ki}\, p(\mathbf{x}|C_k)\, P(C_k) < \sum_{q=1}^{M} L_{qj}\, p(\mathbf{x}|C_q)\, P(C_q), \quad (9.37)$$

$j = 1, 2, \ldots, M,\ j \neq i$.

In most pattern classification problems, the loss is nil for correct decisions. The loss could be assumed to be equal to a certain quantity for all erroneous decisions. Then, $L_{ij} = 1 - \delta_{ij}$, where

$$\delta_{ij} = \begin{cases} 1 & \text{if } i = j \\ 0 & \text{otherwise} \end{cases}, \quad (9.38)$$

and

$$r_j(\mathbf{x}) = \sum_{i=1}^{M} (1 - \delta_{ij})\, p(\mathbf{x}|C_i)\, P(C_i) \quad (9.39)$$
$$= p(\mathbf{x}) - p(\mathbf{x}|C_j)\, P(C_j),$$

since

$$\sum_{i=1}^{M} p(\mathbf{x}|C_i)\, P(C_i) = p(\mathbf{x}).$$

(9.40)

The Bayes classifier will assign a pattern $\mathbf{x}$ to class C_i if

$$p(\mathbf{x}) - p(\mathbf{x}|C_i)P(C_i) < p(\mathbf{x}) - p(\mathbf{x}|C_j)P(C_j), \; j = 1, 2, \ldots, M, \; j \neq i,$$

(9.41)

that is,

$$\mathbf{x} \in C_i \text{ if } p(\mathbf{x}|C_i)P(C_i) > p(\mathbf{x}|C_j)P(C_j), \; j = 1, 2, \ldots, M, \; j \neq i.$$

(9.42)

This is nothing more than using the decision functions

$$d_i(\mathbf{x}) = p(\mathbf{x}|C_i)\, P(C_i), \; i = 1, 2, \ldots, M,$$

(9.43)

where a pattern $\mathbf{x}$ is assigned to class C_i if $d_i(\mathbf{x}) > d_j(\mathbf{x}) \; \forall \; j \neq i$ for that pattern. Using Bayes formula, we get

$$d_i(\mathbf{x}) = P(C_i|\mathbf{x})\, p(\mathbf{x}), \; i = 1, 2, \ldots, M.$$

(9.44)

Since $p(\mathbf{x})$ does not depend upon the class index i, this can be reduced to

$$d_i(\mathbf{x}) = P(C_i|\mathbf{x}), \; i = 1, 2, \ldots, M.$$

(9.45)

The different decision functions given above provide alternative yet equivalent approaches, depending upon whether $p(\mathbf{x}|C_i)$ or $P(C_i|\mathbf{x})$ is used (or available). Estimation of $p(\mathbf{x}|C_i)$ would require a training set for each class C_i. It is common to assume a Gaussian distribution and estimate its mean and variance using the training set.

9.6.2 Bayes classifier for normal patterns

The univariate normal or Gaussian PDF for a single random variable x is given by

$$p(x) = \frac{1}{\sqrt{2\pi}\,\sigma} \exp\left[-\frac{1}{2}\left(\frac{x-m}{\sigma} \right)^2 \right],$$

(9.46)

which is completely specified by two parameters: the mean

$$m = E[x] = \int_{-\infty}^{\infty} x\, p(x)\, dx,$$

(9.47)

and the variance

$$\sigma^2 = E[(x-m)^2] = \int_{-\infty}^{\infty} (x-m)^2\, p(x)\, dx.$$

(9.48)

In the case of M pattern classes and pattern vectors $\mathbf{x}$ of dimension n governed by multivariate normal PDFs, we have

$$p(\mathbf{x}|C_i) = \frac{1}{(2\pi)^{n/2}|\mathbf{C}_i|^{1/2}} \exp\left[-\frac{1}{2}(\mathbf{x} - \mathbf{m}_i)^T \mathbf{C}_i^{-1}(\mathbf{x} - \mathbf{m}_i)\right], \qquad (9.49)$$

$i = 1, 2, \ldots, M$, where each PDF is completely specified by its mean vector $\mathbf{m}_i$ and its $n \times n$ covariance matrix $\mathbf{C}_i$, with

$$\mathbf{m}_i = E_i[\mathbf{x}], \qquad (9.50)$$

and

$$\mathbf{C}_i = E_i[(\mathbf{x} - \mathbf{m}_i)(\mathbf{x} - \mathbf{m}_i)^T]. \qquad (9.51)$$

Here, $E_i[\,]$ denotes the expectation operator over the patterns belonging to class C_i.

Normal distributions occur frequently in nature, and have the advantage of analytical tractability. A multivariate normal PDF reduces to a product of univariate normal PDFs when the elements of $\mathbf{x}$ are mutually independent (then the covariance matrix is a diagonal matrix).

We earlier had formulated the decision functions

$$d_i(\mathbf{x}) = p(\mathbf{x}|C_i)\, P(C_i), \; i = 1, 2, \ldots, M. \qquad (9.52)$$

Given the exponential in the normal PDF, it is convenient to use

$$d_i(\mathbf{x}) = \ln\left[p(\mathbf{x}|C_i)\, P(C_i)\right] = \ln p(\mathbf{x}|C_i) + \ln P(C_i), \qquad (9.53)$$

which is equivalent in terms of classification performance as the natural logarithm ln is a monotonically increasing function. Then [243],

$$d_i(\mathbf{x}) = \ln P(C_i) - \frac{n}{2}\ln 2\pi - \frac{1}{2}\ln|\mathbf{C}_i| - \frac{1}{2}[(\mathbf{x} - \mathbf{m}_i)^T \mathbf{C}_i^{-1}(\mathbf{x} - \mathbf{m}_i)], \quad (9.54)$$

$i = 1, 2, \ldots, M$. The second term does not depend upon i; therefore, we can simplify $d_i(\mathbf{x})$ to

$$d_i(\mathbf{x}) = \ln P(C_i) - \frac{1}{2}\ln|\mathbf{C}_i| - \frac{1}{2}[(\mathbf{x} - \mathbf{m}_i)^T \mathbf{C}_i^{-1}(\mathbf{x} - \mathbf{m}_i)], \; i = 1, 2, \ldots, M. \qquad (9.55)$$

The decision functions above are hyperquadrics; hence the best that a Bayes classifier for normal patterns can do is to place a general second-order decision surface between each pair of pattern classes. In the case of true normal distributions of patterns, the decision functions as above will be optimal on an average basis: they minimize the expected loss with the simplified loss function $L_{ij} = 1 - \delta_{ij}$ [243].

If all the covariance matrices are equal, that is, $\mathbf{C}_i = \mathbf{C}$, $i = 1, 2, \ldots, M$, we get

$$d_i(\mathbf{x}) = \ln P(C_i) + \mathbf{x}^T \mathbf{C}^{-1}\mathbf{m}_i - \frac{1}{2}\mathbf{m}_i^T \mathbf{C}^{-1}\mathbf{m}_i, \; i = 1, 2, \ldots, M, \qquad (9.56)$$

after omitting terms independent of i. The Bayesian classifier is now represented by a set of linear decision functions.

Before one may apply the decision functions as above, it would be appropriate to verify the Gaussian nature of the PDFs of the variables on hand by conducting statistical tests [5, 245]. Furthermore, it would be necessary to derive or estimate the mean vector and covariance matrix for each class; sample statistics computed from a training set may serve this purpose.

9.7 LOGISTIC REGRESSION ANALYSIS

Logistic classification is a statistical technique based on a logistic regression model that estimates the probability of occurrence of an event [250, 251, 252]. The technique is designed for problems where patterns are to be classified into one of two classes. When the response variable is binary, theoretical and empirical considerations indicate that the response function is often curvilinear. The typical response function is shaped as a forward or backward tilted "S", and is known as a sigmoidal function. The function has asymptotes at 0 and 1.

In logistic pattern classification, an event is defined as the membership of a pattern vector in one of the two classes. The method computes a variable that depends upon the given parameters and is constrained to the range $(0, 1)$ so that it may be interpreted as a probability. The probability of the pattern vector belonging to the second class is simply the difference between unity and the estimated value.

For the case of a single feature or parameter, the logistic regression model is given as

$$P(\text{event}) = \frac{\exp(b_0 + b_1 x)}{1 + \exp(b_0 + b_1 x)}, \tag{9.57}$$

or equivalently,

$$P(\text{event}) = \frac{1}{1 + \exp[-(b_0 + b_1 x)]}, \tag{9.58}$$

where b_0 and b_1 are coefficients estimated from the data, and x is the independent (feature) variable. The relationship between the independent variable and the estimated probability is nonlinear, and follows an S-shaped curve that closely resembles the integral of a Gaussian function. In the case of an n-dimensional feature vector $\mathbf{x}$, the model can be written as

$$P(\text{event}) = \frac{1}{1 + \exp(-z)}, \tag{9.59}$$

where z is the linear combination

$$z = b_0 + b_1 x_1 + b_2 x_2 + \cdots + b_n x_n = \langle \mathbf{b}, \mathbf{x} \rangle, \tag{9.60}$$

that is, z is the dot product of the augmented feature vector $\mathbf{x}$ with a coefficient or weight vector $\mathbf{b}$.

In linear regression, the coefficients of the model are estimated using the method of least squares; the selected regression coefficients are those that result in the smallest sum of squared distances between the observed and the predicted values of the dependent variable. In logistic regression, the parameters of the model are estimated using the maximum likelihood method [250, 245]; the coefficients that make the observed results "most likely" are selected. Since the logistic regression model is nonlinear, an iterative algorithm is necessary for estimation of the coefficients [251, 252]. A training set is required to design a classifier based upon logistic regression.

9.8 THE TRAINING AND TEST STEPS

In the situation when a limited number of sample vectors with known classification are available, questions arise as to how many of the samples may be used to design or train a classifier, with the understanding that the classifier designed needs to be tested using an independent set of samples of known classification as well. When a sufficiently large number of samples are available, they may be randomly split into two approximately equal sets, one for use as the training set and the other to be used as the test set. The random-splitting procedure may be repeated a number of times to generate several classifiers. Finally, one of the classifiers so designed may be selected based upon its performance in both the training and test steps.

9.8.1 The leave-one-out method

The leave-one-out method [245] is suitable for the estimation of the classification accuracy of a pattern classification technique, particularly when the number of available samples is small. In this method, one of the available samples is excluded, the classifier is designed with the remaining samples, and then the classifier is applied to the excluded sample. The validity of the classification so performed is noted. This procedure is repeated with each available sample: if N training samples are available, N classifiers are designed and tested. The training and test sets for any one classifier so designed and tested are independent. However, whereas the training set for each classifier has $N - 1$ samples, the test set has only one sample. In the final analysis, every sample will have served $(N - 1)$ times as a training sample, and once as a test sample. An average classification accuracy is then computed using all the test results.

Let us consider a simple case in which the covariances of the sample sets of two classes are equal. Assume that two sample sets, $S_1 = \{\mathbf{x}_1^{(1)}, \ldots, \mathbf{x}_{N_1}^{(1)}\}$ from class C_1, and $S_2 = \{\mathbf{x}_1^{(2)}, \ldots, \mathbf{x}_{N_2}^{(2)}\}$ from class C_2 are given. Here, N_1 and N_2 are the numbers of samples in the sets S_1 and S_2, respectively. Assume also that the prior probabilities of the two classes are equal to each other. Then, according to the Bayes classifier and assuming $\mathbf{x}$ to be governed by a multivariate Gaussian PDF, a sample $\mathbf{x}$ is assigned to class C_1 if

$$(\mathbf{x} - \mathbf{m}_1)^T (\mathbf{x} - \mathbf{m}_1) - (\mathbf{x} - \mathbf{m}_2)^T (\mathbf{x} - \mathbf{m}_2) > \theta, \tag{9.61}$$

where θ is a threshold, and the sample mean $\tilde{\mathbf{m}}_i$ is given by

$$\tilde{\mathbf{m}}_i = \frac{1}{N_i} \sum_{j=1}^{N_i} \mathbf{x}_j^{(i)}. \tag{9.62}$$

In the leave-one-out method, one sample $\mathbf{x}_k^{(i)}$ is excluded from the training set and then used as the test sample. The mean estimate for class C_i without $\mathbf{x}_k^i$, labeled as $\tilde{\mathbf{m}}_{ik}$, may be computed as

$$\tilde{\mathbf{m}}_{ik} = \frac{1}{N_i - 1} \left[\sum_{j=1}^{N_i} \mathbf{x}_j^{(i)} - \mathbf{x}_k^{(i)} \right], \tag{9.63}$$

which leads to

$$\mathbf{x}_k^{(i)} - \tilde{\mathbf{m}}_{ik} = \frac{N_i}{N_i - 1} (\mathbf{x}_k^{(i)} - \tilde{\mathbf{m}}_i). \tag{9.64}$$

Then, testing a sample $\mathbf{x}_k^{(1)}$ from C_1 can be carried out as

$$(\mathbf{x}_k^{(1)} - \tilde{\mathbf{m}}_{1k})^T (\mathbf{x}_k^{(1)} - \tilde{\mathbf{m}}_{1k}) - (\mathbf{x}_k^{(1)} - \tilde{\mathbf{m}}_2)^T (\mathbf{x}_k^{(1)} - \tilde{\mathbf{m}}_2) \tag{9.65}$$

$$= \left(\frac{N_1}{N_1 - 1} \right)^2 (\mathbf{x}_k^{(1)} - \tilde{\mathbf{m}}_1)^T (\mathbf{x}_k^{(1)} - \tilde{\mathbf{m}}_1) - (\mathbf{x}_k^{(1)} - \tilde{\mathbf{m}}_2)^T (\mathbf{x}_k^{(1)} - \tilde{\mathbf{m}}_2) > \theta.$$

Note that when $\mathbf{x}_k^{(1)}$ is tested, only $\tilde{\mathbf{m}}_1$ is changed and $\tilde{\mathbf{m}}_2$ is not changed. Likewise, when a sample $\mathbf{x}_k^{(2)}$ from C_2 is tested, the decision rule is

$$(\mathbf{x}_k^{(2)} - \tilde{\mathbf{m}}_1)^T (\mathbf{x}_k^{(2)} - \tilde{\mathbf{m}}_1) - (\mathbf{x}_k^{(2)} - \tilde{\mathbf{m}}_{2k})^T (\mathbf{x}_k^{(2)} - \tilde{\mathbf{m}}_{2k}) \tag{9.66}$$

$$= (\mathbf{x}_k^{(2)} - \tilde{\mathbf{m}}_1)^T (\mathbf{x}_k^{(2)} - \tilde{\mathbf{m}}_1) - \left(\frac{N_2}{N_2 - 1} \right)^2 (\mathbf{x}_k^{(2)} - \tilde{\mathbf{m}}_2)^T (\mathbf{x}_k^{(2)} - \tilde{\mathbf{m}}_2) < \theta.$$

The leave-one-out method provides the least biased (practically unbiased) estimate of the classification accuracy of a given classification method for a given training set, and is useful when the number of samples available with known classification is small.

9.9   NEURAL NETWORKS

In many practical problems, we may have no knowledge of the prior probabilities of patterns belonging to one class or another. No general classification rules may exist for the patterns on hand. Clinical knowledge may not yield symbolic knowledge bases that could be used to classify patterns that demonstrate exceptional behavior. In such situations, conventional pattern classification methods as described in the preceding sections may not be well-suited for classification of pattern vectors. Artificial neural networks (ANNs), with the properties of experience-based learning and fault tolerance, should be effective in solving such classification problems [247, 248, 249, 253, 254, 255, 256].

Figure 9.2 illustrates a two-layer perceptron with one hidden layer and one output layer for pattern classification. The network learns the similarities among patterns directly from their instances in the training set that is provided initially. Classification rules are inferred from the training data without prior knowledge of the pattern class distributions in the data. Training of an ANN classifier is typically achieved by the *back-propagation* algorithm [247, 248, 249, 253, 254, 255, 256]. The actual output of the ANN y_k is calculated as

$$y_k = f\left(\sum_{j=1}^{J} w_{jk}^{\#} x_j^{\#} - \theta_k^{\#}\right), \quad k = 1, 2, \ldots, K, \tag{9.67}$$

where

$$x_j^{\#} = f\left(\sum_{i=1}^{I} w_{ij} x_i - \theta_j\right), \quad j = 1, 2, \ldots, J, \tag{9.68}$$

and

$$f(\beta) = \frac{1}{1 + \exp(-\beta)}. \tag{9.69}$$

In the above equations, θ_j and $\theta_k^{\#}$ are node offsets, w_{ij} and $w_{jk}^{\#}$ are node weights, x_i are the elements of the pattern vectors (input parameters), and I, J, and K are the numbers of nodes in the input, hidden, and output layers, respectively. The weights and offsets are updated by

$$w_{jk}^{\#}(n+1) = w_{jk}^{\#}(n) + \eta[y_k(1-y_k)(d_k-y_k)]x_j^{\#} + \alpha[w_{jk}^{\#}(n) - w_{jk}^{\#}(n-1)], \tag{9.70}$$

$$\theta_k^{\#}(n+1) = \theta_k^{\#}(n) + \eta[y_k(1-y_k)(d_k-y_k)](-1) + \alpha[\theta_k^{\#}(n) - \theta_k^{\#}(n-1)], \tag{9.71}$$

$$w_{ij}(n+1) = w_{ij}(n) \tag{9.72}$$
$$+ \eta\left[x_j^{\#}(1-x_j^{\#})\sum_{k=1}^{K}\{y_k(1-y_k)(d_k-y_k)w_{jk}^{\#}\}\right]x_i$$
$$+ \alpha[w_{ij}(n) - w_{ij}(n-1)],$$

and

$$\theta_j(n+1) = \theta_j \tag{9.73}$$
$$+ \eta\left[x_j^{\#}(1-x_j^{\#})\sum_{k=1}^{K}\{y_k(1-y_k)(d_k-y_k)w_{jk}^{\#}\}\right](-1)$$
$$+ \alpha[\theta_j(n) - \theta_j(n-1)],$$

where d_k are the desired outputs, α is a momentum term, η is a gain term, and n refers to the iteration number. Equations 9.70 and 9.71 represent the backpropagation steps, with $y_k(1-y_k)x_j^{\#}$ being the sensitivity of y_k to $w_{jk}^{\#}$, that is, $\frac{\partial y_k}{\partial w_{jk}^{\#}}$.

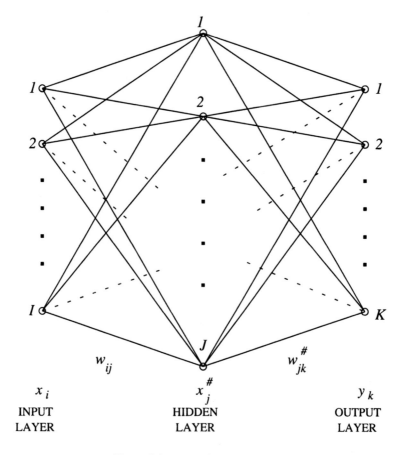

Figure 9.2 A two-layer perceptron.

The classifier training algorithm is repeated until the errors between the desired outputs and actual outputs for the training data are smaller than a predetermined threshold value. Shen et al. [256] present a leave-one-out approach to determine the most suitable values for the parameters J, η, and α.

9.10 MEASURES OF DIAGNOSTIC ACCURACY AND COST

Pattern recognition or classification decisions that are made in the context of medical diagnosis have implications that go beyond statistical measures of accuracy and validity. We need to provide a clinical or diagnostic interpretation of statistical or rule-based decisions made with signal pattern vectors.

Consider the simple situation of *screening*, which represents the use of a test to detect the presence or absence of a specific disease in a certain study population: the decision to be made is binary. Let us represent by A the event that a subject has the particular pathology (or is abnormal), and by N the event that the subject does not have the disease (or is normal). Let the prior probabilities $P(A)$ and $P(N)$ represent the fractions of subjects with the disease and the normal subjects, respectively, in the test population. Let T^+ represent a positive screening test result (indicative of the presence of the disease) and T^- a negative result. The following possibilities arise [257]:

- A *true positive* (TP) is the situation when the test is positive for a subject with the disease (also known as a *hit*). The true-positive fraction (TPF) or *sensitivity* S^+ is given as $P(T^+|A)$ or

$$S^+ = \frac{\text{number of TP decisions}}{\text{number of subjects with the disease}}. \tag{9.74}$$

  The sensitivity of a test represents its capability to detect the presence of the disease of concern.

- A *true negative* (TN) represents the case when the test is negative for a subject who does not have the disease. The true-negative fraction (TNF) or *specificity* S^- is given as $P(T^-|N)$ or

$$S^- = \frac{\text{number of TN decisions}}{\text{number of subjects without the disease}}. \tag{9.75}$$

  The specificity of a test indicates its accuracy in identifying the absence of the disease of concern.

- A *false negative* (FN) is said to occur when the test is negative for a subject who has the disease of concern; that is, the test has missed the case. The probability of this error, known as the false-negative fraction (FNF) is $P(T^-|A)$.

- A *false positive* (FP) is defined as the case where the result of the test is positive when the individual being tested does not have the disease. The probability of this type of error or false alarm, known as the false-positive fraction (FPF) is $P(T^+|N)$.

Table 9.1 summarizes the classification possibilities. Note that

- $FNF + TPF = 1$,

- $FPF + TNF = 1$,

- $S^- = 1 - FPF = TNF$, and

- $S^+ = 1 - FNF = TPF$.

A summary measure of accuracy may be defined as [257]

$$\text{accuracy} = S^+\, P(A) + S^-\, P(N), \tag{9.76}$$

Actual Group	Predicted Group	
	Normal	Abnormal
Normal	$S^- = TNF$	FPF
Abnormal	FNF	$S^+ = TPF$

Table 9.1   Schematic representation of a classification matrix. S^- denotes the specificity (true-negative fraction or TNF), FPF denotes the false-positive fraction, FNF denotes the false-negative fraction, and S^+ denotes the sensitivity (true-positive fraction or TPF).

where $P(A)$ is the fraction of the study population that actually has the disease (that is, the prevalence of the disease) and $P(N)$ is the fraction of the study population that is actually free of the disease.

The efficiency of a test may also be indicated by its predictive values. The *positive predictive value PPV* of a test, defined as

$$PPV = 100 \, \frac{TP}{TP + FP} , \qquad (9.77)$$

represents the percentage of the cases labeled as positive by the test that are actually positive. The *negative predictive value NPV*, defined as

$$NPV = 100 \, \frac{TN}{TN + FN} , \qquad (9.78)$$

represents the percentage of cases labeled as negative by the test that are actually negative.

When a new test or method of diagnosis is being developed and tested, it will be necessary to use another previously established method as a reference to confirm the presence or absence of the disease. Such a reference method is often called the *gold standard*. When computer-based methods need to be tested, it is common practice to use the diagnosis or classification provided by an expert in the field as the gold standard. Results of biopsy, other established laboratory or investigative procedures, or long-term clinical follow-up in the case of normal subjects may also serve this purpose. The term "actual group" in Table 9.1 indicates the result of the gold standard, and the term "predicted group" refers to the result of the test conducted.

Health-care professionals (and the general public) would be interested in knowing the probability that a subject with a positive test result actually has the disease: this is given by the conditional probability $P(A|T^+)$. The question could be answered by using Bayes theorem [245], using which we can obtain

$$P(A|T^+) = \frac{P(A) \, P(T^+|A)}{P(A)P(T^+|A) + P(N)P(T^+|N)} . \qquad (9.79)$$

Note that $P(T^+|A) = S^+$ and $P(T^+|N) = 1 - S^-$. In order to determine the posterior probability as above, the sensitivity and specificity of the test and the prior probabilities of negative cases and positive cases (the rate of prevalence of the disease) should be known.

A cost matrix may be defined, as in Table 9.2, to reflect the overall cost effectiveness of a test or method of diagnosis. The cost of conducting the test and arriving at a TN decision is indicated by C_N: this is the cost of subjecting a normal subject to the test for the purposes of screening for a disease. The cost of the test when a TP is found is shown as C_A: this might include the costs of further tests, treatment, follow-up, and so on, which are secondary to the test itself, but part of the screening and health-care program. The value C^+ indicates the cost of an FP result: this represents the cost of erroneously subjecting an individual without the disease to further tests or therapy. Whereas it may be easy to identify the costs of clinical tests or treatment procedures, it is difficult to quantify the traumatic and psychological effects of an FP result and the consequent procedures on a normal subject. The cost C^- is the cost of an FN result: the presence of the disease in a patient is not diagnosed, the condition worsens with time, the patient faces more complications of the disease, and the health-care system or the patient has to bear the costs of further tests and delayed therapy.

A loss factor due to misclassification may be defined as

$$L = FPF \times C^+ + FNF \times C^-. \tag{9.80}$$

The total cost of the screening program may be computed as

$$C_S = TPF \times C_A + TNF \times C_N + FPF \times C^+ + FNF \times C^-. \tag{9.81}$$

Metz [257] provides more details on the computation of the costs of diagnostic tests.

Actual Group	Predicted Group	
	Normal	Abnormal
Normal	C_N	C^+
Abnormal	C^-	C_A

Table 9.2   Schematic representation of the cost matrix of a diagnostic method.

9.10.1   Receiver operating characteristics

Measures of overall correct classification of patterns as percentages provide limited indications of the accuracy of a diagnostic method. The provision of separate percentage correct classification for each category, such as sensitivity and specificity, can facilitate improved analysis. However, these measures do not indicate the dependence of the results upon the decision threshold. Furthermore, the effect of the rate of incidence or prevalence of the particular disease is not considered.

From another perspective, it is desirable to have a screening or diagnostic test that is both highly sensitive and highly specific. In reality, however, such a test is usually not achievable. Most tests are based on clinical measurements that can assume limited ranges of a variable (or a few variables) with an inherent trade-off between sensitivity and specificity. The relationship between sensitivity and specificity is illustrated by the receiver operating characteristics (ROC) curve, which facilitates improved analysis of the classification accuracy of a diagnostic method [257, 258, 259].

Consider the situation illustrated in Figure 9.3. For a given diagnostic test with the decision variable z, we have predetermined state-conditional PDFs of the decision variable z for actually negative or normal cases indicated as $p(z|N)$, and for actually positive or abnormal cases indicated as $p(z|A)$. As indicated in Figure 9.3, the two PDFs will almost always overlap, given that no method can be perfect. The user or operator needs to determine a decision threshold (indicated by the vertical line) so as to strike a compromise between sensitivity and specificity. Lowering the decision threshold will increase TPF at the cost of increased FPF. (*Note: TNF and FNF* may be derived easily from FPF and TPF, respectively.)

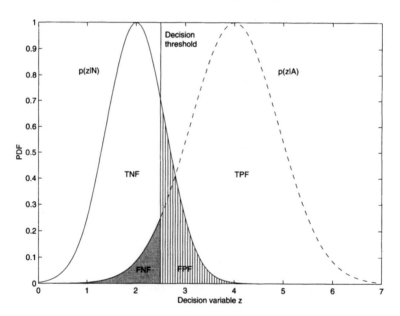

Figure 9.3 State-conditional probability density functions of a diagnostic decision variable z for normal and abnormal cases. The vertical line represents the decision threshold.

An ROC curve is a graph that plots (FPF, TPF) points obtained for a range of decision threshold or cut points of the decision method (see Figure 9.4). The cut point could correspond to the threshold of the probability of prediction. By varying the decision threshold, we get different decision fractions, within the range $(0, 1)$. An ROC curve describes the inherent detection (diagnostic or discriminant) characteristics of a test or method: a receiver (user) may choose to operate at any point along the curve. The ROC curve is independent of the prevalence of the disease or disorder being investigated as it is based upon normalized decision fractions. As all cases may be simply labeled as negative or all may be labeled as positive, an ROC curve has to pass through the points $(0, 0)$ and $(1, 1)$.

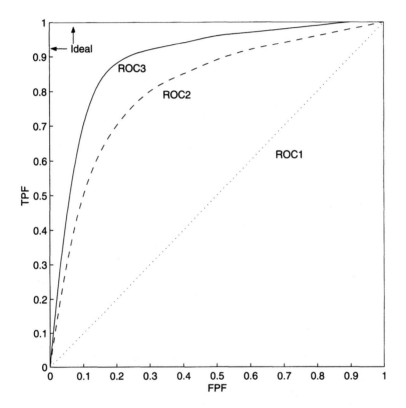

Figure 9.4 Examples of receiver operating characteristic curves.

In a diagnostic situation where a human operator or specialist is required to provide the diagnostic decision, ROC analysis is usually conducted by requiring the specialist to rank each case as one of five possibilities [257]:

1. definitely or almost definitely negative (normal),

2. probably negative,

3. possibly positive,

4. probably positive,

5. definitely or almost definitely positive (abnormal).

Item 3 above may be replaced by "indeterminate" if desired. Various values of TPF and FPF are then calculated by varying the decision threshold from level 5 to level 1 according to the decision items listed above. The resulting (FPF, TPF) points are then plotted to form an ROC curve. The maximum likelihood estimation method [260] is commonly used to fit a binormal ROC curve to data as above.

A summary measure of effectiveness of a test is given by the area under the ROC curve, traditionally labeled as A_z. It is clear from Figure 9.4 that A_z is limited to the range $(0, 1)$. A test that gives a larger area under the ROC curve indicates a better method than one with a smaller area: in Figure 9.4, the method corresponding to ROC3 is better than the method corresponding to ROC2; both are better than the method represented by ROC1 with $A_z = 0.5$. An ideal method will have an ROC curve that follows the vertical line from $(0, 0)$ to $(0, 1)$, and then the horizontal line from $(0, 1)$ to $(1, 1)$, with $A_z = 1$: the method has $TPF = 1$ with $FPF = 0$, which is ideal. (*Note:* This would require the PDFs represented in Figure 9.3 to be non-overlapping.)

9.10.2 McNemar's test of symmetry

Problem: *Suppose we have two methods to perform a certain diagnostic test. How may we compare the classification performance of one against that of the other?*

Solution: Measures of overall classification accuracies such as a percentage of correct classification or the area under the ROC curve provide simple measures to compare two or more diagnostic methods. If more details are required as to how the classifications of groups of cases vary from one method to another, McNemar's test of symmetry [261, 262] would be an appropriate tool.

McNemar's test is based on the construction of contingency tables that compare the results of two classification methods. The rows of a contingency table represent the outcomes of one of the methods used as the reference, possibly a gold standard (labeled as Method A in Table 9.3); the columns represent the outcomes of the other method, which is usually a new method (Method B) to be evaluated against the gold standard. The entries in the table are counts that correspond to particular diagnostic categories, which in Table 9.3 are labeled as normal, indeterminate, and abnormal. A separate contingency table should be prepared for each true category of the patterns; for example, normal and abnormal. (The class "indeterminate" may not be applicable as a true category.) The true category of each case may have to be determined by a third method (for example, biopsy or surgery).

In Table 9.3, the variables a, b, c, d, e, f, g, h, and i denote the counts in each cell, and the numbers in parentheses denote the cell number. The variables $C1$, $C2$, and $C3$ denote the total numbers of counts in the corresponding columns; $R1$, $R2$, and $R3$ denote the total numbers of counts in the corresponding rows. The total number of cases in the true category represented by the table is $N = C1 + C2 + C3 = R1 + R2 + R3$.

| | Method B | | | |
Method A	Normal	Indeterminate	Abnormal	Total
Normal	a (1)	b (2)	c (3)	$R1$
Indeterminate	d (4)	e (5)	f (6)	$R2$
Abnormal	g (7)	h (8)	i (9)	$R3$
Total	$C1$	$C2$	$C3$	N

Table 9.3 Schematic representation of a contingency table for McNemar's test of asymmetry.

Each cell in a contingency table represents a paired outcome. For example, in evaluating the diagnostic efficiency of Method B versus Method A, cell number 3 will contain the number of samples that were classified as normal by Method A but as abnormal by Method B. The row totals $R1$, $R2$, and $R3$, and the column totals $C1$, $C2$, and $C3$ may be used to determine the sensitivity and specificity of the methods.

High values along the main diagonal (a, e, i) of a contingency table (see Table 9.3) indicate no change in diagnostic performance with Method B as compared to Method A. In a contingency table for truly abnormal cases, a high value in the upper-right portion (cell number 3) will indicate an improvement in diagnosis (higher sensitivity) with Method B as compared to Method A. In evaluating a contingency table for truly normal cases, Method B will have a higher specificity than Method A if a large value is found in cell 7. McNemar's method may be used to perform detailed statistical analysis of improvement in performance based upon contingency tables if large numbers of cases are available in each category [261, 262].

9.11   RELIABILITY OF CLASSIFIERS AND DECISIONS

In most practical applications of biomedical signal analysis, the researcher is presented with the problem of designing a pattern classification and decision making system using a small number of training samples (signals), with no knowledge of the distributions of the features or parameters computed from the signals. The size of the training set, relative to the number of features used in the pattern classification system, determines the accuracy and reliability of the decisions made [263, 264]. One should not increase the number of features to be used without a simultaneous increase in the number of training samples, as the two quantities together affect the bias and variance of the classifier. On the other hand, when the training set has a fixed number of samples, the addition of more features beyond a certain limit will lead to poorer performance of the classifier: this is known as the "curse of dimensionality".

It is desirable to be able to analyze the bias and variance of a classification rule while isolating the effects of the functional form of the distributions of the features used.

Raudys and Jain [264] give a rule-of-thumb table for the number of training samples required in relation to the number of features used in order to remain within certain limits of classification errors for five pattern classification methods. When the available features are ordered in terms of their individual classification performance, the optimal number of features to be used with a certain classification method and training set may be determined by obtaining unbiased estimates of the classification accuracy with the number of features increased one at a time in order. A point will be reached when the performance deteriorates, which will indicate the optimal number of features to be used. This method, however, cannot take into account the joint performance of various combinations of features: exhaustive combinations of all features may have to be evaluated to take this aspect into consideration. Software packages such as the Statistical Package for the Social Sciences (SPSS) [251, 252] provide programs to facilitate feature evaluation and selection as well as the estimation of classification accuracies.

Durand et al. [167] reported on the design and evaluation of several pattern classification systems for the assessment of bioprosthetic valves based upon 18 features computed from PCG spectra (see Section 6.6). Based upon the rule of thumb that the number of training samples should be five or more times the number of features used, and with the number of training samples limited to data from 20 normal and 28 degenerated valves, exhaustive combinations of the 18 features taken $2, 3, 4, 5,$ and 6 at a time were used to design and evaluate pattern classification systems. The Bayes method was seen to provide the best performance (98% correct classification) with six features; as many as 511 combinations of the 18 features taken six at a time provided correct classification between 90% and 98%. The nearest-neighbor algorithm with the Mahalanobis distance provided 94% correct classification with only three features, and did not perform any better with more features.

9.12 APPLICATION: NORMAL VERSUS ECTOPIC ECG BEATS

We have seen the distinctions between normal and ectopic (PVC) beats in the ECG in several different contexts (see Sections 1.2.4, 5.4.2, 5.7, and 9.2.2, as well as Figures 5.1 and 5.10). We shall now see how we can put together several of the topics we have studied so far for the purpose of detecting PVCs in an ECG signal.

Training step: Figure 9.5 shows the ECG signal of a patient with several ectopic beats, including episodes of bigeminy (alternating normal beats and PVCs). The beats in the portion of the signal in Figure 9.5 were manually labeled as normals ('o' marks) or PVCs ('x' marks), and used to train a pattern classification system. The training set includes 121 normal beats and 39 PVCs.

The following procedure was applied to the signal to detect each beat, compute features, and develop a pattern classification rule:

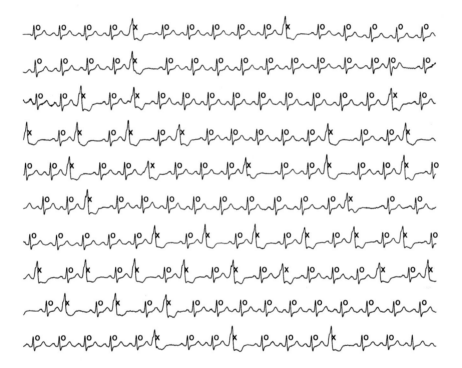

Figure 9.5   The ECG signal of a patient (male, 65 years) with PVCs (training set). Each strip is of duration 10 *s*; the signal continues from top to bottom. The second half of the seventh strip and the first half of the eighth strip illustrate an episode of bigeminy. Each beat was manually labeled as normal ('o') or PVC ('x'). The last beat was not processed.

1. The signal was filtered with a Butterworth lowpass filter of order 8 and cutoff frequency 70 Hz to remove noise (see Section 3.4.1); the sampling rate is 200 Hz.

2. The Pan-Tompkins algorithm was applied to detect each beat (see Section 4.3.2).

3. The QRS – T portion of each beat was segmented by selecting the interval from the sample 160 ms before the peak of the Pan-Tompkins output to the sample 240 ms after the peak (see Figure 5.10).

4. The RR interval and form factor FF were computed for each beat (see Sections 5.6.4 and 5.7, and Figure 5.10). Figure 9.6 illustrates the feature vector plot for the training set.

5. The prototype (mean) feature vectors were computed for the normal and PVC groups in the training set. The prototype vectors are $(RR, FF) = (0.66, 1.58)$ and $(RR, FF) = (0.45, 2.74)$ for the normal and PVC classes, respectively.

6. The equations of the straight line joining the two prototype vectors and its normal bisector were determined; the latter is the optimal linear decision function (see Section 9.4.1 and Figure 9.1). Figure 9.6 illustrates the two lines.

7. The equation of the linear decision function is $RR - 5.56FF + 11.44 = 0$. The decision rule may be stated as

$$\text{if } RR - 5.56FF + 11.44 \begin{cases} > 0 & \text{normal beat} \\ \leq 0 & \text{PVC.} \end{cases} \qquad (9.82)$$

All of the beats in the training set were correctly classified by the decision rule in Equation 9.82.

Observe from Figure 9.6 that a simple threshold on FF alone can effectively separate the PVCs from the normals in the training set. A viable classification rule to detect PVCs may also be stated in a manner similar to that in Section 9.2.2. The example given here is intended to serve as a simple illustration of the design of a 2D linear decision function.

Test step: Figure 9.7 illustrates an ECG segment immediately following that in Figure 9.5. The same procedure as described above was applied to detect the beats in the signal in Figure 9.7 and to compute their features, which were used as the test set. The decision rule in Equation 9.82 was applied to the feature vectors and the beats in the signal were automatically classified as normal or PVC. Figure 9.8 illustrates the feature-vector space of the beats in the test set, along with the decision boundary given by Equation 9.82. Figure 9.7 shows the automatically applied labels of each beat: all the 37 PVCs were correctly classified, and only one of the 120 normal beats was misclassified as a PVC (that is, there was one false positive).

It should be observed that a PVC has, by definition, an RR interval that is less than that for a normal beat (at the same heart rate). However, the heart rate of a subject will vary over time, and the reference RR interval to determine the prematurity of

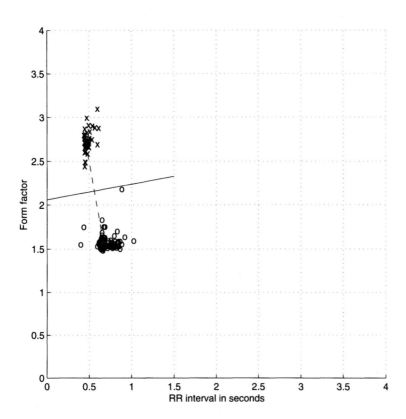

Figure 9.6 (RR, FF) feature-vector space corresponding to the ECG in Figure 9.5 (training set). Normal: 'o', PVC: 'x'. The straight line joining the two prototype vectors (dashed) and its normal bisector (solid) are also shown; the latter is the optimal linear decision function.

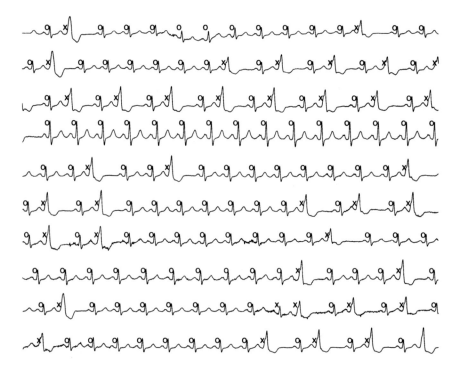

Figure 9.7 The ECG signal of a patient with PVCs (test set); this portion immediately follows that in Figure 9.5. Each strip is of duration 10 s; the signal continues from top to bottom. Each beat was automatically labeled as normal ('o') or PVC ('x') by the decision rule stated in Equation 9.82. The 10th beat in the 9th strip with $(RR, FF) = (0.66, 2.42)$ was misclassified. The last beat was not processed.

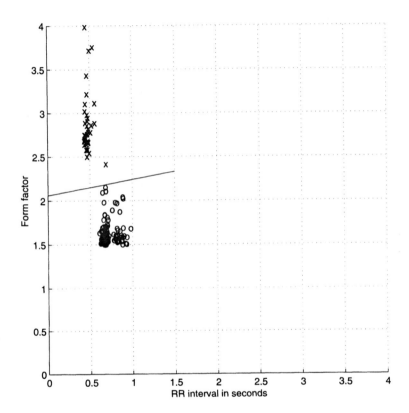

Figure 9.8 (RR, FF) feature-vector space corresponding to the ECG in Figure 9.7 (test set). Normal: 'o', PVC: 'x'. The straight line is the optimal linear decision function given in Equation 9.82. The 'x' mark closest to the decision boundary with $(RR, FF) = (0.66, 2.42)$ corresponds to a false positive classification.

PVCs needs to be updated periodically. A decision rule as in Equation 9.82 cannot be applied on a continuing basis even to the same subject. Note that the proposed method can be extended for the identification of sinus beats (originating from the SA node) that meet the prematurity condition due to sinus arrhythmia but are, nevertheless, normal in waveshape.

The FF values will depend upon the waveshape of each ECG beat, which will vary from one ECG lead to another. Therefore, the same decision rule based upon waveshape cannot be applied to all ECG leads of even the same subject. Furthermore, a given subject may have PVCs originating from various ectopic foci resulting in widely different waveshapes even in the same ECG lead. A shape factor to be used for pattern classification must be capable of maintaining different values between PVCs of various waveshapes as one group, and of normal beats as the other.

The preceding illustration is intended to serve as a simple example of the design of a pattern classification system; in practice, more complex decision rules based upon more than two features will be required. Furthermore, it should be observed that a pattern classification procedure as described above provides beat-by-beat labeling; the overall diagnosis of the patient's condition requires many other items of clinical information and the expertise of a cardiologist.

9.13 APPLICATION: DETECTION OF KNEE-JOINT CARTILAGE PATHOLOGY

Moussavi et al. [56], Krishnan et al. [57], and Rangayyan et al. [58] proposed a series of adaptive segmentation, modeling, and pattern classification techniques for the detection of knee-joint cartilage pathology using VAG signals (see Sections 1.2.13 and 8.2.3). In consideration of the fact that VAG signals are nonstationary, each VAG signal was first divided into locally stationary segments using the RLS or the RLSL algorithm (see Sections 8.6.1 and 8.6.2). Each segment was considered as a separate signal and modeled by the forward-backward linear prediction or the Burg-lattice method (see Section 8.6.2). The model coefficients or poles were used as parameters for pattern classification.

A striking difference that may be observed visually and aurally between normal and abnormal VAG signals is that abnormal signals are much more variable in amplitude across a swing cycle than normal signals. However, this difference is lost in the process of dividing the signals into segments and considering each segment as a separate signal. To overcome this problem, the means (time averages) of the segments of each subject's signal were computed, and then the variance of the means was computed across the various segments of the same signal. The variance of the means represents the above-mentioned difference, and was used as one of the discriminant features.

In addition to quantitative parameters derived from VAG signal analysis, clinical parameters (to be described in the following paragraphs) related to the subjects were also investigated for possible discriminant capabilities. At the outset, as shown in Figure 9.9, knee joints of the subjects in the study were categorized into two groups: normal and abnormal. The normal group was divided into two subgroups: normal-silent and normal-noisy. If no sound was heard during auscultation, a normal knee was considered to be normal-silent; otherwise, it was considered to be normal-noisy. All knees in the abnormal group used were examined by arthroscopy (see Section 8.2.3 and Figure 8.2), and divided into two groups: arthroscopically normal and arthroscopically abnormal.

Labeling of VAG signal segments was achieved by comparing the auscultation and arthroscopy results of each patient with the corresponding segmented VAG and joint angle signals. Localization of the pathology was performed during arthroscopy and the joint angle ranges where the affected areas could come into contact with other joint surfaces were estimated. These results were then compared with the auscultation reports to determine whether the joint angle(s) at which pathology existed corresponded to the joint angle(s) at which sound was heard. For example, if it was found from the arthroscopy report of a patient that the abnormal parts of the patient's knee could cause contact in the range $30° - 90°$, VAG signal segments of the subject corresponding to the angle range of $30° - 90°$ were labeled as arthroscopically abnormal; the rest of the segments of the signal were labeled as arthroscopically normal.

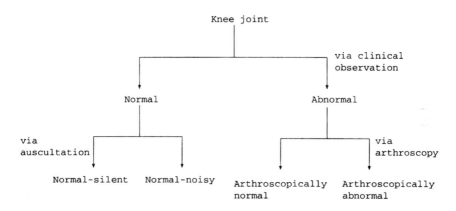

Figure 9.9 Categorization of knee joints based upon auscultation and arthroscopy.

Categorization into four groups as above was done based upon the presumptions that normal-noisy and arthroscopically abnormal signals might be distinguishable in their characteristics, and that normal-silent and arthroscopically normal knees would also be distinguishable. The possibilities of arthroscopically normal knees being associated with sounds, normal-noisy knees not having any associated pathology, and normal-silent knees having undetermined pathologies were also admitted. Krishnan et al. [57] further subdivided the arthroscopically normal and arthroscopically abnormal categories into silent and noisy categories, thereby having a total of six categories; this is not shown in Figure 9.9.

Based on clinical reports and auscultation of knee joints, the following clinical parameters were chosen as features (in addition to AR model parameters) for classification:

Sound: The sound heard by auscultation during flexion and extension movement of the knee joint was coded as:

> 0– silent,
>
> 1– click,
>
> 2– pop,
>
> 3– grinding, or
>
> 4– a mixture of the above-mentioned sounds or other sounds.

Each segment of the VAG signals was labeled with one of the above codes.

Activity level: The activity level of each subject was coded as:

> 1– exercising once per week or less,
>
> 2– exercising two or three times per week, or
>
> 3– exercising more than three times per week.

Age: The age of the subject in years.

Gender: The gender of the subject, which was coded as

> 0– female, or
>
> 1– male.

Among the parameters mentioned above, gender may not be a discriminant parameter; however, it is customary to record gender in clinical analysis. Note that among the four parameters listed above, only the first one can vary between the different segments of a given subject's VAG signal.

Moussavi et al. [56] compared the performance of various sets of features in the classification of VAG signals into two groups and four groups (see Figure 9.9) with random selections of cases. Using a set of 540 segments obtained from 20 normal subjects and 16 subjects with cartilage pathology, different numbers of segments were randomly selected for use in the training step of designing a discriminant function, and finally the selection which provided the best result in the test step was chosen for the final classification system. Two-group classification accuracies in the range $77 - 91\%$ and four-group classification accuracies in the range $65 - 88\%$ were obtained.

By combining the steps of classification into two groups and four groups, a two-step method was proposed by Moussavi et al. [56]; a block diagram of this method is illustrated in Figure 9.10. The algorithm first uses training sets to design classifiers for two and four groups. The resulting discriminant functions are used as Classifier 1 (two groups) and Classifier 2 (four groups), respectively. An unknown signal, which has been adaptively divided into segments, enters Classifier 1. If segments spanning more than 90% of the duration of the signal are classified as being normal, the signal (subject) is considered to be normal. On the other hand, if more than 90% of the duration of the signal is classified as being abnormal, the signal (subject) is considered to be abnormal. If more than 10% but less than 90% of the signal duration is classified as abnormal, the signal goes to Classifier 2, which classifies the signal into four groups (see Figure 9.9). In the second step, if more than 10% of the duration of the signal is classified as being arthroscopically abnormal, the signal is considered to be abnormal; otherwise it is considered to be normal. At this stage, information on the numbers of segments belonging to the four categories shown in Figure 9.9 is available, but the final decision is on the normality of the whole signal (subject or knee joint).

The two-step diagnosis method was trained with 262 segments obtained from 10 normal subjects and eight subjects with cartilage pathology, and was tested with 278 segments obtained from a different set of 10 normal subjects and eight subjects with cartilage pathology but without any restriction on the kind of abnormality. Except for one normal signal which was indicated as being abnormal over 12% of its duration, all of the signals were correctly classified. The results also showed that all of the abnormal signals including signals associated with chondromalacia grades I to IV (see Section 8.2.3 and Figure 8.2) were classified correctly. Based upon this result, it was indicated that the method has the ability to detect chondromalacia patella at its early stages as well as advanced stages. Krishnan et al. [57] and Rangayyan et al. [58] reported on further work along these directions.

9.14 REMARKS

The subject of pattern classification is a vast area by itself. The topics presented in this chapter provide a brief introduction to the subject.

We have now seen how biomedical signals may be processed and analyzed to extract quantitative features that may be used to classify the signals as well as to design diagnostic decision functions. Practical development of such techniques is usually hampered by a number of limitations related to the extent of discriminant information present in the signals selected for analysis, as well as the limitations of the features designed and computed. Artifacts inherent in the signal or caused by the signal acquisition systems impose further limitations.

A pattern classification system that is designed with limited data and information about the chosen signals and features will provide results that should be interpreted with due care. Above all, it should be borne in mind that the final diagnostic decision requires far more information than that provided by signal analysis: this aspect is best left to the physician or health-care specialist in the spirit of computer-*aided* diagnosis.

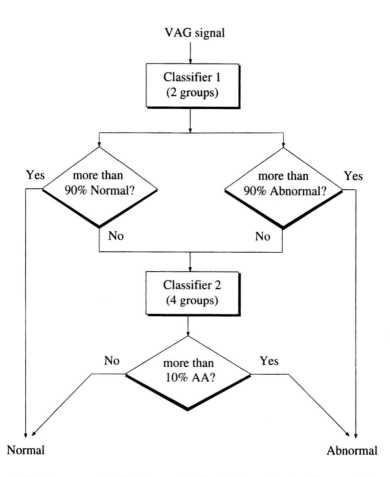

Figure 9.10 A two-step classification method for the diagnosis of cartilage pathology. AA – Arthroscopically abnormal. See also Figure 9.9. Reproduced with permission from Z.M.K. Moussavi, R.M. Rangayyan, G.D. Bell, C.B. Frank, K.O. Ladly, and Y.T. Zhang, Screening of vibroarthrographic signals via adaptive segmentation and linear prediction modeling, *IEEE Transactions on Biomedical Engineering*, 43(1):15–23, 1996. ©IEEE.

9.15   STUDY QUESTIONS AND PROBLEMS

1. The prototype vectors of two classes of signals are specified as Class 1 : $(1, 0.5)$, and Class 2 : $(3, 3)$. A new sample vector is given as $(2, 1)$. Give the equations for two measures of similarity or dissimilarity, compute the measures for the sample vector, and classify the sample as Class 1 or Class 2 using each measure.

2. In a three-class pattern classification problem, the three decision boundaries are $d_1(\mathbf{x}) = -x_1 + x_2$, $d_2(\mathbf{x}) = x_1 + x_2 - 5$, and $d_3(\mathbf{x}) = -x_2 + 1$.
   Draw the decision boundaries on a sheet of graph paper.
   Classify the sample pattern vector $\mathbf{x} = (6, 5)$ using the decision functions.

3. Two pattern class prototype vectors are given to you as $\mathbf{z}_1 = (3, 4)$ and $\mathbf{z}_2 = (10, 2)$. Classify the sample pattern vector $\mathbf{x} = (4, 5)$ using (a) the normalized dot product, and (b) the Euclidean distance.

4. A researcher makes two measurements per sample on a set of 10 normal and 10 abnormal samples.
   The set of feature vectors for the normal samples is
   $\{(2, 6), (22, 20), (10, 14), (10, 10), (24, 24), (8, 10), (8, 8), (6, 10), (8, 12), (6, 12)\}$.
   The set of feature vectors for the abnormal samples is
   $\{(4, 10), (24, 16), (16, 18), (18, 20), (14, 20), (20, 22), (18, 16), (20, 20), (18, 18), (20, 18)\}$.
   Plot the scatter diagram of the samples in both classes in the feature-vector space (on a sheet of graph paper). Draw a linear decision function to classify the samples with the least error of misclassification. Write the decision function as a mathematical rule.
   How many (if any) samples are misclassified by your decision function? Mark the misclassified samples on the plot.
   Two new observation sample vectors are provided to you as $\mathbf{x}_1 = (12, 15)$ and $\mathbf{x}_2 = (14, 15)$. Classify the samples using your decision rule.
   Now, classify the samples $\mathbf{x}_1$ and $\mathbf{x}_2$ using the k-nearest-neighbor method, with $k = 7$. Measure distances graphically on your graph paper plot and mark the neighbors used in this decision process for each sample.
   Comment upon the results — whether the two methods resulted in the same classification result or not — and provide reasons.

5. A researcher makes measurements of RR intervals (in seconds) and form factor (FF) for a number of ECG beats including (i) normal beats, (ii) premature ventricular contractions (PVC), and (iii) normal beats with a compensatory pause (NBCP). The values (training set) are given in Table 9.4.
   (a) Plot the (RR, FF) feature-vector points for the three classes of beats on a graph paper.
   (b) Compute the prototype vectors for each class as the class means. Indicate the prototypes on the plot.
   (c) Derive the optimal linear discriminant functions (or decision functions) as the perpendicular bisectors of the straight lines joining the prototypes. State the decision rule(s) for each type of beat.
   (d) Three new beats are observed to have the parameters listed in Table 9.5. Classify each beat using the decision functions derived in part (c).

6. For the training data given in the preceding problem, compute the mean and covariance matrices of the feature vectors for each class, as well as the pooled covariance matrix.

Normal Beats		PVCs		NBCPs	
RR	FF	RR	FF	RR	FF
0.700	1.5	0.600	5.5	0.800	1.2
0.720	1.0	0.580	6.1	0.805	1.1
0.710	1.2	0.560	6.4	0.810	1.6
0.705	1.3	0.570	5.9	0.815	1.3
0.725	1.4	0.610	6.3	0.790	1.4

Table 9.4   Training set of (RR, FF) feature vectors.

Beat No.	RR	FF
1	0.650	5.5
2	0.680	1.9
3	0.820	1.8

Table 9.5   Test set of (RR, FF) feature vectors.

Design a classifier based upon the Mahalanobis distance using the pooled covariance matrix.

7. You have won a contract to design a software package for computer-aided diagnosis of cardiovascular diseases using the heart sound signal (PCG) as the main source of information. The main task is to identify the presence of murmurs in systole and/or diastole. You may use other signals for reference.

Propose a signal processing system to

(i) acquire the required signals;

(ii) preprocess them as required;

(iii) extract at least two features for classification; and

(iv) classify the PCG signals as:

class 1 – Normal (no murmurs),

class 2 – Systolic murmur,

class 3 – Diastolic murmur, or

class 4 – Systolic and diastolic murmur.

Provide a block diagram of the complete procedure. Explain the reason behind the application of each step and state the expected results or benefits. Provide algorithmic details and/or mathematical definitions for at least two major steps in your procedure.

Draw a schematic plot of the feature-vector space and indicate where samples from the four classes listed above would fall. Propose a framework of decision rules to classify an incoming signal as belonging to one of the four classes.

9.16   LABORATORY EXERCISES AND PROJECTS

Note: Data files related to the exercises are available at the site
ftp://ftp.ieee.org/uploads/press/rangayyan/

1. The data file ecgpvc.dat contains the ECG signal of a patient with PVCs (see Figures 9.5 and 9.7). Refer to the file ecgpvc.m for details. Use the first 40% of the signal as training data to develop a PVC detection system (see Section 9.12). Develop code to segment the QRS – T portion of each beat using the Pan-Tompkins method (see Section 4.3.2), and compute the RR interval, QRS width (see Figure 4.5), and form factor FF for each beat (see Section 5.6.4). Design linear discriminant functions using (i) RR and QRS width, and (ii) RR and FF as the features; see Figure 9.6. Analyze the results in terms of TPF and FPF.

   Code the decision function into your program as a classification rule. Test the pattern classifier program with the remaining 60% of the signal as the test signal. Compute the test-stage classification accuracy in terms of TPF and FPF.

2. Repeat the previous exercise replacing the linear discriminant function with the k-nearest-neighbor method, with $k = 1, 3, 5,$ and 7. Evaluate the method with feature sets composed as

   - RR and QRS width,
   - RR and FF, and
   - RR, FF, and QRS width.

   Compare the performances of the three classifiers and provide reasons for any differences between them.

References

1. Lathi BP. *Signal Processing and Linear Systems*. Berkeley-Cambridge, Carmichael, CA, 1998.

2. Oppenheim AV, Willsky AS, and Nawab SH. *Signals and Systems*. Prentice-Hall, Englewood Cliffs, NJ, 2nd edition, 1997.

3. Papoulis A. *Signal Analysis*. McGraw-Hill, New York, NY, 1977.

4. Papoulis A. *Probability, Random Variables, and Stochastic Processes*. McGraw-Hill, New York, NY, 1965.

5. Bendat JS and Piersol AG. *Random Data: Analysis and Measurement Procedures*. Wiley, New York, NY, 2nd edition, 1986.

6. Auñón JI and Chandrasekar V. *Introduction to Probability and Random Processes*. McGraw-Hill, New York, NY, 1997.

7. Ramsey FL and Schafer DW. *The Statistical Sleuth — A Course in Methods of Data Analysis*. Wadsworth Publishing Company, Belmont, CA, 1997.

8. Riffenburgh RH. *Statistics in Medicine*. Academic, San Diego, CA, 1993.

9. Bailar III JC and Mosteller F, editors. *Medical Uses of Statistics*. NEJM Books, Boston, MA, 2nd edition, 1992.

10. Webster JG, editor. *Medical Instrumentation: Application and Design*. Wiley, New York, NY, 3rd edition, 1998.

11. Bronzino JD. *Biomedical Engineering and Instrumentation*. PWS Engineering, Boston, MA, 1986.

12. Bronzino JD, editor. *The Biomedical Engineering Handbook*. CRC and IEEE, Boca Raton, FL, 1995.

13. Aston R. *Principles of Biomedical Instrumentation and Measurement*. Merrill, Columbus, OH, 1990.

14. Oppenheim AV and Schafer RW. *Discrete-time Signal Processing*. Prentice-Hall, Englewood Cliffs, NJ, 1989.

15. Cooper KE, Cranston WI, and Snell ES. Temperature regulation during fever in man. *Clinical Science*, 27(3):345–356, 1964.

16. Cooper KE. Body temperature and its regulation. In *Encyclopedia of Human Biology*, volume 2, pages 73–83. Academic, New York, NY, 1997.

17. Cromwell L, Weibell FJ, and Pfeiffer EA. *Biomedical Instrumentation and Measurements*. Prentice-Hall, Englewood Cliffs, NJ, 2nd edition, 1980.

18. Plonsey R. Action potential sources and their volume conductor fields. *Proceedings of the IEEE*, 65(5):601–611, 1977.

19. Clark R. *Action potentials (Personal Communication)*. University of Calgary, Calgary, Alberta, Canada, 1999.

20. Hille B. Membrane excitability: Action potential propagation in axons. In Patton H, Fuchs A, Hille B, Scher A, and Steiner R, editors, *Textbook of Physiology*, pages 49–79. WB Saunders, Philadelphia, PA, 21st edition, 1989.

21. Koester J. Action conductances underlying the action potential. In Kandel E and Schwartz J, editors, *Principles of Neural Science*, pages 53–62. Elsevier - North Holland, New York, NY, 1981.

22. Goodgold J and Eberstein A. *Electrodiagnosis of Neuromuscular Diseases*. Williams and Wilkins, Baltimore, MD, 3rd edition, 1983.

23. Rushmer RF. *Cardiovascular Dynamics*. WB Saunders, Philadelphia, PA, 4th edition, 1976.

24. de Luca CJ. Physiology and mathematics of myoelectric signals. *IEEE Transactions on Biomedical Engineering*, 26:313–325, 1979.

25. Mambrito B and de Luca CJ. Acquisition and decomposition of the EMG signal. In Desmedt JE, editor, *Progress in Clinical Neurophysiology, Volume 10: Computer-aided Electromyography*, pages 52–72. S. Karger AG, Basel, Switzerland, 1983.

26. Platt RS, Hajduk EA, Hulliger M, and Easton PA. A modified Bessel filter for amplitude demodulation of respiratory electromyograms. *Journal of Applied Physiology*, 84(1):378–388, 1998.

27. Tompkins WJ. *Biomedical Digital Signal Processing*. Prentice-Hall, Upper Saddle River, NJ, 1995.

28. Goldberger E. *Unipolar Lead Electrocardiography and Vectorcardiography*. Lea & Febiger, Philadelphia, PA, 3rd edition, 1954.

29. Jenkins JM. Computerized electrocadiography. *CRC Critical Reviews in Bio-engineering*, pages 307–350, November 1981.

30. Jenkins JM. Automated electrocardiography and arrythmia monitoring. *Progress in Cardiovascular Disease*, 25(5):367–408, 1983.

31. Cox Jr. JR, Nolle FM, and Arthur RM. Digital analysis of the electroencephalo-gram, the blood pressure wave, and the electrocardiogram. *Proceedings of the IEEE*, 60(10):1137–1164, 1972.

32. Cooper R, Osselton JW, and Shaw JC. *EEG Technology*. Butterworths, London, UK, 3rd edition, 1980.

33. Kooi KA, Tucker RP, and Marshall RE. *Fundamentals of Electroencephalog-raphy*. Harper & Row, Hagerstown, MD, 2nd edition, 1978.

34. Hughes JR. *EEG in Clinical Practice*. Butterworth, Woburn, MA, 1982.

35. Verhagen MAMT, van Schelven LJ, Samsom M, and Smout AJPM. Pitfalls in the analysis of electrogastrographic recordings. *Gastroenterology*, 117:453–460, 1999.

36. Mintchev MP and Bowes KL. Capabilities and limitations of electrogastro-grams. In Chen JDZ and McCallum RW, editors, *Electrogastrography: Principles and Applications*, pages 155–169. Raven, New York, NY, 1994.

37. Mintchev MP and Bowes KL. Extracting quantitative information from digi-tal electrogastrograms. *Medical and Biological Engineering and Computing*, 34:244–248, 1996.

38. Chen JDZ, Stewart Jr. WR, and McCallum RW. Spectral analysis of episodic rhythmic variations in the cutaneous electrogastrogram. *IEEE Transactions on Biomedical Engineering*, 40(2):128–135, 1993.

39. Mintchev MP, Stickel A, and Bowes KL. Dynamics of the level of randomness in gastric electrical activity. *Digestive Diseases and Sciences*, 43(5):953–956, 1998.

40. Rangayyan RM and Lehner RJ. Phonocardiogram signal processing: A review. *CRC Critical Reviews in Biomedical Engineering*, 15(3):211–236, 1988.

41. Tavel ME. *Clinical Phonocardiography and External Pulse Recording.* Year Book Medical, Chicago, IL, 3rd edition, 1978.

42. Luisada AA and Portaluppi F. *The Heart Sounds — New Facts and Their Clinical Implications.* Praeger, New York, NY, 1982.

43. Shaver JA, Salerni R, and Reddy PS. Normal and abnormal heart sounds in cardiac diagnosis, Part I: Systolic sounds. *Current Problems in Cardiology,* 10(3):1–68, 1985.

44. Reddy PS, Salerni R, and Shaver JA. Normal and abnormal heart sounds in cardiac diagnosis, Part II: Diastolic sounds. *Current Problems in Cardiology,* 10(4):1–55, 1985.

45. Childers DG and Bae KS. Detection of laryngeal function using speech and electroglottographic data. *IEEE Transactions on Biomedical Engineering,* 39(1):19–25, 1992.

46. Rabiner LR and Schafer RW. *Digital Processing of Speech Signals.* Prentice-Hall, Englewood Cliffs, NJ, 1978.

47. Zhang YT, Frank CB, Rangayyan RM, and Bell GD. A comparative study of vibromyography and electromyography obtained simultaneously from active human quadriceps. *IEEE Transactions on Biomedical Engineering,* 39(10):1045–1052, 1992.

48. Zhang YT, Frank CB, Rangayyan RM, and Bell GD. Relationships of the vibromyogram to the surface electromyogram of the human rectus femoris muscle during voluntary isometric contraction. *Journal of Rehabilitation Research and Development,* 33(4):395–403, October 1996.

49. Ellison AE. *Athletic Training and Sports Medicine.* American Academy of Orthopaedic Surgeons, Chicago, IL, 1984.

50. Moore KL. *Clinically Oriented Anatomy.* Williams/Wilkins, Baltimore, MD, 1984.

51. Tortora GJ. Articulations. In Wilson CM and Helfgott N, editors, *Principles of Human Anatomy,* pages 167–203. Harper and Row, New York, NY, 1986.

52. Frankel VH and Nordin M, editors. *Basic Biomechanics of the Skeletal System.* Lea and Febiger, Philadelphia, PA, 1980.

53. Nicholas JA and Hershman EB, editors. *The Lower Extremity and Spine in Sports Medicine.* CV Mosby, Missouri, KS, 1986.

54. Frank CB, Rangayyan RM, and Bell GD. Analysis of knee sound signals for non-invasive diagnosis of cartilage pathology. *IEEE Engineering in Medicine and Biology Magazine,* pages 65–68, March 1990.

55. Tavathia S, Rangayyan RM, Frank CB, Bell GD, Ladly KO, and Zhang YT. Analysis of knee vibration signals using linear prediction. *IEEE Transactions on Biomedical Engineering*, 39(9):959–970, 1992.

56. Moussavi ZMK, Rangayyan RM, Bell GD, Frank CB, Ladly KO, and Zhang YT. Screening of vibroarthrographic signals via adaptive segmentation and linear prediction modeling. *IEEE Transactions on Biomedical Engineering*, 43(1):15–23, 1996.

57. Krishnan S, Rangayyan RM, Bell GD, Frank CB, and Ladly KO. Adaptive filtering, modelling, and classification of knee joint vibroarthrographic signals for non-invasive diagnosis of articular cartilage pathology. *Medical and Biological Engineering and Computing*, 35(6):677–684, 1997.

58. Rangayyan RM, Krishnan S, Bell GD, Frank CB, and Ladly KO. Parametric representation and screening of knee joint vibroarthrographic signals. *IEEE Transactions on Biomedical Engineering*, 44(11):1068–1074, 1997.

59. Kernohan WG, Beverland DE, McCoy GF, Hamilton A, Watson P, and Mollan RAB. Vibration arthrometry. *Acta Orthopedica Scandinavia*, 61(1):70–79, 1990.

60. Chu ML, Gradisar IA, and Mostardi R. A noninvasive electroacoustical evaluation technique of cartilage damage in pathological knee joints. *Medical and Biological Engineering and Computing*, 16:437–442, 1978.

61. Probst R, Lonsbury-Martin B, and Martin GK. A review of otoacoustic emissions. *Journal of the Acoustical Society of America*, 89(5):2027–2067, 1991.

62. Widrow B, Glover Jr. JR, McCool JM, Kaunitz J, Williams CS, Hearn RH, Zeidler JR, Dong Jr. E, and Goodlin RC. Adaptive noise cancelling: Principles and applications. *Proceedings of the IEEE*, 63(12):1692–1716, 1975.

63. Zhang YT, Rangayyan RM, Frank CB, and Bell GD. Adaptive cancellation of muscle contraction interference from knee joint vibration signals. *IEEE Transactions on Biomedical Engineering*, 41(2):181–191, 1994.

64. Bartlett J. *Familiar Quotations*. Little, Brown and Co., Boston, MA, 15th edition, 1980.

65. Akay AM, Semmlow JL, Welkowitz W, Bauer MD, and Kostis JB. Detection of coronary occlusions using autoregressive modeling of diastolic heart sounds. *IEEE Transactions on Biomedical Engineering*, 37(4):366–373, 1990.

66. Lehner RJ and Rangayyan RM. A three-channel microcomputer system for segmentation and characterization of the phonocardiogram. *IEEE Transactions on Biomedical Engineering*, 34:485–489, 1987.

67. Jenkins JM, Wu D, and Arzbaecher RC. Computer diagnosis of abnormal cardiac rhythms employing a new P-wave detector for interval measurement. *Computers and Biomedical Research*, 11:17–33, 1978.

68. Jenkins JM, Wu D, and Arzbaecher RC. Computer diagnosis of supraventricular and ventricular arrhythmias. *Circulation*, 60(5):977–987, 1979.

69. Sayers B.McA. Analysis of heart rate variability. *Ergonomics*, 16(1):17–32, 1973.

70. Kobayashi M and Musha T. 1/f fluctuation of heartbeat period. *IEEE Transactions on Biomedical Engineering*, 29(6):456–457, 1982.

71. Rompelman O, Snijders JBIM, and van Spronsen CJ. The measurement of heart rate variability spectra with the help of a personal computer. *IEEE Transactions on Biomedical Engineering*, 29(7):503–510, 1982.

72. deBoer RW, Karemaker JM, and Strackee J. Comparing spectra of a series of point events particularly for heart rate variability studies. *IEEE Transactions on Biomedical Engineering*, 31(4):384–387, 1984.

73. Rosenblum MG, Kurths J, Pikovsky A, Schäfer C, Tass P, and Abel HH. Synchronization in noisy systems and cardiorespiratory interaction. *IEEE Engineering in Medicine and Biology Magazine*, 17(6):46–53, 1998.

74. Pompe B, Blidh P, Hoyer D, and Eiselt M. Using mutual information to measure coupling in the cardiorespiratory system. *IEEE Engineering in Medicine and Biology Magazine*, 17(6):32–39, 1998.

75. Durand LG, Genest Jr. J, and Guardo R. Modeling of the transfer function of the heart-thorax acoustic system in dogs. *IEEE Transactions on Biomedical Engineering*, 32(8):592–601, 1985.

76. Kantz H, Kurtis J, and Mayer-Kress G, editors. *Nonlinear Analysis of Physiological Data*. Springer-Verlag, Berlin, Germany, 1998.

77. Haykin S. *Adaptive Filter Theory*. Prentice-Hall, Upper Saddle River, NJ, 3rd edition, 1996.

78. Kendall M. *Time-Series*. Charles Griffin, London, UK, 2nd edition, 1976.

79. Challis RE and Kitney RI. Biomedical signal processing (in four parts): Part 1. Time-domain methods. *Medical and Biological Engineering and Computing*, 28:509–524, 1990.

80. Shanks JL. Recursion filters for digital processing. *Geophysics*, 32(1):33–51, 1967.

81. Rabiner LR and Gold B. *Theory and Application of Digital Signal Processing*. Prentice-Hall, Englewood Cliffs, NJ, 1975.

82. Hamming RW. *Digital Filters*. Prentice-Hall, Englewood Cliffs, NJ, 2nd edition, 1983.

83. Antoniou A. *Digital Filters: Analysis, Design, and Applications*. McGraw-Hill, New York, NY, 2nd edition, 1993.

84. Williams CS. *Designing Digital Filters*. Prentice-Hall, Englewood Cliffs, NJ, 1986.

85. Haykin S. *Modern Filters*. Macmillan, New York, NY, 1989.

86. Oppenheim AV and Schafer RW. *Digital Signal Processing*. Prentice-Hall, Englewood Cliffs, NJ, 1975.

87. Little JN and Shure L. *Signal Processing Toolbox for Use with MATLAB*. The MathWorks, Inc., Natick, MA, 1992.

88. Krishnan S. Adaptive filtering, modeling, and classification of knee joint vibroarthrographic signals. Master's thesis, Department of Electrical and Computer Engineering, University of Calgary, Calgary, AB, Canada, April 1996.

89. Riegler R and Compton Jr. R. An adaptive array for interference rejection. *Proceedings of the IEEE*, 61(6):748–758, 1973.

90. Sesay AB. *ENEL 671: Adaptive Signal Processing*. Unpublished lecture notes, Department of Electrical and Computer Engineering, University of Calgary, Calgary, Alberta, Canada, 1995.

91. Ferrara ER and Widrow B. Fetal electrocardiogram enhancement by time-sequenced adaptive filtering. *IEEE Transactions on Biomedical Engineering*, 29(6):458–460, 1982.

92. Rangayyan RM, Krishnan S, Bell GD, Frank CB, and Ladly KO. Impact of muscle contraction interference cancellation on vibroarthrographic screening. In *Proceedings of the International Conference on Biomedical Engineering*, pages 16–19, Kowloon, Hong Kong, June 1996.

93. Krishnan S and Rangayyan RM. Automatic denoising of knee joint vibration signals using adaptive time-frequency representations. *Medical and Biological Engineering and Computing*, 38(1):2–8, 2000.

94. Maragos P and Schafer RW. Morphological filters – Part I: Their set-theoretic analysis and relations to linear shift-invariant filters. *IEEE Transactions on Acoustics, Speech, and Signal Processing*, 35(8):1153–1169, 1987.

95. Maragos P and Schafer RW. Morphological filters – Part II: Their relations to median, order-statistic, and stack filters. *IEEE Transactions on Acoustics, Speech, and Signal Processing*, 35(8):1170–1184, 1987.

96. Dumermuth G, Huber PJ, Kleiner B, and Gasser T. Numerical analysis of electroencephalographic data. *IEEE Transactions on Audio and Electroacoustics*, 18(4):404–411, 1970.

97. Barlow JS. Computerized clinical electroencephalography in perspective. *IEEE Transactions on Biomedical Engineering*, 26(7):377–391, 1979.

98. Bodenstein G and Praetorius HM. Feature extraction from the electroencephalogram by adaptive segmentation. *Proceedings of the IEEE*, 65(5):642–652, 1977.

99. Balda RA, Diller G, Deardorff E, Doue J, and Hsieh P. The HP ECG analysis program. In van Bemmel JH and Willems JL, editors, *Trends in Computer-processed Electrocardiograms*, pages 197–205. North Holland, Amsterdam, The Netherlands, 1977.

100. Ahlstrom ML and Tompkins WJ. Digital filters for real-time ECG signal processing using microprocessors. *IEEE Transactions on Biomedical Engineering*, 32:708–713, 1985.

101. Friesen GM, Jannett TC, Jadallah MA, Yates SL, Quint SR, and Nagle HT. A comparison of the noise sensitivity of nine QRS detection algorithms. *IEEE Transactions on Biomedical Engineering*, 37(1):85–97, 1990.

102. Murthy ISN and Rangaraj MR. New concepts for PVC detection. *IEEE Transactions on Biomedical Engineering*, 26(7):409–416, 1979.

103. Pan J and Tompkins WJ. A real-time QRS detection algorithm. *IEEE Transactions on Biomedical Engineering*, 32:230–236, 1985.

104. Starmer CF, McHale PA, and Greenfield Jr. JC. Processing of arterial pressure waves with a digital computer. *Computers and Biomedical Research*, 6:90–96, 1973.

105. Schwartz M. *Information Transmission, Modulation, and Noise*. McGraw-Hill, New York, NY, 3rd edition, 1980.

106. Wade JG. *Signal Coding and Processing: An introduction based on video systems*. Ellis Horwood, Chichester, England, 1987.

107. Hengeveld SJ and van Bemmel JH. Computer detection of P waves. *Computers and Biomedical Research*, 9:125–132, 1976.

108. Gritzali F, Frangakis G, and Papakonstantinou G. Detection of the P and T waves in an ECG. *Computers and Biomedical Research*, 22:83–91, 1989.

109. Willems JL, Arnaud P, van Bemmel JH, Bourdillon PJ, Brohet C, Volta SD, Andersen JD, Degani R, Denis B, Demeester M, Dudeck J, Harms FMA, Macfarlane PW, Mazzocca G, Meyer J, Michaelis J, Pardaens J, Pöppl SJ, Reardon BC, van Eck HJR, de Medina EOR, Rubel P, Talmon JL, and Zywietz

C. Assessment of the performance of electrocardiographic computer programs with the use of a reference data base. *Circulation*, 71(3):523–534, 1985.

110. Willems JL, Arnaud P, van Bemmel JH, Bourdillon PJ, Degani R, Denis B, Harms FMA, Macfarlane PW, Mazzocca G, Meyer J, van Eck HJR, de Medina EOR, and Zywietz C. Establishment of a reference library for evaluating computer ECG measurement programs. *Computers and Biomedical Research*, 18:439–457, 1985.

111. Bogert BP, Healy MJR, and Tukey JW. The quefrency alanysis of time series for echoes: Cepstrum, pseudo-autocovariance, cross-cepstrum, and saphe cracking. In Rosenblatt M, editor, *Proceedings of the Symposium on Time Series Analysis*, pages 209–243. Wiley, New York, NY, 1963.

112. Oppenheim AV, Schafer RW, and Stockham Jr. TG. Nonlinear filtering of multiplied and convolved signals. *Proceedings of the IEEE*, 56(8):1264–1291, 1968.

113. Oppenheim AV and Schafer RW. Homomorphic analysis of speech. *IEEE Transactions on Audio and Electroacoustics*, AU-16(2):221–226, 1968.

114. Gonzalez RC and Woods RE. *Digital Image Processing*. Addison-Wesley, Reading, MA, 1992.

115. Childers DG, Skinner DP, and Kemerait RC. The cepstrum: A guide to processing. *Proceedings of the IEEE*, 65(10):1428–1443, 1977.

116. MacCanon DM, Arevalo F, and Meyer EC. Direct detection and timing of aortic valve closure. *Circulation Research*, 14:387–391, 1964.

117. Stein PD, Sabbah HN, Anbe DT, and Khaja F. Hemodynamic and anatomic determinants of relative differences in amplitude of the aortic and pulmonary components of the second heart sound. *American Journal of Cardiology*, 42:539–544, 1978.

118. Stein PD and Sabbah H. Intensity of the second heart sound: Relation of physical, physiological and anatomic factors to auscultatory evaluation. *Henry Ford Hospital Medical Journal*, 28(4):205–209, 1980.

119. Sarkady AA, Clark RR, and Williams R. Computer analysis techniques for phonocardiogram diagnosis. *Computers and Biomedical Research*, 9:349–363, 1976.

120. Baranek HL, Lee HC, Cloutier G, and Durand LG. Automatic detection of sounds and murmurs in patients with Ionescu-Shiley aortic bioprostheses. *Medical and Biological Engineering and Computing*, 27:449–455, 1989.

121. Durand LG, de Guise J, Cloutier G, Guardo R, and Brais M. Evaluation of FFT-based and modern parametric methods for the spectral analysis of bioprosthetic

valve sounds. *IEEE Transactions on Biomedical Engineering*, 33(6):572–578, 1986.

122. Wallace AG. Electrophysiology of the myocardium. In *Clinical Cardiopulmonary Physiology*. Grune & Stratton, New York, NY, 3rd edition, 1969.

123. Berkhout AJ. On the minimum phase criterion of sampled signals. *IEEE Transactions on Geoscience Electronics*, 11:186–198, 1973.

124. Berkhout AJ. On the minimum-length property of one-sided signals. *Geophysics*, 38:657–672, 1978.

125. Amazeen RL, Moruzzi RL, and Feldman CL. Phase detection of R-waves in noisy electrocardiograms. *IEEE Transactions on Biomedical Engineering*, 19(1):63–66, 1972.

126. Ulrych TJ and Lasserre M. Minimum-phase. *Canadian Journal of Exploration Geophysicists*, 2:22–32, 1966.

127. Treitel S and Robinson EA. The stability of digital filters. *IEEE Transactions on Geoscience Electronics*, 2:6–18, 1964.

128. Oppenheim AV, Kopec GE, and Tribolet JM. Signal analysis by homomorphic prediction. *IEEE Transactions on Acoustics, Speech, and Signal Processing*, 24(4):327–332, 1976.

129. Nolle F. *Argus, A Clinical Computer System for Monitoring Electrocardiographic Rhythms*. PhD thesis, Washington University School of Medicine, Saint Louis, MO, December 1972.

130. Shin SJ, Tapp WN, Reisman SS, and Natelson BH. Assessment of autonomic regulation of heart rate variability by the method of complex demodulation. *IEEE Transactions on Biomedical Engineering*, 36(2):274–283, 1989.

131. Hayano J, Taylor JA, Yamada A, Mukai S, Hori R, Asakawa T, Yokoyama K, Watanabe Y, Takata K, and Fujinami T. Continuous assessment of hemodynamic control by complex demodulation of cardiovascular variability. *American Journal of Physiology*, 264:H1229–H1238, 1993.

132. Bloomfield P. *Fourier Analysis of Time Series: An Introduction*. Wiley, New York, NY, 1976.

133. Karpman L, Cage J, Hill C, Forbes AD, Karpman V, and Cohn K. Sound envelope averaging and the differential diagnosis of systolic murmurs. *American Heart Journal*, 90(5):600–606, 1975.

134. Gerbarg DS, Holcomb Jr. FW, Hofler JJ, Bading CE, Schultz GL, and Sears RE. Analysis of phonocardiogram by a digital computer. *Circulation Research*, 11:569–576, 1962.

135. Gerbarg DS, Taranta A, Spagnuolo M, and Hofler JJ. Computer analysis of phonocardiograms. *Progress in Cardiovascular Diseases*, 5(4):393–405, 1963.

136. Saltzberg B and Burch NR. Period analytic estimates of moments of the power spectrum: A simplified EEG time domain procedure. *Electroencephalography and Clinical Neurophysiology*, 30:568–570, 1971.

137. Jacobs JE, Horikoshi K, and Petrovick MA. Feasibility of automated analysis of phonocardiograms. *Journal of the Audio Engineering Society*, 17(1):49–54, 1969.

138. Yokoi M, Uozumi Z, Okamoto N, Mizuno Y, Iwatsuka T, Takahashi H, Watanabe Y, and Yasui S. Clinical evaluation on 5 years' experience of automated phonocardiographic analysis. *Japanese Heart Journal*, 18(4):482–490, 1977.

139. Willison RG. Analysis of electrical activity in health and dystrophic muscle in man. *Journal of Neurology, Neurosurgery, and Psychiatry*, 27:386–394, 1964.

140. Fuglsang-Frederiksen A and Månsson A. Analysis of electrical activity of normal muscle in man at different degrees of voluntary effort. *Journal of Neurology, Neurosurgery, and Psychiatry*, 38:683–694, 1975.

141. Dowling MH, Fitch P, and Willison RG. A special purpose digital computer (BIOMAC 500) used in the analysis of the human electromyogram. *Electroencephalography and Clinical Neurophysiology*, 25:570–573, 1968.

142. Hjorth B. EEG analysis based on time domain properties. *Electroencephalography and Clinical Neurophysiology*, 29:306–310, 1970.

143. Hjorth B. The physical significance of time domain descriptors in EEG analysis. *Electroencephalography and Clinical Neurophysiology*, 34:321–325, 1973.

144. Hjorth B. Time domain descriptors and their relation to a particular model for generation of EEG activity. In Dolce G and Künkel H, editors, *CEAN: Computerised EEG Analysis*, pages 3–8. Gustav Fischer, Stuttgart, Germany, 1975.

145. Binnie CD, Batchelor BG, Bowring PA, Darby CE, Herbert L, Lloyd DSL, Smith DM, Smith GF, and Smith M. Computer-assisted interpretation of clinical EEGs. *Electroencephalography and Clinical Neurophysiology*, 44:575–585, 1978.

146. Binnie CD, Batchelor BG, Gainsborough AJ, Lloyd DSL, Smith DM, and Smith GF. Visual and computer-assisted assessment of the EEG in epilepsy of late onset. *Electroencephalography and Clinical Neurophysiology*, 47:102–107, 1979.

147. Hornero R, Espino P, Alonso A, and López M. Estimating complexity from EEG background activity of epileptic patients. *IEEE Engineering in Medicine and Biology Magazine*, 18(6):73–79, November/December 1999.

148. Celka P, Mesbah M, Keir M, Boashash B, and Colditz P. Time-varying dimension analysis of EEG using adaptive principal component analysis and model selection. In *World Congress on Medical Physics and Biomedical Engineering*, page 4 pages on CDROM. IFMBE/IEEE, Chicago, IL, 2000.

149. Hsia PW, Jenkins JM, Shimoni Y, Gage KP, Santinga JT, and Pitt B. An automated system for ST segment and arrhythmia analysis in exercise radionuclide ventriculography. *IEEE Transactions on Biomedical Engineering*, 33(6):585–593, 1986.

150. Lawrence JH and de Luca CJ. Myoelectric signal versus force relationship in different human muscles. *Journal of Applied Physiology*, 54(6):1653–1659, 1983.

151. Sakai A, Feigen LP, and Luisada AA. Frequency distribution of heart sounds in normal man. *Cardiovascular Research*, 5:358–363, 1971.

152. Frome EL and Frederickson EL. Digital spectrum analysis of the first and second heart sounds. *Computers and Biomedical Research*, 7:421–431, 1974.

153. Yoganathan AP, Gupta R, Udwadia FE, Miller JW, Corcoran WH, Sarma R, Johnson JL, and Bing RJ. Use of the fast Fourier transform for frequency analysis of the first heart sound in normal man. *Medical and Biological Engineering*, 14:69–73, 1976.

154. Yoganathan AP, Gupta R, Udwadia FE, Corcoran WH, Sarma R, and Bing RJ. Use of the fast Fourier transform in the frequency analysis of the second heart sound in normal man. *Medical and Biological Engineering*, 14:455–459, 1976.

155. Adolph RJ, Stephens JF, and Tanaka K. The clinical value of frequency analysis of the first heart sound in myocardial infarction. *Circulation*, 41:1003–1014, 1970.

156. Clarke WB, Austin SM, Shah PM, Griffen PM, Dove JT, McCullough J, and Schreiner BF. Spectral energy of the first heart sound in acute myocardial ischemia. *Circulation*, 57(3):593–598, 1978.

157. Geckeler GD, Likoff W, Mason D, Riesz RR, and Wirth CH. Cardiospectrograms: A preliminary report. *American Heart Journal*, 48:189–196, 1954.

158. McKusick VA, Talbot SA, and Webb GN. Spectral phonocardiography: Problems and prospects in the application of the Bell sound spectrograph to phonocardiography. *Bulletin of the Johns Hopkins Hospital*, 94:187–198, 1954.

159. McKusick VA, Webb GN, Humphries JO, and Reid JA. On cardiovascular sound: Further observations by means of spectral phonocardiography. *Circulation*, 11:849–870, 1955.

160. Winer DE, Perry LW, and Caceres CA. Heart sound analysis: A three dimensional approach. Contour plotting of sound for study of cardiovascular acoustics. *American Journal of Cardiology*, 16:547–551, 1965.

161. Yoshimura S. Principle and practice of phonocardiography in reference to frequency intensity characteristics of heart sounds and murmurs. *Japanese Circulation Journal*, 24:921–931, 1960.

162. van Vollenhoven E, van Rotterdam A, Dorenbos T, and Schlesinger FG. Frequency analysis of heart murmurs. *Medical and Biological Engineering*, 7:227–231, 1969.

163. Johnson GR, Adolph RJ, and Campbell DJ. Estimation of the severity of aortic valve stenosis by frequency analysis of the murmur. *Journal of the American College of Cardiology*, 1(5):1315–1323, 1983.

164. Johnson GR, Myers GS, and Lees RS. Evaluation of aortis stenosis by spectral analysis of the murmur. *Journal of the American College of Cardiology*, 6(1):55–63, 1985.

165. Welch PD. The use of fast Fourier transform for the estimation of power spectra: A method based on time averaging over short, modified periodograms. *IEEE Transactions on Audio and Electroacoustics*, 15:70–73, 1967.

166. Harris FJ. On the use of windows for harmonic analysis with the discrete Fourier transform. *Proceedings of the IEEE*, 66(1):51–83, 1978.

167. Durand LG, Blanchard M, Cloutier G, Sabbah HN, and Stein PD. Comparison of pattern recognition methods for computer-assisted classification of spectra of heart sounds in patients with a porcine bioprosthetic valve implanted in the mitral position. *IEEE Transactions on Biomedical Engineering*, 37(12):1121–1129, 1990.

168. Cloutier G, Durand LG, Guardo R, Sabbah HN, and Stein PD. Bias and variability of diagnostic spectral parameters extracted from closing sounds produced by bioprosthetic valves implanted in the mitral position. *IEEE Transactions on Biomedical Engineering*, 36(8):815–825, 1989.

169. Agarwal GC and Gottlieb GL. An analysis of the electromyogram by Fourier, simulation and experimental techniques. *IEEE Transactions on Biomedical Engineering*, 22(3):225–229, 1975.

170. Abeles M and Goldstein Jr. MH. Multispike train analysis. *Proceedings of the IEEE*, 65(5):762–773, 1977.

171. Landolt JP and Correia MJ. Neuromathematical concepts of point process theory. *IEEE Transactions on Biomedical Engineering*, 25(1):1–12, 1978.

172. Anderson DJ and Correia MJ. The detection and analysis of point processes in biological signals. *Proceedings of the IEEE*, 65(5):773–780, 1977.

173. Cohen A. *Biomedical Signal Processing*. CRC Press, Boca Raton, FL, 1986.

174. Zhang YT, Frank CB, Rangayyan RM, and Bell GD. Mathematical modeling and spectrum analysis of the physiological patello-femoral pulse train produced by slow knee movement. *IEEE Transactions on Biomedical Engineering*, 39(9):971–979, 1992.

175. Beverland DE, Kernohan WG, and Mollan RAB. Analysis of physiological patello-femoral crepitus. In Byford GH, editor, *Technology in Health Care*, pages 137–138. Biological Engineering Society, London, UK, 1985.

176. Beverland DE, Kernohan WG, McCoy GF, and Mollan RAB. What is physiological patellofemoral crepitus? In *Proceedings of the XIV International Conference on Medical and Biological Engineering and VII International Conference on Medical Physics*, pages 1249–1250. IFMBE, Espoo, Finland, 1985.

177. Beverland DE, McCoy GF, Kernohan WG, and Mollan RAB. What is patellofemoral crepitus? *Journal of Bone and Joint Surgery*, 68-B:496, 1986.

178. Parker PA, Stuller JA, and Scott RN. Signal processing for the multistate myoelectric channel. *Proceedings of the IEEE*, 65(5):662–674, 1977.

179. Lindström LH and Magnusson RI. Interpretation of myoelectric power spectra: A model and its applications. *Proceedings of the IEEE*, 65(5):653–662, 1977.

180. Zhang YT, Parker PA, and Scott RN. Study of the effects of motor unit recruitment and firing statistics on the signal-to-noise ratio of a myoelectric control channel. *Medical and Biological Engineering and Computing*, 28:225–231, 1990.

181. Parker PA and Scott RN. Statistics of the myoelectric signal from monopolar and bipolar electrodes. *Medical and Biological Engineering*, 11:591–596, 1973.

182. Shwedyk E, Balasubramanian R, and Scott RN. A nonstationary model for the electromyogram. *IEEE Transactions on Biomedical Engineering*, 24(5):417–424, 1977.

183. Person RS and Libkind MS. Simulation of electromyograms showing interference patterns. *Electroencephalography and Clinical Neurophysiology*, 28:625–632, 1970.

184. Person RS and Kudina LP. Cross-correlation of electromyograms showing interference patterns. *Electroencephalography and Clinical Neurophysiology*, 25:58–68, 1968.

185. de Luca CJ. A model for a motor unit train recorded during constant force isometric contractions. *Biological Cybernetics*, 19:159–167, 1975.

186. de Luca CJ and van Dyk EJ. Derivation of some parameters of myoelectric signals recorded during sustained constant force isometric contractions. *Biophysical Journal*, 15:1167–1180, 1975.

187. Makhoul J. Linear prediction: A tutorial. *Proceedings of the IEEE*, 63(4):561–580, 1975.

188. Durbin J. *The fitting of time-series models*. Mimeograph Series No. 244, Institute of Statistics, University of North Carolina, Chapel Hill, NC, 1959.

189. Durbin J. Estimation of parameters in time-series regression models. *Journal of the Royal Statistical Society, Series B (Methodological)*, 22(1):139–153, 1960.

190. Akaike H. A new look at the statistical model identification. *IEEE Transactions on Automatic Control*, 19:716–723, 1974.

191. Atal BS. Effectiveness of linear prediction characteristics of the speech wave for automatic speaker identification and verification. *Journal of the Acoustical Society of America*, 55(6):1304–1313, June 1974.

192. Kang WJ, Shiu JR, Cheng CK, Lai JS, Tsao HW, and Kuo TS. The application of cepstral coefficients and maximum likelihood method in EMG pattern recognition. *IEEE Transactions on Biomedical Engineering*, 42(8):777–785, 1995.

193. Kopec GE, Oppenheim AV, and Tribolet JM. Speech analysis by homomorphic prediction. *IEEE Transactions on Acoustics, Speech, and Signal Processing*, 25(1):40–49, 1977.

194. Steiglitz K and McBride LE. A technique for the identification of linear systems. *IEEE Transactions on Automatic Control*, 10:461–464, 1965.

195. Steiglitz K. On the simultaneous estimation of poles and zeros in speech analysis. *IEEE Transactions on Acoustics, Speech, and Signal Processing*, 25(3):229–234, 1977.

196. Kalman RE. Design of a self-optimizing control system. *Transactions of the ASME*, 80:468–478, 1958.

197. Joo TH, McClellan JH, Foale RA, Myers GS, and Lees RA. Pole-zero modeling and classification of phonocardiograms. *IEEE Transactions on Biomedical Engineering*, 30(2):110–118, 1983.

198. Murthy ISN and Prasad GSSD. Analysis of ECG from pole-zero models. *IEEE Transactions on Biomedical Engineering*, 39(7):741–751, 1992.

199. Murthy ISN, Rangaraj MR, Udupa KJ, and Goyal AK. Homomorphic analysis and modeling of ECG signals. *IEEE Transactions on Biomedical Engineering*, 26(6):330–344, 1979.

200. Akay AM, Welkowitz W, Semmlow JL, and Kostis JB. Application of the ARMA method to acoustic detection of coronary artery disease. *Medical and Biological Engineering and Computing*, 29:365–372, 1991.

201. Sikarskie DL, Stein PD, and Vable M. A mathematical model of aortic valve vibration. *Journal of Biomechanics*, 17(11):831–837, 1984.

202. Wang JZ, Tie B, Welkowitz W, Semmlow JL, and Kostis JB. Modeling sound generation in stenosed coronary arteries. *IEEE Transactions on Biomedical Engineering*, 37(11):1087–1094, 1990.

203. Wang JZ, Tie B, Welkowitz W, Kostis J, and Semmlow J. Incremental network analogue model of the coronary artery. *Medical and Biological Engineering and Computing*, 27:416–422, 1989.

204. Fredberg JJ. Origin and character of vascular murmurs: Model studies. *Journal of the Acoustical Society of America*, 61(4):1077–1085, 1977.

205. Akselrod S, Gordon D, Ubel FA, Shannon DC, Barger AC, and Cohen RJ. Power spectrum analysis of heart rate fluctuation: A quantitative probe of beat-to-beat cardiovascular control. *Science*, 213:220–222, 10 July 1981.

206. Iwata A, Suzumara N, and Ikegaya K. Pattern classification of the phonocardiogram using linear prediction analysis. *Medical and Biological Engineering and Computing*, 15:407–412, 1977.

207. Iwata A, Ishii N, Suzumara N, and Ikegaya K. Algorithm for detecting the first and the second heart sounds by spectral tracking. *Medical and Biological Engineering and Computing*, 18:19–26, 1980.

208. Akay AM, Semmlow JL, Welkowitz W, Bauer MD, and Kostis JB. Noninvasive detection of coronary stenoses before and after angioplasty using eigenvector methods. *IEEE Transactions on Biomedical Engineering*, 37(11):1095–1104, 1990.

209. Goodfellow J, Hungerford DS, and Woods C. Patellofemoral joint mechanics and pathology. *Journal of Bone and Joint Surgery*, 58B:921, 1976.

210. Woo SLY and Buckwalter JA, editors. *Injury and Repair of the Musculoskeletal Soft Tissues*. American Academy of Orthopaedic Surgeons, Park Ridge, IL, 1987.

211. Hwang WS, Li B, and Jin LH. Collagen fibril structure of normal, aging, and osteoarthritic cartilage. *Journal of Pathology*, 167:425–433, 1992.

212. Fulkerson JP and Hungerford DS, editors. *Disorders of the Patello-femoral Joint*. Williams/Wilkins, Baltimore, MD, 1990.

213. Noyes FR and Stabler CL. A system for grading articular cartilage lesions at arthroscopy. *American Journal of Sports Medicine*, 17(4):505–513, 1989.

214. Kulund DN, editor. *The Injured Athlete*. Lippincott, Philadelphia, PA, 2nd edition, 1988.

215. Meisel AD and Bullough PG. Osteoarthritis of the knee. In Krieger A, editor, *Atlas of Osteoarthritis*, pages 5.1–5.19. Gower Medical Publishing, New York, NY, 1984.

216. Smillie IS. *Injuries of the Knee Joint*. Churchill Livingstone, Edinburgh, Scotland, 5th edition, 1978.

217. Mankin HJ. The articular cartilages, cartilage healing, and osteoarthritis. In Cruess RL and Rennie WRJ, editors, *Adult Orthopaedics*, pages 163–270. Churchill Livingstone, New York, NY, 1984.

218. McCoy GF, McCrea JD, Beverland DE, Kernohan WG, and Mollan RAB. Vibration arthrography as a diagnostic aid in disease of the knee. *Journal of Bone and Joint Surgery*, 69-B(2):288–293, 1987.

219. Appel U and v. Brandt A. Adaptive sequential segmentation of piecewise stationary time series. *Information Sciences*, 29:27–56, 1983.

220. Appel U and v. Brandt A. A comparative analysis of three sequential time series segmentation algorithms. *Signal Processing*, 6:45–60, 1984.

221. Arnold M, Witte H, Leger P, Boccalon H, Bertuglia S, and Colantuoni A. Time-variant spectral analysis of LDF signals on the basis of multivariate autoregressive modelling. *Technology and Health Care*, 7:103–112, 1999.

222. Arnold M, Miltner WHR, Witte H, Bauer R, and Braun C. Adaptive AR modeling of nonstationary time series by means of Kalman filtering. *IEEE Transactions on Biomedical Engineering*, 45(5):553–562, 1998.

223. Bohlin T. Analysis of EEG signals with changing spectra using a short-word Kalman estimator. *Mathematical Biosciences*, 35:221–259, 1977.

224. Gath I, Feuerstein C, Pham DT, and Rondouin G. On the tracking of rapid dynamic changes in seizure EEG. *IEEE Transactions on Biomedical Engineering*, 39(9):952–958, 1992.

225. Bianchi AM, Mainardi L, Petrucci E, Signorini MG, Mainardi M, and Cerutti S. Time-variant power spectrum analysis for the detection of transient episodes in HRV signal. *IEEE Transactions on Biomedical Engineering*, 40(2):136–144, 1993.

226. Oppenheim AV and Lim JS. The importance of phase in signals. *Proceedings of the IEEE*, 69(5):529–541, 1981.

227. Hayes MH and Oppenheim AV. Signal reconstruction from phase or magnitude. *IEEE Transactions on Acoustics, Speech, and Signal Processing*, 28(6):672–680, 1980.

228. Nikias CL and Mendel JM. Signal processing with higher-order spectra. In Ackenhusen JG, editor, *Signal Processing Technology and Applications*, pages 7–34. IEEE Technology Update Series, New York, NY, 1995.

229. Nikias CL and Raghuveer MR. Bispectrum estimation — A digital signal processing framework. *Proceedings of the IEEE*, 75:869–891, 1987.

230. Hlawatsch F and Boudreaux-Bartels GF. Linear and quadratic time-frequency signal representations. *IEEE Signal Processing Magazine*, pages 21–67, April 1992.

231. Cohen L. Time-frequency distributions — A review. *Proceedings of the IEEE*, 77:941–981, 1989.

232. Boashash B, editor. *Time-Frequency Signal Analysis*. Wiley, New York, NY, 1992.

233. Akay M, editor. *Time Frequency and Wavelets in Biomedical Signal Processing*. IEEE, New York, NY, 1998.

234. Praetorius HM, Bodenstein G, and Creutzfeldt OD. Adaptive segmentation of EEG records: A new approach to automatic EEG analysis. *Electroencephalography and Clinical Neurophysiology*, 42:84–94, 1977.

235. Ferber G. Treatment of some nonstationarities in the EEG. *Neuropsychobiology*, 17:100–104, 1987.

236. Bodenstein G, Schneider W, and Malsburg CVD. Computerized EEG pattern classification by adaptive segmentation and probability-density-function classification. Description of the method. *Computers in Biology and Medicine*, 15(5):297–313, 1985.

237. Creutzfeldt OD, Bodenstein G, and Barlow JS. Computerized EEG pattern classification by adaptive segmentation and probability density function classification. Clinical evaluation. *Electroencephalography and Clinical Neurophysiology*, 60:373–393, 1985.

238. Michael D and Houchin J. Automatic EEG analysis: A segmentation procedure based on the autocorrelation function. *Electroencephalography and Clinical Neurophysiology*, 46:232–235, 1979.

239. Barlow JS, Creutzfeldt OD, Michael D, Houchin J, and Epelbaum H. Automatic adaptive segmentation of clinical EEGs. *Electroencephalography and Clinical Neurophysiology*, 51:512–525, 1981.

240. Willsky AS and Jones HL. A generalized likelihood ratio approach to the detection and estimation of jumps in linear systems. *IEEE Transactions on Automatic Control*, 21:108–112, February 1976.

241. Basseville M and Benveniste A. Sequential segmentation of nonstationary digital signals using spectral analysis. *Information Sciences*, 29:57–73, 1983.

242. GE-Marquette Medical Systems, Inc., Milwaukee, WI. *Physician's Guide to Resting ECG Analysis Program, 12SL-tm*, 1991.

243. Tou JT and Gonzalez RC. *Pattern Recognition Principles*. Addison-Wesley, Reading, MA, 1974.

244. Duda RO and Hart PE. *Pattern Classification and Scene Analysis*. Wiley, New York, NY, 1973.

245. Fukunaga K. *Introduction to Statistical Pattern Recognition*. Academic, San Diego, CA, 2nd edition, 1990.

246. Johnson RA and Wichern DW. *Applied Multivariate Statistical Analysis*. Prentice-Hall, Englewood Cliffs, NJ, 3rd edition, 1992.

247. Schürmann J. *Pattern Classification — A unified view of statistical and neural approaches*. Wiley, New York, NY, 1996.

248. Duda RO, Hart PE, and Stork DG. *Pattern Classification*. Wiley, New York, NY, 2nd edition, 2001.

249. Micheli-Tzanakou E. *Supervised and Unsupervised Pattern Recognition*. CRC Press, Boca Raton, FL, 2000.

250. Neter J, Kutner MH, Nachtsheim CJ, and Wasserman W. *Applied Linear Statistical Models*. Irwin, Chicago, IL, 4th edition, 1990.

251. SPSS Inc., Chicago, IL. *SPSS Advanced Statistics User's Guide*, 1990.

252. SPSS Inc., Chicago, IL. *SPSS Base System User's Guide*, 1990.

253. Pao YH. *Adaptive Pattern Recognition and Neural Networks*. Addison-Wesley, Reading, MA, 1989.

254. Lippmann RP. An introduction to computing with neural nets. *IEEE Signal Processing Magazine*, pages 4–22, April 1987.

255. Nigrin A. *Neural Networks for Pattern Recognition*. MIT Press, Cambridge, MA, 1993.

256. Shen L, Rangayyan RM, and Desautels JEL. Detection and classification of mammographic calcifications. *International Journal of Pattern Recognition and Artificial Intelligence*, 7(6):1403–1416, 1993.

257. Metz CE. Basic principles of ROC analysis. *Seminars in Nuclear Medicine*, VIII(4):283–298, 1978.

258. Metz CE. ROC methodology in radiologic imaging. *Investigative Radiology*, 21:720–733, 1986.

259. Swets JA and Pickett RM. *Evaluation of diagnostic systems: Methods from signal detection theory*. Academic, New York, NY, 1982.

260. Dorfman DD and Alf E. Maximum likelihood estimation of parameters of signal detection theory and determination of confidence intervals — rating method data. *Journal of Mathematical Psychology*, 6:487–496, 1969.

261. Fleiss JL. *Statistical Methods for Rates and Proportions*. Wiley, New York, NY, 2nd edition, 1981.

262. Zar JH. *Biostatistical Analysis*. Prentice-Hall, Englewood Cliffs, NJ, 2nd edition, 1984.

263. Fukunaga K and Hayes RR. Effects of sample size in classifier design. *IEEE Transactions on Pattern Analysis and Machine Intelligence*, 11(8):873–885, 1989.

264. Raudys SJ and Jain AK. Small sample size effects in statistical pattern recognition: Recommendations for practitioners. *IEEE Transactions on Pattern Analysis and Machine Intelligence*, 13(3):252–264, 1991.

Index

509

CPSIA information can be obtained at www.ICGtesting.com
Printed in the USA
BVOW070739301111

276760BV00007B/12/A